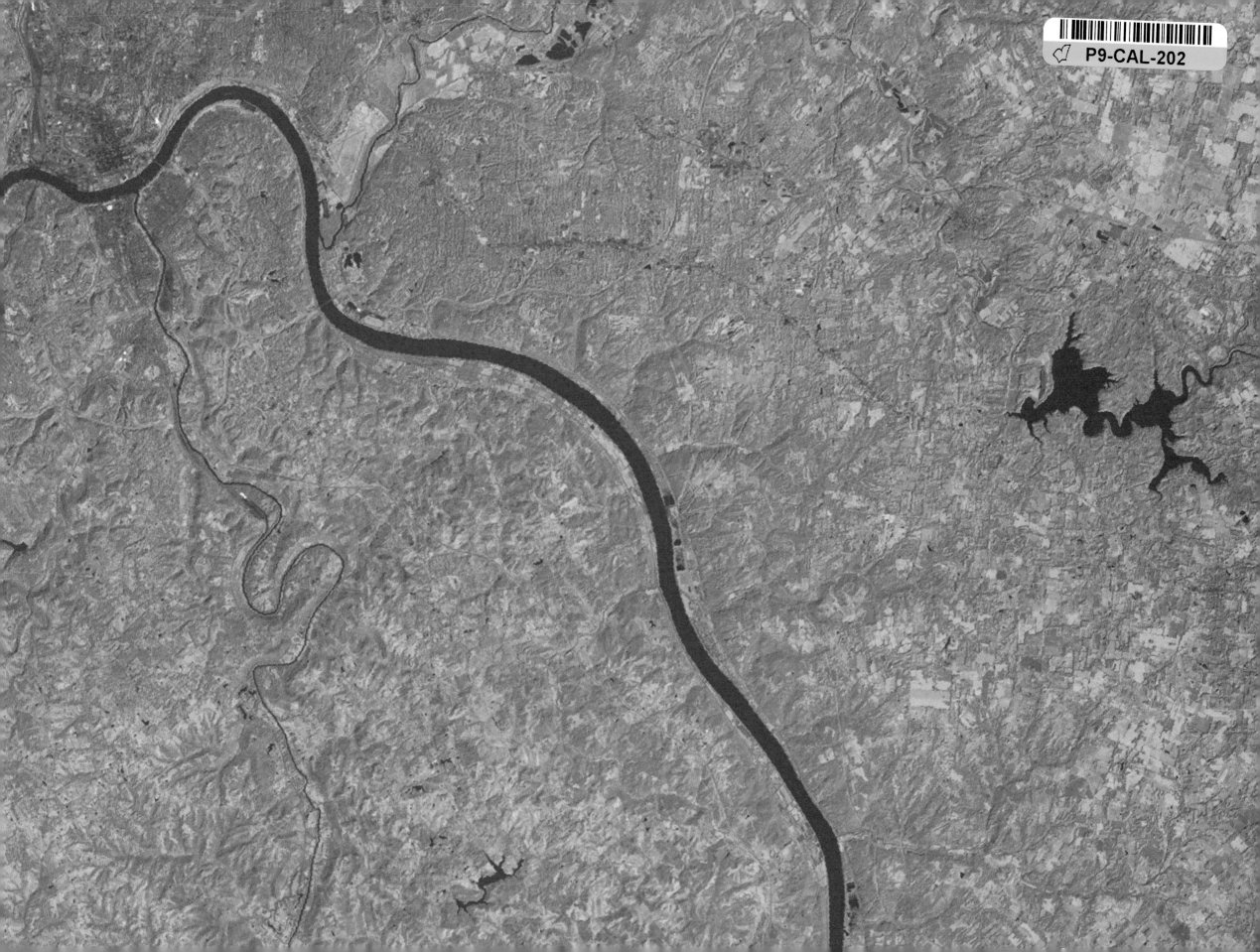
P9-CAL-202

Atlas of Kentucky

Atlas of Kentucky

EDITOR-IN-CHIEF

Richard Ulack, University of Kentucky

CO-EDITOR

Karl Raitz, University of Kentucky

CARTOGRAPHIC EDITOR

Gyula Pauer, University of Kentucky

MAJOR PARTICIPANTS
David A. Howarth, University of Louisville
Ronald L. Mitchelson, Morehead State University
John F. Watkins, University of Kentucky

CARTOGRAPHERS
Richard Gilbreath, University of Kentucky
Donna Gilbreath, University of Kentucky

THE UNIVERSITY PRESS OF KENTUCKY

Copyright © 1998 by The University Press of Kentucky

Scholarly publisher for the Commonwealth,
serving Bellarmine College, Berea College, Centre
College of Kentucky, Eastern Kentucky University,
The Filson Club Historical Society, Georgetown College,
Kentucky Historical Society, Kentucky State University,
Morehead State University, Murray State University,
Northern Kentucky University, Transylvania University,
University of Kentucky, University of Louisville,
and Western Kentucky University.

Editorial and Sales Offices: The University Press of Kentucky
663 South Limestone Street, Lexington, Kentucky 40508-4008

02 01 00 99 98   5 4 3 2 1

Front endpaper: Northern Kentucky. An enhanced Landsat Thematic Mapper (TM) image,
acquired in September 1994, using bands 5 (middle infrared), 4 (near infrared), and 3 (red).
This false-color rendition depicts forest as medium green and actively growing cropland/
grassland as lighter greens. Cropland with standing mature crops and fields that have been
harvested are shades of purple and magenta. Water bodies such as the Ohio River are dark
blue. Roads (purple) and commercial/industrial areas (light to dark purple) are visible in the
densely populated urban area. Covington, Kentucky, is located in the upper center of the
image where the Licking River meets the Ohio. The Cincinnati central business district is
across the Ohio River to the north; the Cincinnati/Northern Kentucky International Airport is
visible to the west of Covington, south of the Ohio River.

Frontispiece: Morning fog along the Pottsville Escarpment, Estill County.

Library of Congress Cataloging-in-Publication Data

Atlas of Kentucky / Richard Ulack, editor-in-chief, Karl Raitz,
    co-editor ; Gyula Pauer, cartographic editor.
        p.   cm.
        Includes bibliographical references and index.
        ISBN 0-8131-2005-5 (cloth : alk. paper)
    1. Kentucky—Maps.   2. Kentucky—Economic conditions—Maps.
3. Kentucky—Social conditions—Maps.   4. Kentucky—History.
I. Ulack, Richard, 1942–   . II. Pauer, Gyula.   III. Raitz, Karl B.
G1330 .A7 1998 <G&M>
912.769—DC21   97-38042
                                         MAPS

Financial Supporters

James Graham Brown Foundation, Inc.
Mary and Barry Bingham, Sr. Fund
The University of Kentucky
Kentucky Economic Development Corporation
Kentucky Cabinet for Economic Development
Morehead State University
Columbia Gas of Kentucky, Inc.
Western Kentucky Gas Company
Kentucky Utilities Company
Hazard Compensation Agency, Inc.
William A. Withington
Landmark Community Newspapers, Inc.
Kentucky Post
Paducah Sun
Bicentennial Commission of Boyd and Greenup Counties
Kentucky-American Water Company
FIVCO Area Development District
Pennyrile Area Development District
Barren River Area Development District
Gateway Area Development District
Big Sandy Area Development District
Bluegrass Area Development District
Buffalo Trace Area Development District
Green River Area Development District
Kentucky River Area Development District
Purchase Area Development District
Lake Cumberland Area Development District, Inc.
Kentuckiana Regional Planning and Development Agency

Contributors

Text and Research Contributors

Stanley D. Brunn, University of Kentucky
Michael T. Childress, Kentucky Long-Term Policy Research Center
Berle Clay, University of Kentucky
D. Glen Conner, Western Kentucky University
Wayne L. Hoffman, Western Kentucky University
Clara A. Leuthart, University of Louisville
Yu Luo, Indiana State University
Conrad T. Moore, Western Kentucky University
Nancy O'Malley, University of Kentucky
Albert J. Petersen, Western Kentucky University
Richard Schein, University of Kentucky
Peter Schirmer, Kentucky Long-Term Policy Research Center
Michal Smith-Mello, Kentucky Long-Term Policy Research Center
L. Michael Trapasso, Western Kentucky University

Cartographic Contributors

Kent Anness, Bluegrass Area Development District
David Elbon, University of Kentucky
Terry Hounshell, Kentucky Geological Survey
Gary O'Dell, Kentucky Division of Water

Major Photographic Contributors

Thomas Barnes, University of Kentucky
John McGregor, U.S. Forest Service
Keith Mountain, University of Louisville
Karl Raitz, University of Kentucky
Neil Weber, Murray State University
Kentucky Department of Travel

Contents

Continues on next page

Foreword

KENTUCKIANS HAVE ALWAYS taken great pride in their land. Over the past two centuries it has nourished the hopes and dreams of the people of the Commonwealth. Kentucky's resources and natural wonders have shaped the lives of its inhabitants and fostered extraordinary growth.

Now, as we are poised at the opening of a new millennium, the time is ripe for the publication of a new resource detailing the many facets of our state. The *Atlas of Kentucky* offers a wealth of information on every aspect of Kentucky's people, land, economy, and cultural life.

Never before has so much effort gone into upgrading our educational system or attracting new business and industry to improve the state's economy and the quality of life for our citizens. The information in this volume makes clear the importance of our educational institutions, our transportation systems, our communication networks, and the natural and human resources that make Kentucky an ideal location for modern manufacturing and service industries.

A model of intercollegiate cooperation, this new publication brings together the work of geographers and cartographers at five of our state's great universities. The result is an *Atlas of Kentucky* that documents the sweeping changes that have taken place over the past several years and their impact on the Commonwealth as we move into the twenty-first century.

Governor of the Commonwealth of Kentucky

Acknowledgments

During the five years we have been working on this project, numerous agencies and individuals have given us invaluable advice and assistance, as well as the data, photographs, and other materials necessary to produce the graphics and text herein. Clearly, without such help the *Atlas* could not have been produced. At the risk of omitting an agency or individual, what follows is our record of those who helped in producing the *Atlas of Kentucky*. Any errors and omissions, and all interpretations, are, however, our responsibility alone.

Employees of various state government and other agencies gave much of their time, as well as information, in finding and providing answers to our questions. Included are the Natural Resources and Environmental Protection Cabinet, where Julie Smithers, Kenneth Bates, and Jeno Balassa provided maps, data, and other information. In the Cabinet for Economic Development, Ronald Morgan and Ronald Decker were both most helpful in providing the data that was necessary for producing many of the graphics, primarily in Chapters 5, 6, and 8. Gwyn Boyd of the Kentucky Economic Development Finance Authority also provided data for Chapter 8. J. Dan Guffey of the Kentucky Coal Marketing and Export Council and John K. Hiett of the Department of Mines and Minerals assisted by providing data and other materials for the coal section of Chapter 6. From the Public Service Commission, Kyle Willard, Elie El-Rouaiheb, Ralph Dennis, Eddie Smith, Martha Morton, and Bill Feldman were most helpful in providing materials and helping us to understand the distribution of public services available in the state and graphically illustrated in Chapters 6 and 9.

Several individuals at the Kentucky Geological Survey, including James Cobb, Garland Dever, Warren Anderson, Bart Davidson, Donald Chesnut, and Brandon Nuttal, also supplied us with the data necessary to produce the graphics on Kentucky's coal and other minerals for Chapter 6, and with information and maps on Kentucky's geology for Chapter 2. State and other resource agencies were also contacted to assist in the completion of Chapter 2, which depicts the natural environment. Bill Craddock, State Soil Scientist at the Natural Resources Conservation Service, assisted by providing valuable information on Kentucky's soils. Jafar Hadizadeh of the University of Louisville provided reference material on the New Madrid fault zone; David Foster of the American Cave Association helped with information on caves and karst; as did Angelo George, who provided maps and other information on the location of caves throughout the state. Doug Jackson of the State Department for Health Services, Cabinet for Human Resources, provided the data to construct the maps on radon levels, and Hugh Spencer of the University of Louisville provided the information for the maps on greenhouse gas emissions. Thanks also to H. Scott Hankla, Ted Stumbur, Tom Van Arsdall, Joe Ray, and Gary O'Dell of the Division of Water of the Natural Resources and Environmental Protection Cabinet, who provided much insight on the state's water resources.

Information on wildlife and other aspects of the natural environment were graciously provided by Lynn Garrison, Pam Renner, Bob Berry, Lauren Schaaf, and Peter Pfeiffer of the Department of Fish and Wildlife Resources; special thanks to Lynn Garrison for the time he spent answering many questions and providing contacts. Lew Kornman, fisheries biologist with the Department of Fish and Wildlife Resources, supplied information on the game fishes of Kentucky as well as a number of photographic slides from his extensive personal collection. Julian Campbell of The Nature Conservancy provided the information necessary for an understanding of the state's natural regions, as well as nature photographs. Robert McCance Jr. and Marc Evans of the Kentucky State Nature Preserves Commission were most helpful in providing information on the state's endangered plant and animal species.

Much of the data and other information, as well as many photographs, in Chapter 10 were furnished by the Kentucky Department of Travel Development. We wish to extend special thanks to Larry Southard, Don Cecil, Barbara Atwood, Deborah Weis, Dana Stratton, Beth Evans Cooke, and Bruce Brooks II for their assistance in the completion of Chapter 10. Larry Newby of the Kentucky Lottery Commission provided the data for those maps depicting participation in the state's lottery. To our knowledge no other state atlas has a chapter devoted solely to recreation and tourism.

Karla Arnold of the Kentucky State Board of Elections provided us with the most recent official general election results. These data were used in the preparation of several maps and graphs that appear in Chapter 11.

Librarians at the University of Kentucky and Morehead State University researched, located, and collected for us much of the data, maps, and historic photographs and other information necessary for the production of this atlas. At the University of Kentucky, special thanks to Bill Marshall of Special Collections; Gwen Curtis from the Map Library; Diane Brunn of the Agricultural Library; Brad Grissom of the Biological Sciences Library; and Kandace Rogers, Shawn Livingston, Sandee McAninch, and Roxanna Jones from Research and Information Services. Clara Keyes and Julia Lewis from Morehead State University's Camden Carroll Library spent much time finding information from special collections and government documents. Laura Whayne at UK's Transportation Research Center Library was also most helpful in gathering information for Chapter 9. Thanks also to Leanne S. Brehob, research assistant at Mid-America Remote Sensing Center at Murray State University, who helped in the production of the satellite imagery.

Thanks are due also to Tom Woods of the *Lexington Herald-Leader,* who assisted in identifying and providing several photographs.

At the University of Kentucky, special thanks are due to President Charles T. Wethington Jr., who provided support to launch the project, and to former Campus Chancellor Robert Hemenway, Vice Chancellor for Research and Graduate Studies David Watt, and former College of Arts and Sciences Dean Richard Edwards for providing funding to support cartographic production through UK's Center for Cartography and Geographic Information. At Morehead State University, Vice Presidents Keith Kappes and John Philley provided significant monetary support and release time for Morehead State University atlas contributors. One of those supported by Morehead, Kevin Calhoun, assembled and mapped many graphics that were adapted for use in Chapter 9.

Thanks also to Professors Matt Pelkki and Jim Ringe of UK's College of Forestry, who reviewed the forestry section of Chapter 6 at our request, and to the anonymous reviewers whose comments were also most helpful. Tom Hodler and Howard Schretter of the University of Georgia and editors of the *Atlas of Georgia,* as well as Will Fontanez of the University of Tennessee, should be thanked for their guidance, support, and advice at a very early stage in the project.

All of the maps and other graphics for the *Atlas of Kentucky* were proofed on a color ink jet printer made available by the University of Kentucky's FACTS Center. We appreciate their support of this project.

We could not end this section without acknowledging several other individuals who were critical to the completion of this project: Glenda King, who assisted and advised in all facets of the design and layout of the atlas; Angela G. Ray, who provided most careful editorial assistance until other duties necessitated her leaving the project; Paula Call Wadlington, who was the final proofreader; and Martin L. White, who indexed the work. Finally, the University Press of Kentucky staff was most helpful and supportive throughout the duration of the project.

Kentucky: Its Setting

CHAPTER ONE

Kentucky: Its Setting

Downtown Louisville stands beside the Falls of the Ohio.

THE COMMONWEALTH OF KENTUCKY is one of the most accessible states in the nation. Centered in the eastern half of the United States, Kentucky is equidistant from Los Angeles and Panama, and from Havana and Winnipeg. In driving time, Lexington is about the same distance from Detroit, Atlanta, and St. Louis; from New York, New Orleans, and Minneapolis; or from Miami, San Antonio, Denver, and Portland, Maine. The state, located within one day's drive of nearly 70 percent of the nation's population, is served by an excellent interstate highway and parkway system that includes the east–west routes I-64 and I-24 and the north–south routes I-65 and I-75. Since Kentucky has about eleven hundred miles of commercially navigable waterways, much of its territory is also accessible through the nation's inland waterway system via the Mississippi-Ohio and Tennessee-Tombigbee systems. Indeed, Kentucky's waterways have more miles of running water than any other state except Alaska. Kentucky is also well served by rail, with nearly twenty-five hundred miles of mainline track used by two major trunk carriers, CSX Transportation and Norfolk Southern, and fourteen regional and shortline railroads. There are two major international airports, Louisville's International Airport at Standiford Field and the Cincinnati/Northern Kentucky International Airport, and several regional airports, most notably Blue Grass Field in Lexington, which serves Kentuckians in the central and eastern parts of the state.

In short, Kentucky's highway, waterway, rail, and air accessibility means that the state is an ideal location for manufacturing, corporate headquarters, and distribution centers. It is not surprising that corporations such as Toyota, Ford, and General Electric are among the state's major industrial employers, or that United Parcel Service has designated Louisville as its national air hub. This accessibility, coupled with the state's scenic attractions and historic landscapes, also makes Kentucky an attractive destination for visitors and new residents, young and old alike.

A GLOBAL PERSPECTIVE

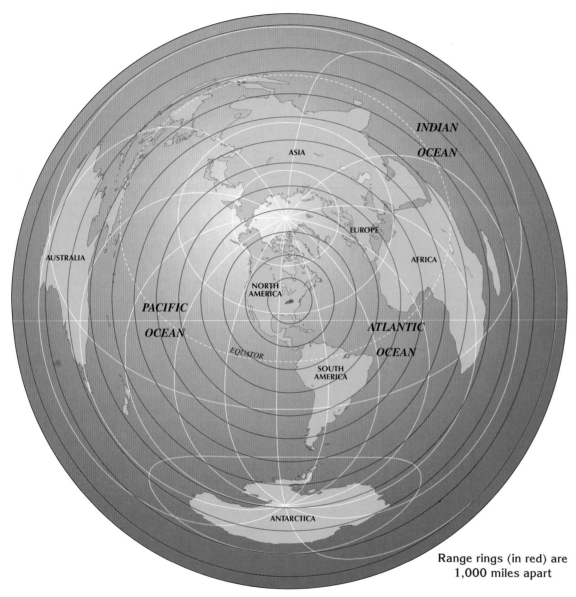

ASIA

INDIAN OCEAN

EUROPE

AUSTRALIA

AFRICA

NORTH AMERICA

PACIFIC OCEAN

ATLANTIC OCEAN

EQUATOR

SOUTH AMERICA

ANTARCTICA

Range rings (in red) are
1,000 miles apart

**Azimuthal Equidistant Projection
Centered on Frankfort, Kentucky
38°12'N., 84°51'W.**

The Army Corp of Engineers' Wolf Creek Dam on the Cumberland River was completed in 1952, impounding Lake Cumberland, the state's largest lake.

*A Bluegrass horse farm,
Fayette County.*

KENTUCKY'S PLACE IN THE WESTERN HEMISPHERE

Orthographic Projection
Centered on 25°00'N., 75°00'W.

TRAVEL TIME FROM LEXINGTON, KENTUCKY

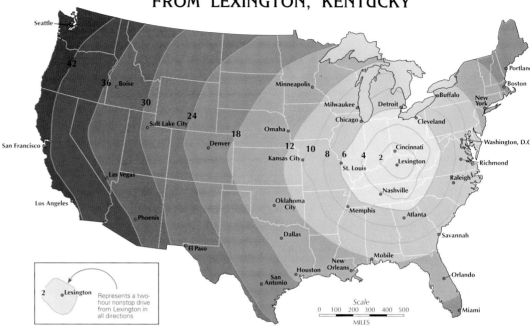

2 ⊙ Lexington   Represents a two-hour nonstop drive from Lexington in all directions

Scale
0 100 200 300 400 500
MILES

The Kentucky River floodplain in Owen County.

PHYSICAL FRAMEWORK

ELEVATION IN FEET

3,001 and above
2,001 – 3,000
1,001 – 2,000
401 – 1,000
0 – 400

Kentucky is a land of diversity. Within its boundaries one can find forested highlands of the Appalachian Mountain chain, rocky gorges and tumbling waterfalls, broad valleys framing several major rivers, and fertile plains that support an array of agricultural activities. Five major physiographic regions constitute the state's terrain: the Eastern Kentucky Coal Field, the Bluegrass, the Pennyroyal (or Mississippian Plateaus), the Western Kentucky Coal Field, and the Jackson Purchase (or Mississippi Embayment). Perhaps the best-known region is the Bluegrass, comprising the Outer and Inner Bluegrass Regions. A place-name recognizable to people far and wide, "Bluegrass" conjures images of grassy rolling hills, Thoroughbred horses, picturesque farmland, and miles of wooden or rock fences.

(Text continues on page 10)

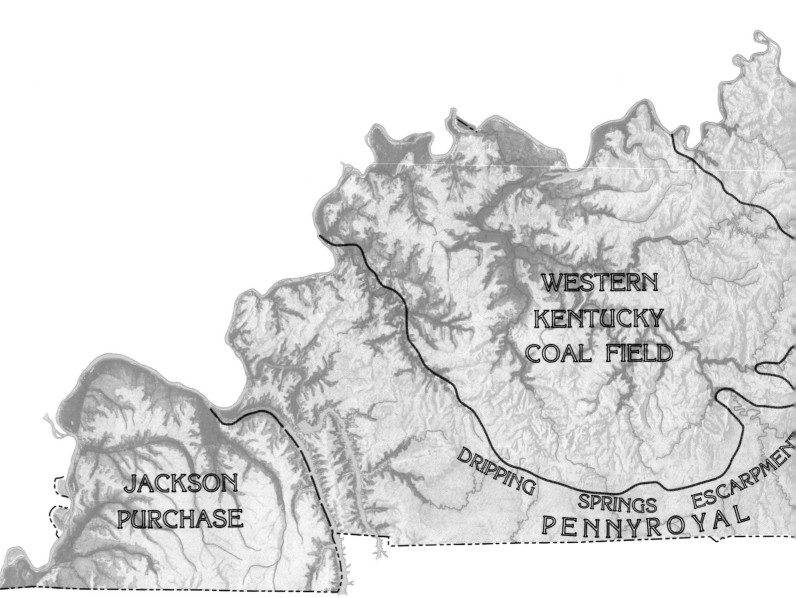

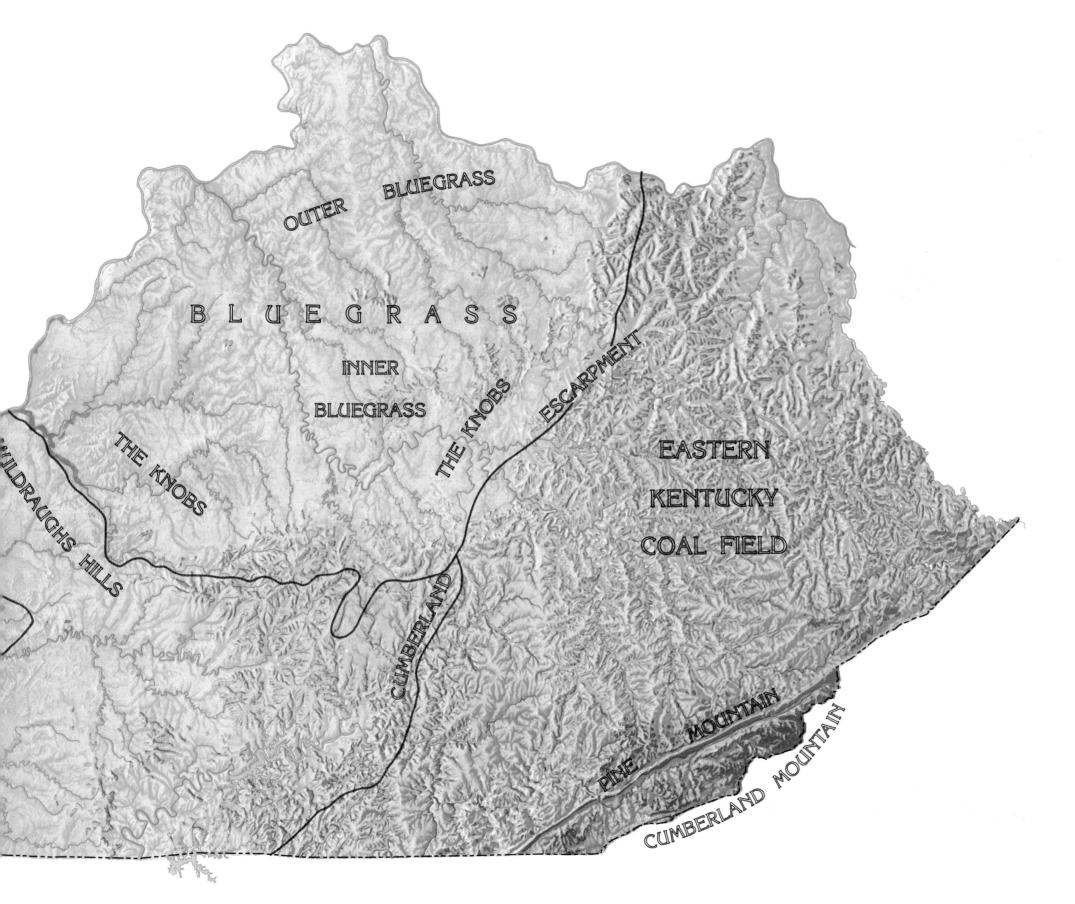

Kentucky's state flag.

Goldenrod, Kentucky's state flower.

The cardinal, Kentucky's state bird.

Kentucky's capitol building in Frankfort, designed by architect Frank M. Andrews and dedicated in 1910.

COMMONWEALTH OF KENTUCKY

20	0	20	40 MILES
20	0	20	40 KILOMETERS

POPULATION

Bardstown	0 – 5,000
Williamsburg	5,000 – 25,000
Covington	25,000 – 55,000
LEXINGTON	Above 200,000

★ State Capitol
⊚ County Seat
● Town or City

——————— Limited Access Highway
——————— Major Highway
——————— Other Highway
—+—+—+— Railway
— — — — State Boundary
——————— County Boundary
——————— Time Zone

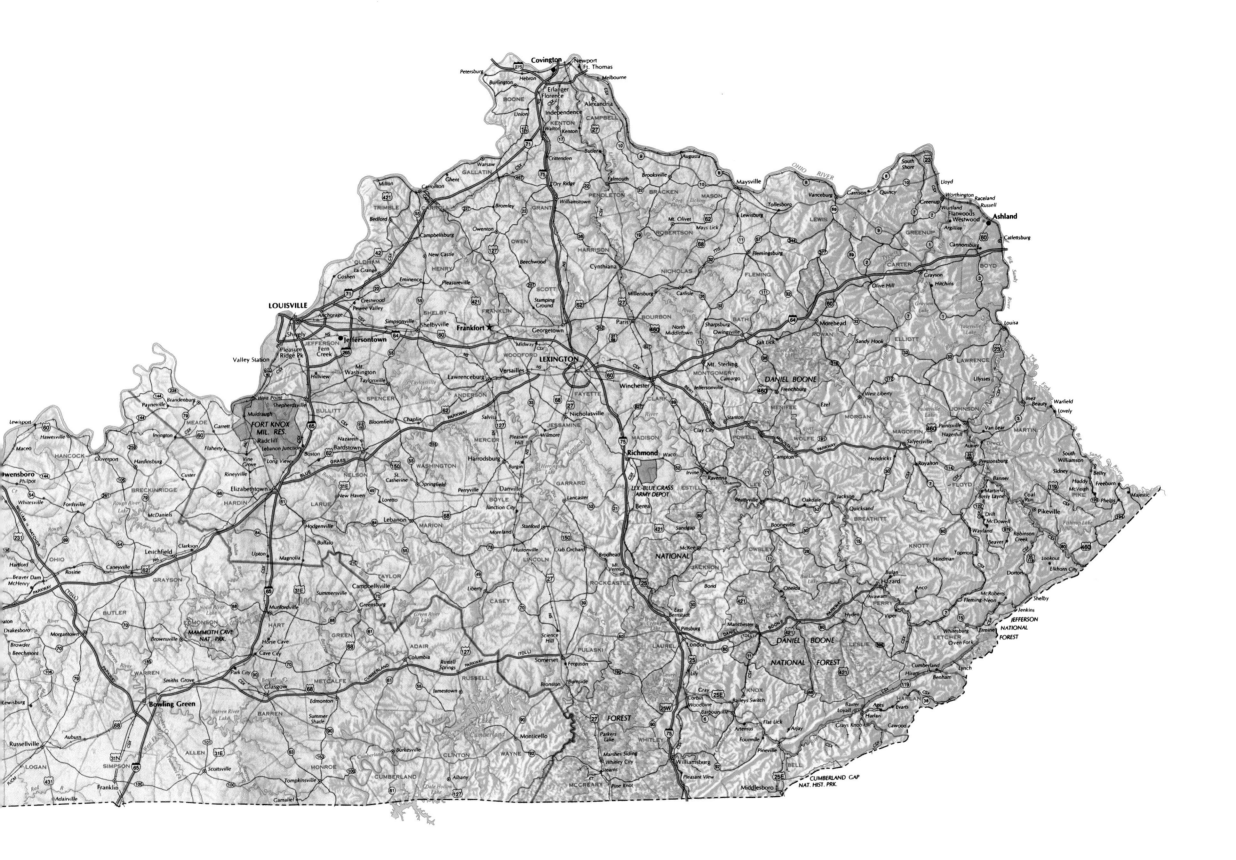

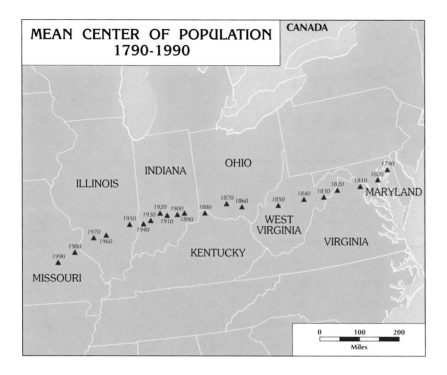

MEAN CENTER OF POPULATION
1790-1990

Miles
0   100   200

Kentucky's humid climate, landform and drainage patterns, and location mean that flooding, tornadoes, and earthquakes are the most significant environmental hazards in the state. The potential for disaster from these hazards varies across the state. Historically, flooding has been most serious in the Eastern Kentucky Coal Field Region, tornadoes in the central portion of the state, and earthquake activity in the west because of the nearby New Madrid fault. Flooding is of course brought on by the sometimes heavy precipitation associated with the humid climate in a hilly region like eastern Kentucky. The deforestation and strip mining that have characterized economic activity in the region, however, have also promoted floods.

Kentuckians, like people across the nation, are concerned about the environment. A 1994 statewide poll conducted by the University of Kentucky revealed that 91 percent of respondents expressed medium to high concern about environmental issues in their communities. The growing population in some places, the expansion of urban and agricultural lands, continued mining, and increasing industrialization have all added to these concerns.

The first settlers of European ancestry probably entered Kentucky in the early eighteenth century, and in 1774 Kentucky's first permanent settlement, Harrodsburg, was established. In the spring of 1775 Daniel Boone and other pioneers established

Kentucky's second settlement at Boonesborough, near the banks of the Kentucky River, about fifteen miles southeast of present-day downtown Lexington. Kentucky joined the nation as its fifteenth state in 1792, and Frankfort became the capital. By the early nineteenth century, the Wilderness Road from the Cumberland Gap, the Limestone Road from Maysville (formerly Limestone), and the Midland Trail, which linked Ashland (formerly Poage Settlement) and Louisville, all passed through or terminated in Lexington. Kentucky's Bluegrass Region became a prime agricultural destination for early settlers.

In 1820, Lexington, at the heart of the Bluegrass, was still the largest settlement in Kentucky, twice as large as Louisville; indeed, even Frankfort was larger than Louisville in these early years. As the home of Transylvania University, the oldest college west of the Alleghenies, and other cultural and educational influences, Lexington laid claim to being the "Athens of the West." Other activities were centered on agriculture: wealthy estate owners raised high-quality livestock, including Thoroughbred horses, and in 1828 the first track for horse racing in Kentucky was built in Lexington.

By the end of the 1820s, however, the river port of Louisville became the premier city in the state, and it has remained so since then. Louisville's emergence as the state's largest city can

be attributed to its site and situation. The location of the Falls of the Ohio and the break in navigation caused by this physical feature, together with the invention of the steamboat, shifted the focus of urban life in Kentucky to Louisville. Before the steamboat, captains of keelboats needed three to four months to bring their cargo upstream from New Orleans to Louisville; the steamboat made the same trip in only one week. The first steamboat from New Orleans arrived at Louisville in 1815, and by 1830 the city boasted a population of more than ten thousand citizens, which made it the state's largest urban place. In 1830 Lexington's population numbered just over six thousand, and Frankfort's was less than seventeen hundred.

The mean center of the population of the United States has shifted westward since the first national census of 1790; in 1880 the mean center was located in extreme northern Kentucky, near Covington. In 1990 the state's population numbered 3,685,296, placing it twenty-third among the states, between Alabama and Arizona. By 1993 the state ranked twenty-fourth in population. It is projected that by the year 2020 Kentucky will have 4.5 million persons, but its ranking will have dropped to twenty-seventh. Kentucky experienced net out-migration during the intercensal years of 1980–90. The state's population increased less than 1 percent during the decade, compared with a national average increase of almost 10 percent. Lack of economic opportunities in the state relative to some other states in the 1980s contributed to Kentucky's slow growth, as people sought jobs elsewhere, most notably in the Sun Belt.

Kentucky ranks twenty-second in population density, with ninety-three persons per square mile. This density is higher than one would expect, given that the eastern third of Kentucky is rugged and less than half of the state's population resides in counties that are part of the U.S. Census Bureau-defined Metropolitan Areas (48.5 percent in 1992). The national average is nearly 80 percent. Kentucky contains twenty-two counties that are part of seven such Metropolitan Areas (MAs) as defined in mid-1993 (see Chapter 12). The largest such areas within the state are Louisville, Lexington, and northern Kentucky. The Kentucky portion of the Louisville Metropolitan Statistical Area (MSA)—four of its seven counties are in Indiana—had a population of 745,953 in 1990. The Lexington MSA, comprising seven counties, contained 405,936 people in 1990; and the Kentucky part of the Cincinnati Primary Metropolitan Statistical Area (PMSA)—six of the twelve counties are in Kentucky—had a 1990 population of 316,652, or only one-fifth of the more than 1.5 million who lived in the Cincinnati PMSA.

STATE POPULATIONS, 1990

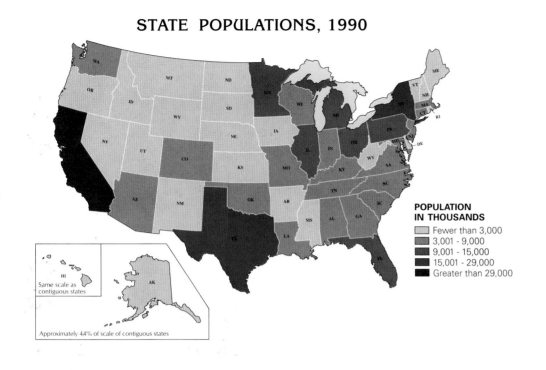

POPULATION IN THOUSANDS
- Fewer than 3,000
- 3,001 - 9,000
- 9,001 - 15,000
- 15,001 - 29,000
- Greater than 29,000

Same scale as contiguous states

Approximately 44% of scale of contiguous states

POPULATION CHANGE, 1980-1990

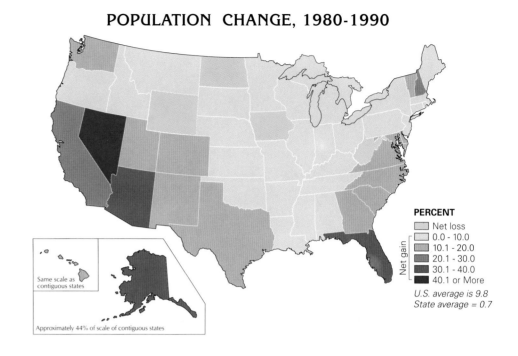

PERCENT
- Net loss
- 0.0 - 10.0
- 10.1 - 20.0
- 20.1 - 30.0
- 30.1 - 40.0
- 40.1 or More

U.S. average is 9.8
State average = 0.7

Net loss / Net gain

Same scale as contiguous states

Approximately 44% of scale of contiguous states

POPULATION DENSITY, 1990

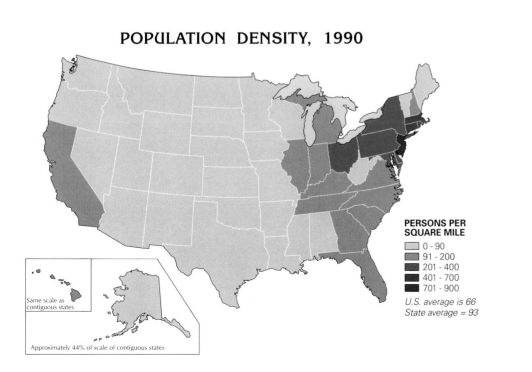

PERSONS PER SQUARE MILE
- 0 - 90
- 91 - 200
- 201 - 400
- 401 - 700
- 701 - 900

U.S. average is 66
State average = 93

Same scale as contiguous states

Approximately 44% of scale of contiguous states

POPULATION LIVING IN METROPOLITAN AREAS, 1992

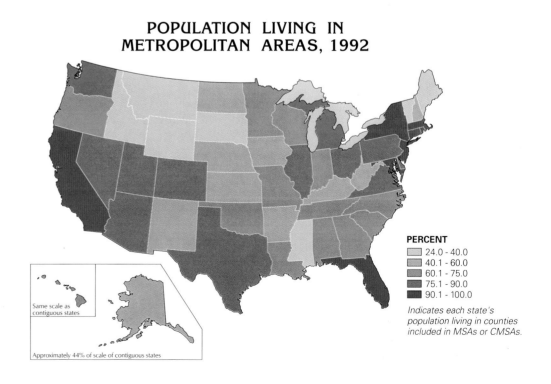

PERCENT
- 24.0 - 40.0
- 40.1 - 60.0
- 60.1 - 75.0
- 75.1 - 90.0
- 90.1 - 100.0

Indicates each state's population living in counties included in MSAs or CMSAs.

Same scale as contiguous states

Approximately 44% of scale of contiguous states

COUNTY SEATS

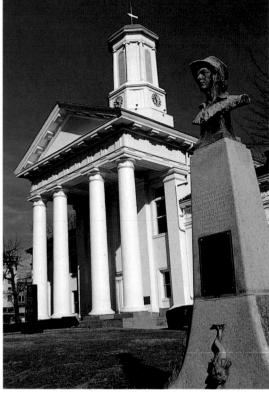

The Madison County
Courthouse, Richmond.

Six other Kentucky counties are included in MSAs that are partly or wholly in Kentucky: the Huntington-Ashland MSA includes three Kentucky counties (with 112,232 people in 1990); the Owensboro MSA comprises Daviess County (68,941); and the Evansville-Henderson MSA includes Henderson County (43,044).

If cities are ranked according to the population within their incorporated limits, then the state's seven largest cities in 1990 were Louisville (269,063), Lexington-Fayette Urban County (225,366), Owensboro (53,549), Covington (43,264), Bowling Green (40,641), Hopkinsville (29,809), and Paducah (27,256). Frankfort, one of the nation's smallest state capitals, ranked eighth among Kentucky's incorporated cities, with a population of 25,968. Frankfort would not have emerged as such a significant place except for its early selection as the state capital, its location on the Kentucky River, and its situation between Louisville and Lexington. Its role as state capital is crucial to the city's economy today: in 1990, Franklin County, where Frankfort is located, ranked second among the nation's 3,141 counties in the concentration of people employed in state and local government per ten thousand residents. For every ten thousand people living in Franklin County, more than thirty-three hundred employees worked in state and local government.

Kentucky is well known for its large number of counties,

most of which were created during the nineteenth century. Today the state has 120 counties. Only Texas and Georgia have more. Some people criticize the large number of counties in Kentucky, saying that the numerous political subdivisions necessarily compromise educational opportunities and governmental services. Nonetheless, loyalty to the home county remains a significant feature of the life of Kentuckians. Writing in *The Kentucky Encyclopedia,* Robert M. Ireland says that "because of their small size" the counties "are like hometowns to many."

In addition to counties and numerous other regions such as travel and tourism regions, congressional districts, and MSAs (see Chapters 10, 11, and 12, respectively), the state is subdivided into fifteen administrative regions known as Area Development Districts (ADDs). These regions comprise counties that have been grouped together based on their similarity in terms of social, economic, and geographic characteristics. ADDs sometimes cross state boundaries as is exhibited by Clark and Floyd Counties, Indiana (not to be confused with the two in Kentucky with the same names). Both have recently joined the KIPDA (Kentucky-Indiana Planning Development Agency) ADD because their characteristics are similar to those of other counties in the Louisville metropolitan region. ADDs emerged from federal government initiatives including the Economic Development Administration

and the Appalachian Regional Commission in the mid-1960s. Kentucky was the first state to establish a statewide system of regional organizations when Governor Edward Breathitt signed an Executive Order dividing the state into fifteen multi-county regions for planning purposes; four years later, in 1971, Governor Louie Nunn signed the Executive Order that established the ADDs as the official regional planning and development agencies for their respective areas. In 1972, the Kentucky General Assembly enacted legislation that set up the ADDs as public agencies and provided for their basic organizational structure; however, the ADDs are not state agencies nor are they another level of government regulation. Locally elected officials and citizen members comprise each ADD board of directors and include county judge-executives, mayors, and private citizens. The ADD staffs are made up of professionals with a wide range of backgrounds in such areas as economic development, human services, management, and planning. The idea is to share the technical and management expertise found on the ADD boards and staff so that local governments are collectively able to provide for the planned growth of each multi-county area.

The income and educational levels in Kentucky are among the lowest in the nation. Whereas recent years have witnessed improvements, Kentucky still does not fare well relative to other states on many common measures of socioeconomic well-being, and many of the state's eastern counties remain among the most impoverished in the nation. For example, with median household income calculated on a three-year average for 1991–93, only three states—Arkansas, West Virginia, and Mississippi—ranked below Kentucky. The national average was $31,585; Kentucky's was $24,563. According to the *County and City Data Book* for 1994, seven of the nation's twenty-five lowest-ranking counties in personal income were in Kentucky (six of them in eastern Kentucky). Again using a three-year average for 1991–93, only four states—New Mexico, West Virginia, Louisiana, and Mississippi—ranked below Kentucky in the percentage of the population living below the poverty level. The official poverty threshold for a family of four in 1993 was $14,654, and 19.6 percent of Kentucky households had incomes less than that sum.

On the other hand, Kentucky's cost of living is also below the national average. The cost of living in Louisville and Lexington is among the highest in the state, and yet in 1992 the cost of living in these two cities was only 92 percent and 98 percent of the national average, respectively. In general, many consider Kentucky to be an attractive place in which to live. Kentucky is known for its moderate climate, and the state has one of the lowest rates of serious crime (violent crime plus burglary, larceny, theft, and arson) in the nation. Only four states—New Hampshire, South Dakota, North Dakota, and West Virginia—have serious crime rates lower than that of Kentucky.

AREA DEVELOPMENT DISTRICTS

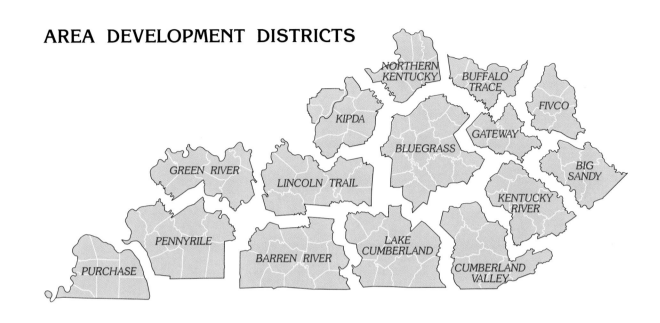

POPULATION BELOW POVERTY LEVEL, 1989

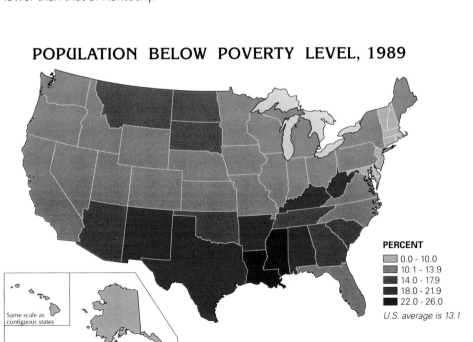

PERCENT
- 0.0 - 10.0
- 10.1 - 13.9
- 14.0 - 17.9
- 18.0 - 21.9
- 22.0 - 26.0

U.S. average is 13.1

POPULATION BELOW POVERTY LEVEL
(KENTUCKY AND ADJACENT STATES)

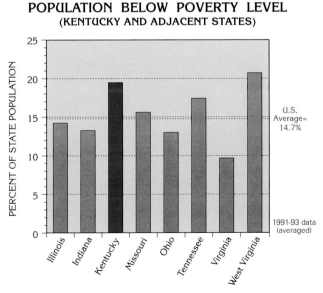

U.S. Average= 14.7%

1991-93 data (averaged)

MEDIAN HOUSEHOLD INCOME
(KENTUCKY AND ADJACENT STATES)

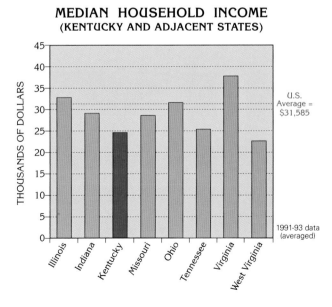

U.S. Average = $31,585

1991-93 data (averaged)

COMPARATIVE COST OF LIVING, 1992

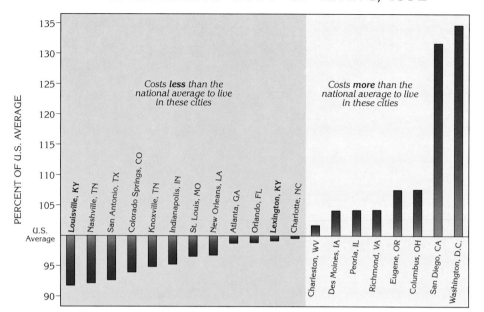

PERCENT OF U.S. AVERAGE

Costs *less* than the national average to live in these cities

Costs *more* than the national average to live in these cities

U.S. Average

Louisville, KY
Nashville, TN
San Antonio, TX
Colorado Springs, CO
Knoxville, TN
Indianapolis, IN
St. Louis, MO
New Orleans, LA
Atlanta, GA
Orlando, FL
Lexington, KY
Charlotte, NC

Charleston, WV
Des Moines, IA
Peoria, IL
Richmond, VA
Eugene, OR
Columbus, OH
San Diego, CA
Washington, D.C.

SERIOUS CRIMES
(KENTUCKY AND ADJACENT STATES)

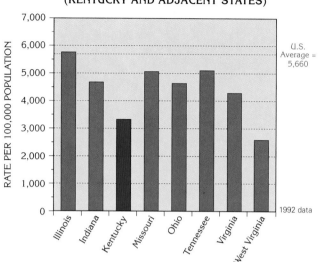

RATE PER 100,000 POPULATION

U.S. Average = 5,660

Illinois
Indiana
Kentucky
Missouri
Ohio
Tennessee
Virginia
West Virginia

1992 data

STATE AND LOCAL TAXES PAID
(FAMILY OF FOUR IN SELECTED CITIES)

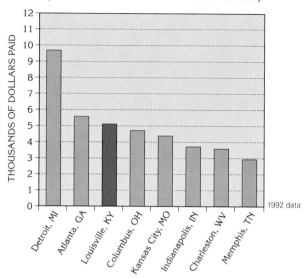

THOUSANDS OF DOLLARS PAID

Detroit, MI
Atlanta, GA
Louisville, KY
Columbus, OH
Kansas City, MO
Indianapolis, IN
Charleston, WV
Memphis, TN

1992 data

GROSS STATE PRODUCT FOR MAJOR INDUSTRIES
1963-1990

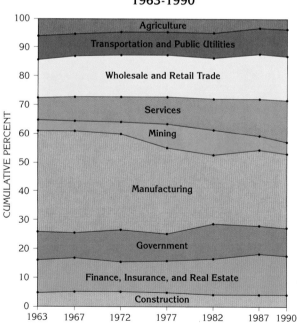

CUMULATIVE PERCENT

Agriculture
Transportation and Public Utilities
Wholesale and Retail Trade
Services
Mining
Manufacturing
Government
Finance, Insurance, and Real Estate
Construction

1963 1967 1972 1977 1982 1987 1990

A new subdivision in Woodford County.

In 1993, *Places Rated Almanac* ranked Louisville the tenth-best metropolitan area in which to live among 343 metropolitan areas in the United States and Canada. The ranking was based on a composite score derived from evaluating such characteristics as education, health care, crime, climate, recreational opportunities, housing, transportation, and the arts. The Cincinnati–Northern Kentucky metropolitan area was ranked first among all such areas, and the Lexington metropolitan area was ranked thirty-sixth.

Kentucky has long been known for its farm crops and livestock, including tobacco, corn, horses, and beef cattle, and for its coal, produced from strip mines and underground mines in the Eastern and Western Kentucky Coal Field Regions. Indeed, these commodities are still important to the state's economy and employment, but their significance is decreasing relative to other commodities and services. As measured by share of Gross Domestic Product, for example, manufacturing is far more important to the state's economy, constituting one-quarter of the Gross State Product in 1990. The importance of manufacturing, however, as well as that of agriculture and mining, has declined in recent years, at least as measured by the Gross State Product. While new manufacturing plants such as Toyota and scores of automobile-related production industries continue to be attracted to the Commonwealth, the service, financial, and wholesale and retail trade sectors are becoming the state's more important economic activities. Like most of the nation, Kentucky is moving away from an economy that was first dominated by primary activities—agriculture, forestry, and mining—and then by secondary activities (manufacturing), to one that will be dominated by tertiary (service-related) and quaternary (information) economic activities and jobs. The state's likely future will involve tertiary and quaternary activities, and that is the type of economy for which the state's residents must prepare.

The Natural Environment

CHAPTER TWO

The Natural Environment

Valley farming in the Knobs Region south of Louisville.

THE LIVES AND LIVELIHOODS of Kentuckians, and in fact all of the world's populations, are closely tied to the earth's natural environment. Many cities, including Ashland, Covington, Louisville, Paducah, and Frankfort, were located to take advantage of the water and transportation potential of major streams and rivers. Kentucky's great coalfields are a consequence of prehistoric carbon-rich swamps in the east and west, and the famous horse industry is purported to have emerged in part because of ancient limestone deposits that are now exposed in the Bluegrass and have decomposed into rich soils. Agriculture is found along both valley bottoms where weathered and eroded particles of rock and organic debris collect and deepen, and in more rolling, upland areas such as the Bluegrass and Pennyroyal Regions. Forestry is supported by a climate favorable to a mix of hardwood species that are in demand for home interior finishing and furniture building. Kentucky's biodiversity reflects its topographic conditions and its weather and climate, as well as its central continental location. We must also look to the natural environment to understand the peril to Kentuckians from flooding, landslides, earthquakes, ice storms, tornadoes, and both surface and groundwater pollution.

The physical landscapes now visible in the state, as well as the underground resources, are the evolving product of many forces operating over a variety of time and space scales. Hundreds of millions of years ago, powerful forces at the continental and global scale gradually began to establish the bedrock foundations of Kentucky, from the types and depths of rock layers to the warping and dipping of certain rock layers in specific locations. Both physical and chemical weathering and erosional forces at work over the past several million years and still at work today have continuously sculpted and reshaped the bedrock foundations into rolling hills, sinuous cave systems, pock-marked sinkhole plains, and majestic palisades. These forces tend to be local or regional in scale and

GENERALIZED GEOLOGY

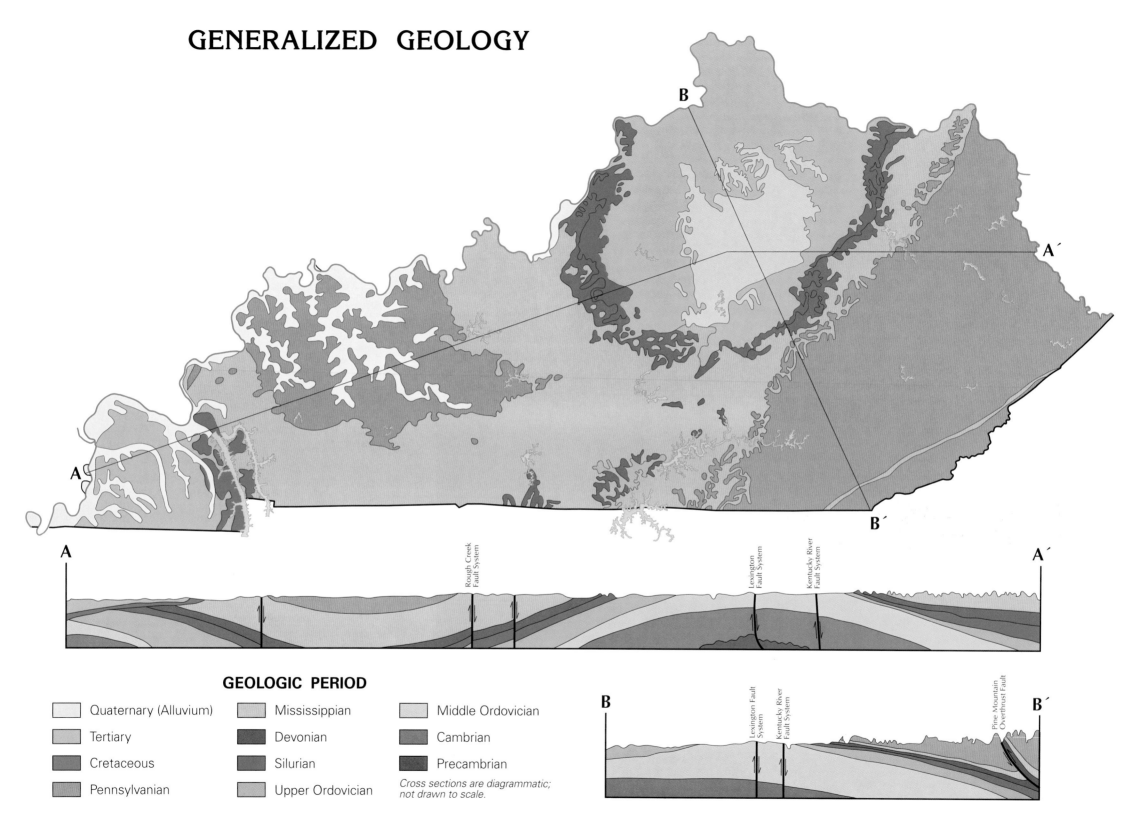

GEOLOGIC PERIOD

- Quaternary (Alluvium)
- Tertiary
- Cretaceous
- Pennsylvanian
- Mississippian
- Devonian
- Silurian
- Upper Ordovician
- Middle Ordovician
- Cambrian
- Precambrian

Cross sections are diagrammatic; not drawn to scale.

Rough Creek Fault System

Lexington Fault System

Kentucky River Fault System

Pine Mountain Overthrust Fault

GEOLOGIC TIMETABLE

ERA	ROCKS EXPOSED IN KENTUCKY	PERIOD (DURATION IN MILLIONS OF YEARS)		AGE (MILLIONS OF YEARS)
CENOZOIC		QUATERNARY	1.6	1.6
		TERTIARY	64.4	66
MESOZOIC		CRETACEOUS	78	144
		JURASSIC	64	208
		TRIASSIC	37	245
PALEOZOIC		PERMIAN	41	286
		PENNSYLVANIAN	34	320
		MISSISSIPPIAN	40	360
		DEVONIAN	48	408
		SILURIAN	30	438
		ORDOVICIAN	67	505
		CAMBRIAN	65	570
PRECAMBRIAN		PRECAMBRIAN	4000	

are strongly linked to the atmospheric variables of temperature and precipitation. Living organisms, both plant and animal, have become established on this dynamic physical landscape as part of an ongoing search for life-supporting environments, and these organisms also exert a force capable of changing the landscape. Humans in particular have modified the physical earth in a short period by cutting trees, tilling soils or covering them with impervious materials, channeling streams and rivers, building reservoirs, laying railroads and highways, removing entire hills, and building new hills from debris.

The natural environment is indeed complex, and investigations of how it links with the human environment (including demographic, socioeconomic, political, and recreational patterns and processes) can increase the complexity by several orders of magnitude. To promote a basic understanding of this complexity, we present in this chapter a general survey of Kentucky's more important physical elements, which are ordered roughly according to the broad classes of forces introduced: *geologic* (evolution and structure); *erosional* (physiographic regions, weather and climate); and *biotic* (flora and fauna, human/environmental hazards).

Geologic Foundations

Traversing the state on an interstate highway provides significant clues to Kentucky's geologic past. Wherever the highway cuts through a hill or ridge or bridges a deep river gorge, one can see exposed rock arranged in distinct layers, or strata. The layering is one indication that bedrock in Kentucky is sedimentary in nature, having been formed in one or more of the following ways: from the deposition and solidification of small mineral and rock fragments (often derived from sandstones and shales), from minerals that once were dissolved in water (limestones), or from organic materials (coal). It is possible in several areas of the state (for example, at Louisville's Falls of the Ohio) to find fossilized remains of marine animals embedded in limestone sedimentary rock. Most strata are oriented horizontally, or nearly so, although a trip through any number of areas in Kentucky may reveal layers that are tilted, in some cases with an almost vertical orientation. The presence of sedimentary bedrock suggests that the land now at or near the surface must have been under water at some time during the geologic past. Furthermore, the tilting of rock layers in some areas of the state provides evidence that strong internal forces slowly raised, lowered, and warped the earth's crust.

Geologists now know that most of the state's rocks were formed during the Paleozoic Era, which dates from about 570 million to about 245 million years ago. During the early portion of this era, Kentucky was completely submerged under varying depths of water. Concurrently, deposition of material occurred in the shallow inland seas or along the continental margins of the region that was to become Kentucky. The oldest exposed rock formations in the state are from the Ordovician Period and date back to about 500 million years ago; they originated from the deposition, compaction, and hardening into rock of the deposited materials. This exposed Ordovician rock covers an area that extends from Cincinnati south to Berea and from Louisville east

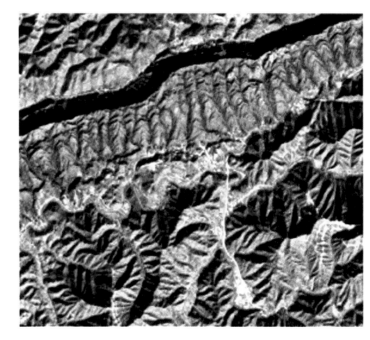

Pine Mountain, as shown in a satellite image: The linear feature trending east northeast–west southwest in the upper portion of the image represents part of a major geologic feature, the Pine Mountain Thrust Fault. The low sun altitude in the winter (December 26, 1991) emphasizes the relief in the area as a result of extended shadowing. Most of the land cover is deciduous forest (grey tones); coniferous forest is green. Harlan, Kentucky, is in the center of the image, in the valley of Martin's Fork. The Landsat TM rendition is natural color, bands 3,2,1.

to Mt. Sterling and Flemingsburg; it is composed primarily of limestones and some shales, and today defines the physiographic region known as the Bluegrass. In fact, rocks of Ordovician age underlie virtually all of the state.

Interestingly, during part of the Paleozoic Era, Kentucky was located *south* of the equator in tropical latitudes, a consequence of "continental drift" and the location of large crustal plates during that era. During the second half of the Paleozoic, water levels over the area fluctuated substantially, and portions of the state began to emerge from the shallow seas. As the Ordovician Period ended, tectonic forces began to bend or warp existing layers of sedimentary rock. Up-warping took place primarily during the Silurian and Devonian Periods and gave rise to the Cincinnati Arch, a long, gentle, ridgelike feature with a crest that runs between Cincinnati and Nashville. This arch effectively divided the state

MAJOR TECTONIC FEATURES

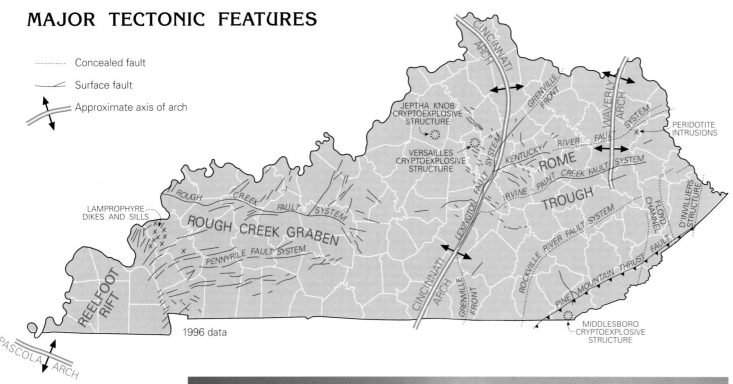

.... Concealed fault

⌒⌒ Surface fault

⤒ Approximate axis of arch

LAMPROPHYRE
DIKES AND SILLS

ROUGH CREEK FAULT SYSTEM

ROUGH CREEK GRABEN

PENNYRILE FAULT SYSTEM

REELFOOT RIFT

PASCOLA ARCH

1996 data

CINCINNATI ARCH

JEPTHA KNOB
CRYPTOEXPLOSIVE
STRUCTURE

VERSAILLES
CRYPTOEXPLOSIVE
STRUCTURE

GRENVILLE FRONT

WAVERLY ARCH

PERIDOTITE
INTRUSIONS

KENTUCKY RIVER FAULT SYSTEM

LEXINGTON FAULT SYSTEM

IRVINE PAINT CREEK FAULT SYSTEM

ROME TROUGH

ROCKVILLE RIVER FAULT SYSTEM

FLOYD CHANNEL

D'INVILLIERS STRUCTURE

CINCINNATI ARCH

GRENVILLE FRONT

PINE MOUNTAIN THRUST FAULT

MIDDLESBORO
CRYPTOEXPLOSIVE
STRUCTURE

Cumberland Mountain near Middlesboro.

into two basins, one on either side of the arch. Within these basins, additional limestone beds continued to form.

Still later in the Paleozoic Era, as climatic conditions continued to favor a less variable environment, additional limestone beds, sometimes interspersed with shales, were formed that are up to several hundred feet thick in places. A warm and humid tropical environment supported vast populations of organisms with calcium carbonate shells. The remains of dead organisms settled and accumulated on the sea floor and were mixed with the chemical precipitation of calcium carbonate to form beds of limestone. In these limestone beds, originally deposited some 350 million years ago during the Mississippian Period, were later formed most of Kentucky's myriad cave systems and sinkhole plains. Additionally, and more significantly for Kentuckians today, the warm tropical conditions of the Pennsylvanian Period (320 million years ago) permitted abundant vegetation to flourish in expansive forests and swamps. Sands and silts, eroded from surrounding uplands, buried the Pennsylvanian vegetation in a series of cycles that produced a sandwich of sedimentary rock layers. Compressed under the weight of the sand and silt sediments, and in the absence of oxygen, the vegetative matter slowly transformed into coal, forming the beds for which the Eastern and Western Kentucky Coal Fields are named.

The end of the Paleozoic Era (about 245 million years ago) was marked by a second major tectonic phenomenon, called the Appalachian Orogeny, that would have further effects on Kentucky's landscape. Horizontal compression of the North American crustal plate helped to create the Appalachian Mountains, which extend from Canada's Maritime provinces through New England to Mississippi. One important result of this mountain-building period for Kentucky was the redirection of surface water and sediment flow westward away from the mountain core toward what is now eastern Kentucky. This period of mountain building increased the elevation of eastern Kentucky, forming a plateau that has evolved into a distinctive local pattern of relief as a consequence of rapid downcutting of streams through the surface rocks. What has evolved is a characteristic hill-valley-hollow landscape. Additionally, forces along the Pine Mountain fault in extreme southeastern Kentucky forced the rock strata to move horizontally and slightly upward toward the northwest, producing a huge overthrust sheet demarcated by Cumberland Mountain on the east and Pine Mountain on the west.

PHYSIOGRAPHIC REGIONS

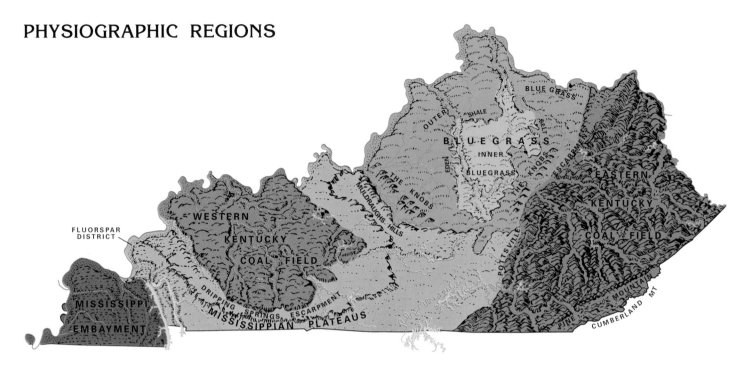

The boundary between the Knobs and Outer Bluegrass Regions, Hardin County.

Physiography

Kentucky's land surface, formed by tectonic, erosional, and depositional actions operating over hundreds of million of years, presents a unique physiography. Careful examination of the accompanying maps demonstrates that subsurface geology, landforms or physiography, and soils are closely interrelated. The composition and distribution of sedimentary rock strata determine, in part, the rate at which degradational and erosional forces can proceed. In general, sandstones and conglomerates are more resistant to weathering and erosion, while limestones and shales can be broken down and removed at faster rates. Consequently, resistant sandstones cap higher topography whereas the limestones and shales weather more rapidly to yield surfaces of lower elevation.

Kentucky's topographic surface, if observed in cross-section from the Mississippi River in the west to the Appalachian Mountains in the east, is wedge-shaped. The thin western edge is approximately 400 feet above sea level and the thick eastern upland exceeds 4,000 feet in elevation at Black Mountain. Surface streams flow generally toward the west and north down the surface of the wedge, emptying primarily into the Ohio River. No major streams flow eastward. Stream erosion and other weathering and erosional processes have further sculpted the physiographic regions that were fundamentally defined by the underlying bedrock and sediments.

The Mississippi Embayment Region (Jackson Purchase) in extreme western Kentucky is a low-lying area formed when a tongue of the ancestral Gulf of Mexico extended north along what is now the Mississippi Valley, covering the area with shallow waters. Since that time some 100 million years ago, streams have deposited sand, gravel, and silt in layers across the surface; more recently winds have deposited loess in layers of varying thickness. Silty soils have subsequently developed on this surface although the sand and clay content of these soils increases in the immediate vicinity of the Mississippi River.

The Bluegrass Region is a broad upland area in north central Kentucky that developed on the weathered Ordovician limestone and shale associated with the Cincinnati Arch. The topographic surface varies from very gently rolling in the Inner Bluegrass Region to more pronounced undulations in the Outer Bluegrass. The Inner Bluegrass limestones are especially rich in calcium and phosphate. Soils here are deep and well-drained brownish silt loams overlying silty clays. As one moves out from the Inner Bluegrass, clay soils predominate and soil fertility decreases. The fertile Inner Bluegrass soils, especially those high in phosphate, are especially suited to production of tobacco, hay, and alfalfa. The area's rich pasturelands provide a principal resource necessary for the development of the nation's largest concentration of horse farms, surrounding Lexington.

An area known as the Knobs, which consists of a series of low but well-defined hills, acts as a transition zone of the Bluegrass Region on the west, south, and east. These remnant landforms (originally part of the Muldraughs and Pottsville Escarpments) have a distinctive structure—shale overlain by a more resistant caprock. The blocky caprock erodes slowly, thus protecting the underlying rock and slowing the rate of weathering and erosion immediately beneath. The resulting "knobs" are cone-shaped hills that form a narrow band of rugged topography around the Bluegrass Region. The Knobs area is broadest south of Louisville, where it resembles a staggered line of pyramids cut free from the Muldraughs Escarpment in the drainage basins of the Salt and Rolling Fork Rivers.

The Mississippian Plateaus Region (Pennyroyal) extends from the Mississippi Embayment in the west to the Bluegrass

GENERAL SOILS

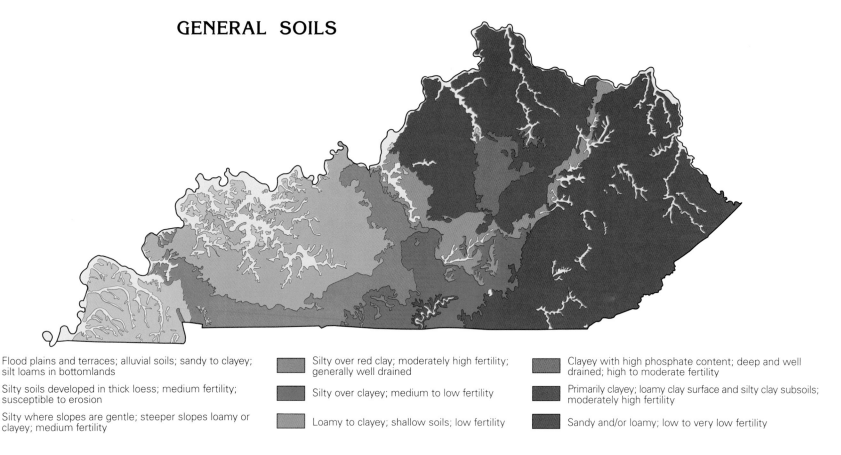

- Flood plains and terraces; alluvial soils; sandy to clayey; silt loams in bottomlands
- Silty soils developed in thick loess; medium fertility; susceptible to erosion
- Silty where slopes are gentle; steeper slopes loamy or clayey; medium fertility
- Silty over red clay; moderately high fertility; generally well drained
- Silty over clayey; medium to low fertility
- Loamy to clayey; shallow soils; low fertility
- Clayey with high phosphate content; deep and well drained; high to moderate fertility
- Primarily clayey; loamy clay surface and silty clay subsoils; moderately high fertility
- Sandy and/or loamy; low to very low fertility

Harlan lies between Pine Mountain, in the background, and Cumberland Mountain. The city is linked to the north by the Daniel Boone Parkway and south to Virginia by U.S. 421, one of the region's major highways.

and Eastern Kentucky Coal Field Regions and surrounds much of the Western Kentucky Coal Field. It is a physiographically diverse region. The Mammoth Cave area, for example, borders the Western Kentucky Coal Field and is bounded on the outer edge by the Dripping Springs Escarpment. This area's soils are primarily derived from sandstones, shales, and ancient loess deposits that cover the deep limestones in which the extensive Mammoth Cave system was formed. The balance of the Mississippian Plateaus Region extends outward from the Dripping Springs Escarpment to the Knobs and the Pottsville Escarpment. Here soils are predominantly limestone-based; they are primarily silty over red clay and provide some of the state's most fertile soils. This area also contains abundant karst terrain with associated sinkholes, disappearing streams, and subsurface cave systems. Indeed, the southern portion of the Mississippian Plateaus is often highlighted in earth science textbooks as a classic example of karst processes and landscapes.

The Western and Eastern Kentucky Coal Fields are structurally very similar from a geologic standpoint, both having been formed as a consequence of the uplift of the Cincinnati Arch. Their respective topographies are quite different, however. The Western Kentucky Coal Field is lower in elevation and exhibits less relief; the Eastern Kentucky Coal Field, by contrast, is much higher in elevation and has been so deeply dissected by erosional forces (primarily running water) that the region has little level land. The Cumberland, Kentucky, and Licking Rivers and their tributaries have shaped the surface of the Eastern Kentucky Coal Field Region into a landscape of rugged topography that severely limits agricultural potential and forces highway engineers to construct roads along the narrow valley bottoms. Some of the state's most spectacular scenery—gorges, arches (found mainly within the Cumberland Escarpment bordering the Eastern Kentucky Coal Field), and Pine and Cumberland Mountains—can be found in this eastern Kentucky area.

AREAS WITH POTENTIAL FOR KARSTIC DEVELOPMENT

■ Karst or potentially karst

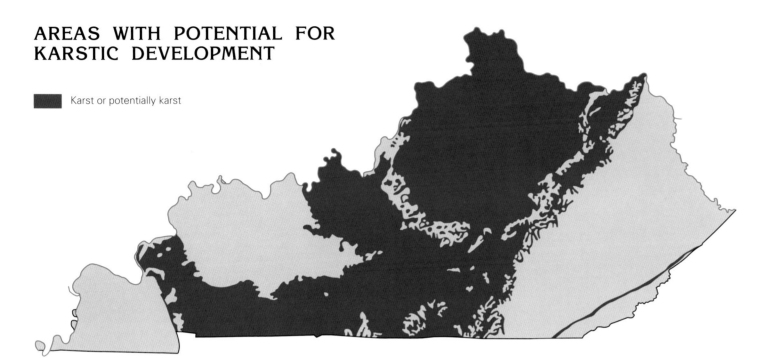

A karst sinkhole plain in Barren County.

Karst Topography and the Mammoth Cave System

The term *karst* (derived from a Slavic word) refers both to a distinctive landscape and to the processes that produce features such as sinkholes, caves, springs, and disappearing streams. Many of the unique features found in karst landscapes are formed not by the actions of surface erosional forces but by the dissolution of limestone bedrock through chemical reaction with groundwater. Approximately 55 percent of Kentucky is underlain by soluble rocks and about 25 percent of the state has intensive karst development.

The formation of a karst landscape occurs as a consequence of interactions among components of the gaseous, liquid, and solid portions of the environment. Rain falling through the atmosphere naturally absorbs carbon dioxide, producing a weak solution of carbonic acid. As the rainwater is absorbed and moves downward through the soil, it often collects additional carbon dioxide from decaying organic matter, which slightly increases the acidity. The groundwater in effect becomes a weak acidic solution capable of dissolving limestones (carbonate rocks) with which it comes in contact. Thus the underlying limestone strata are locally dissolved to form caves below the ground. As caves grow larger, collapse sinks occasionally result from the structural failure of cave roofs. Most sinks, however, are formed in the soil and broken rock layer overlying the bedrock as soil particles are washed underground through gradually enlarging cracks and fissures.

Premier among the state's karst features is the 350-mile-long Mammoth-Flint-Roppel Cave System—the longest cave system in the world. Mammoth Cave

KARST AREAS IN THE UNITED STATES

CAVE DISTRIBUTION

NUMBER OF CAVES

- · 1
- · 5
- · 10
- ● 25
- ● 50

1985 data

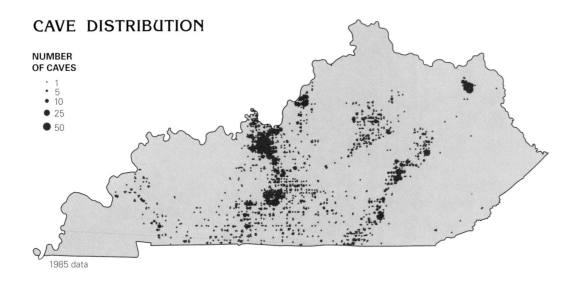

GENERALIZED KARST LANDSCAPE

Spring

Sinking Creek

Swallow Hole

Blind Valley

Subsidence Sinkholes

Collapse Sinkhole

Entrenched Stream

Water Table

Dry Cave

Water-filled Cave

Limestone

Limestone Palisades

Sandstone

Limestone

Shale

Breakdown

Sink Pond

The Bridal Altar (so named for the numerous wedding ceremonies performed there) consists of a cluster of five columns 9-10 feet high that form a tight semicircle within the Mammoth Cave system. It is nicely framed by the twin columns "Caesar" and "Pompeii."

National Park is now recognized by the United Nations Educational, Scientific and Cultural Organization (UNESCO) as a World Heritage Site and is also an International Biosphere Reserve.

Special geologic circumstances set the stage for Mammoth Cave's formation. As southeastern North America began to rise at the end of the Paleozoic Era, the thick limestone beds in the Mammoth Cave region were lifted up from beneath shallow seas. Ultimately the limestone strata dipped northwestward toward the Green River, which is the longest river within the state's boundaries and which drains much of the Mississippian Plateaus and Western Kentucky Coal Field Regions. As some of the abundant, weakly acidic precipitation was absorbed by the surface and entered the groundwater contained within the limestone, that water flowed slowly down-gradient toward the Green River (which had begun to incise itself into the surface), carrying with it dissolved limestone. Precipitation recharged the groundwater which, over millions of years, gradually dissolved more and more limestone. Bedding planes (the boundaries between different rock layers) and joints (cracks that develop as a result of stresses) within the limestone layers offered natural pathways for water to flow, and they gradually enlarged into cave passages. As the Green River cut deeper into the surface, gradually lowering the water table, numerous cave levels developed at successively lower elevations. The oldest, and now dry, cave passageways are nearest to the surface. Deeper passages are more recent and are still in the process of being formed.

MAMMOTH CAVE SYSTEM

Pond

Intermediate contour
(represents elevation change in 20' increments)

Depression contour
(represents elevation decline in 20' increments)

Index contour
(represents elevation change in 100' increments)

Cave passage

Intermittent stream

0   1/2 mile

Scale of main map

ENTIRE CAVE SYSTEM

AREA
SHOWN
IN
DETAIL

N

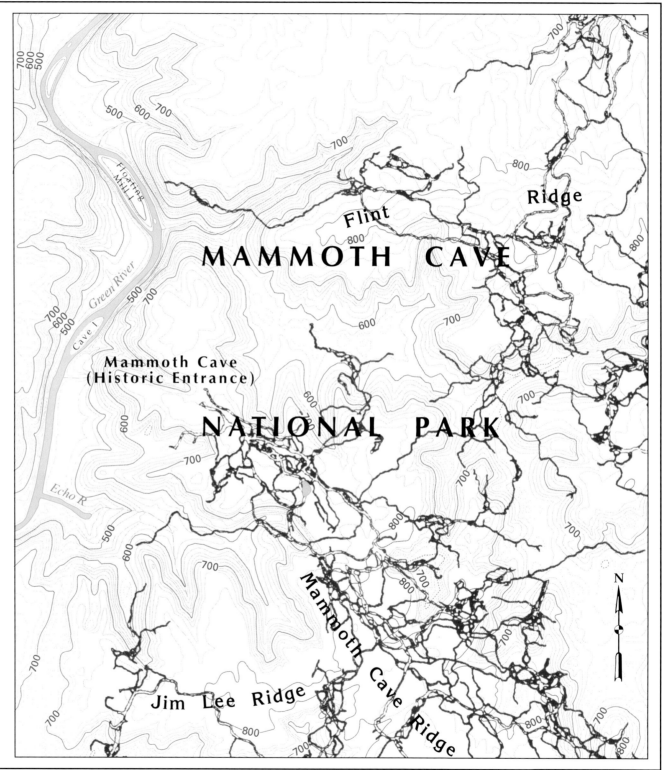

Flint   Ridge

MAMMOTH CAVE

Mammoth Cave
(Historic Entrance)

NATIONAL PARK

Green River

Cave I

Echo R

Floating Mill

Mammoth Cave Ridge

Jim Lee Ridge

N

DRAINAGE BASINS

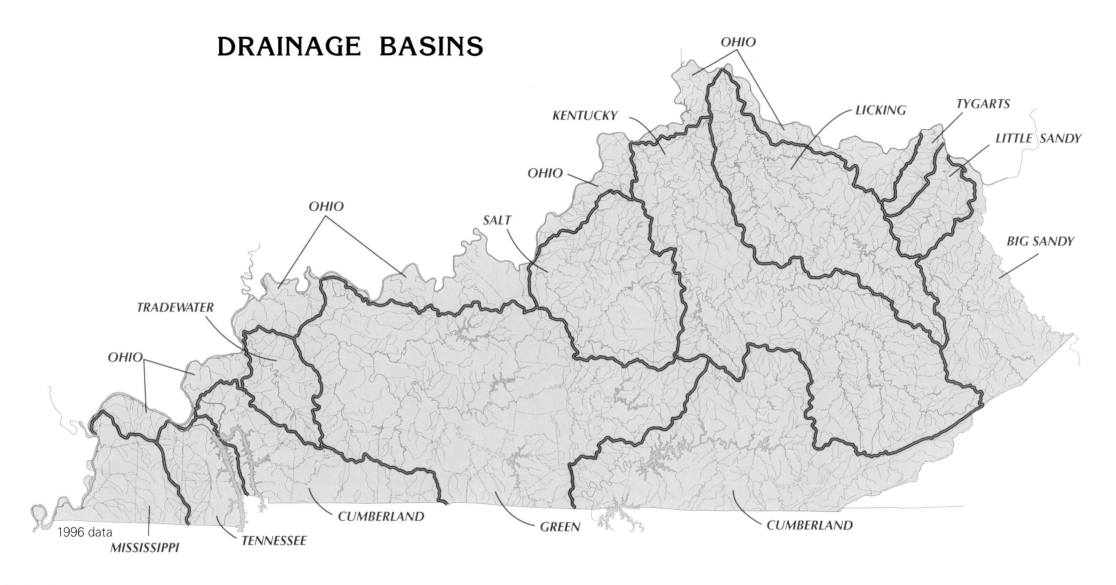

1996 data

Hydrology

Kentucky is blessed with abundant water in the form of precipitation, surface water (lakes and other impoundments, streams, and rivers), and groundwater. Located in a humid subtropical climatic zone, Kentucky receives annually about 46 inches of precipitation. This precipitation is distributed fairly evenly throughout the year, although a spring maximum and a fall minimum are evident across much of the state. The water that falls as precipitation can follow many paths. Some may infiltrate the soil, where a portion is used seasonally by plants; there it is incorporated and stored in plant cells or released back to the atmosphere through the transpiration process. Water may also penetrate downward to be-

come part of the groundwater supply. Alternatively, precipitation may run across the surface and become part of the state's extensive drainage network of streams and rivers. About 60 percent of all water that falls on the state as precipitation is ultimately evaporated or transpired by vegetation back into the atmosphere; the remaining 40 percent is eventually captured by streams or lakes or becomes part of groundwater storage. A full and accurate accounting of a region's water budget is very difficult to determine because of great variations that occur over both short periods of time and small distances. Despite these variations, a basic understanding of the water cycle and local water budget can pro-

mote better resource management, from planning for public water supply demands to individual household use.

Runoff from precipitation is effectively accomplished by twelve major river basins that contain nearly 90,000 miles of streams. Virtually all of these streams form part of the larger Ohio River basin. Of the 90,000 miles of streams in the state, approximately 700 miles have been designated as Outstanding Resource Waters, and 114 miles—primarily in southern and southeastern Kentucky—have been preserved as Wild Rivers under the Wild River Act of 1972.

Kentucky's surface water includes more than 2,700 natural

GROUNDWATER SENSITIVITY REGIONS

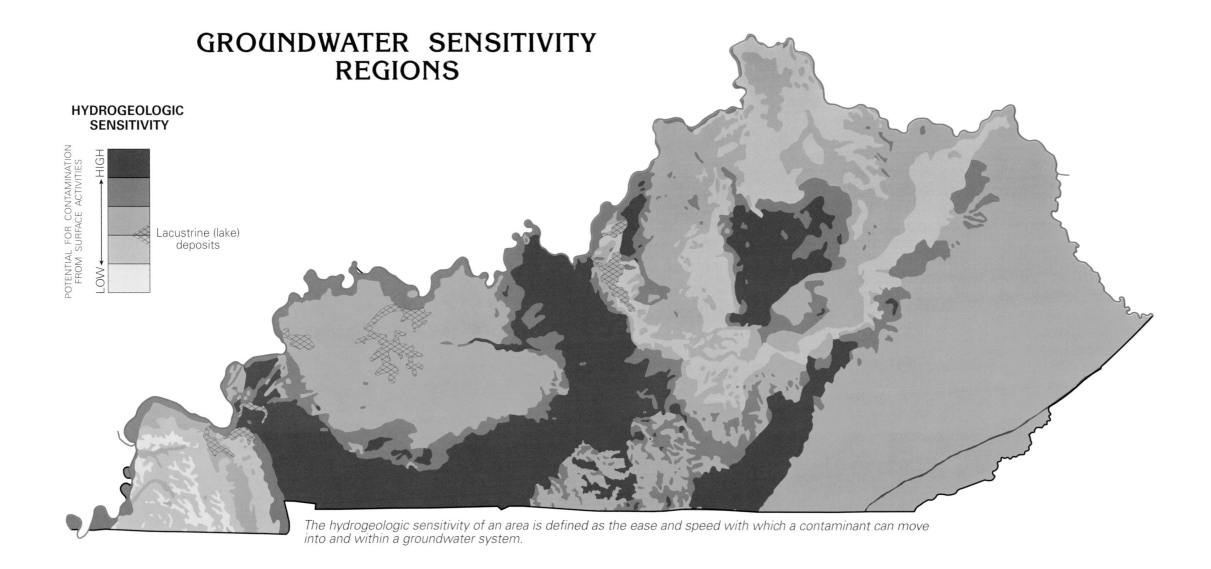

HYDROGEOLOGIC SENSITIVITY

POTENTIAL FOR CONTAMINATION FROM SURFACE ACTIVITIES

HIGH

LOW

Lacustrine (lake) deposits

The hydrogeologic sensitivity of an area is defined as the ease and speed with which a contaminant can move into and within a groundwater system.

and artificial impoundments, of which roughly one-third are larger than ten acres in size. The state's reservoir system, which comprises the vast majority of surface water impoundments, was planned as a way to increase the available surface storage of water flowing in the stream and river networks as well as to reduce the risk of flooding along these networks. Kentucky's reservoirs contain, on average, about three trillion gallons of water, with a storage minimum experienced in the autumn and a maximum usually recorded in the spring season. This average amount, if it could be perfectly allocated, would sustain all current users for two years at current rates of usage.

Wetlands constitute an additional class of surface water storage. Wetlands once comprised 1.5 million acres in the state, but today it is estimated that less than 300,000 acres remain. The value of wetlands in the natural environment has only recently been brought to the public's attention, and efforts are being made nationwide to protect remaining areas and increase their size and number. In the past wetlands were viewed as perfect locations for dumps or for filling and development. Their existence, however, is now known to be essential in maintaining ecosystem quality; they serve as valuable wildlife habitat and low impact recreation areas, and they play an important role in nutri-

ent cycling, flood mitigation, erosion control, and natural water filtration. Although there is at least a small area of wetland in nearly every Kentucky county, the vast majority of the state's wetlands are located in its western portion, in the low-lying flatlands that border the Mississippi and lower Ohio Rivers in the Mississippi Embayment Region. This area, which includes the Reelfoot National Wildlife Refuge (located predominantly in Tennessee) and the proposed Clarks River National Wildlife Refuge, is a major migratory stopover or nesting area for a number of species of waterfowl, including snow geese, Canada geese, and several species of ducks, herons, and egrets (Aldrich, 1996). It is critical to maintain and improve the state's wetlands for several reasons, especially since aquatic ecosystems provide habitat for about 55 percent of all species listed by federal and state government agencies as threatened, endangered, or of special concern.

Contamination of Kentucky's rivers, streams, reservoirs, and lakes continues to be a problem, although improvement has occurred over the last several years. Of the total stream miles in the state, about 20 percent are monitored on a regular basis to determine whether they meet standards designed to protect aquatic life, public water supply uses, and swimming; nearly 75 percent of these assessed streams do in fact meet the established standards. Sixty-five percent of the roughly one hundred lakes that are monitored in the state meet safe standards for drinking water use, swimming, and fishing.

While some surface waters can be contaminated by natural elements such as iron, salts, or hydrogen sulfide, the primary cause of water pollution problems in the state is related to human activities. Runoff from agricultural lands, coal mines, and urban areas is the principal means by which contaminants enter Kentucky's surface waters. Bacteria from both human and animal waste and sediment eroding from farmlands (which causes siltation in streams, lakes, and reservoirs) are the leading causes of water pollution. More than 50 percent of the lakes assessed in the state showed high levels of eutrophication caused by nutrients associated with agricultural runoff and human and animal waste.

Although surface waters supply 95 percent of the total water used in Kentucky, about 25 percent of the population rely on groundwater for household use. Wells, both public and private, are the primary means by which groundwater is extracted. More than 200,000 wells now supply potable water needs, and since the issuance of permits and registration was implemented in 1985, more than 20,000 new wells have been drilled. Wells provide direct access to groundwater supplies, given that median depths to the water table vary from only 20 feet in the Bluegrass Region to a little more than 60 feet in the Mississippi Embayment Region. Potential and identifiable contributors to pollution of this source of drinking water include septic systems, underground storage tanks, oil and gas wells, dumps and landfills, and hazardous waste sites. Private wells drilled to provide drinking water are not required to be tested for quality (except for bacteria at the time of drilling), yet experts estimate that about half have high bacteria counts.

As noted previously, approximately one-fourth of the state exhibits karst features associated with limestone bedrock. Karst topography offers innumerable points where surface water can rap-

idly enter the groundwater supply and, in turn, move quickly into streams and lakes. The quality of the groundwater supply is especially important, yet much remains unknown about groundwater quality and the potential sources of pollution of this important water resource. Increased awareness of the need to maintain water availability and quality and rigorous monitoring practices will help maintain Kentucky's surface and groundwater resources for future generations.

WETLANDS

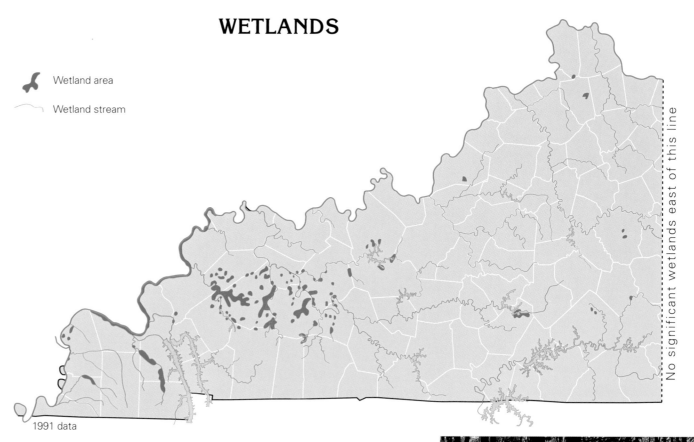

🦅 Wetland area

〰️ Wetland stream

No significant wetlands east of this line

1991 data

Cypress Creek Nature Preserve in Marshall County is one of the best remnant examples of old tupelo swamp forest in Kentucky. A heron rookery is nearby.

DOMESTIC WATER WELL LOCATIONS

One dot · represents one water well

Kentucky total = 16,538

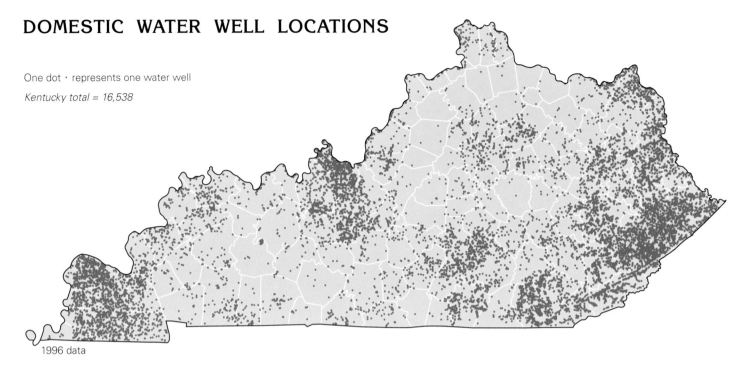

1996 data

Orphaned strip mines in the Western Kentucky Coal Field often trap pools of water that become highly acidified, causing the reddish color shown here.

DOMESTIC WATER WELL FLOW RATE

GALLONS PER MINUTE

· 0 - 20
· 20 - 50
· 50 - 100
· More than 100

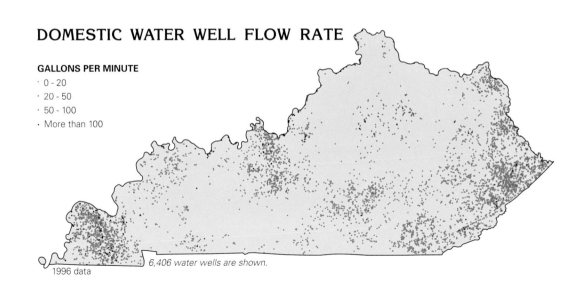

6,406 water wells are shown.

1996 data

DEPTH TO GROUNDWATER

FEET

25
50
75
100
125
150
175
200
225
250
275
300

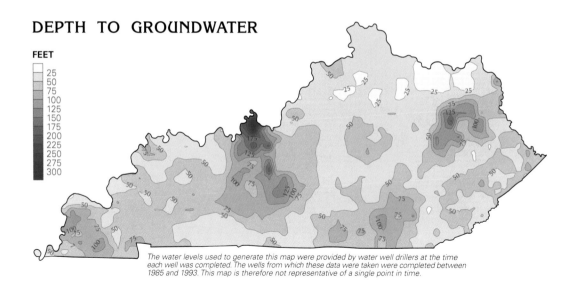

The water levels used to generate this map were provided by water well drillers at the time each well was completed. The wells from which these data were taken were completed between 1985 and 1993. This map is therefore not representative of a single point in time.

A late April snow dusts the trees along Chandler Creek in Raven Run Nature Sanctuary in Fayette County.

UNITED STATES CLIMATIC SYSTEM

Same scale as contiguous states

Approximately 44% of scale of contiguous states

After Köppen-Geiger System of Climate Classification.

First and Second Letter:
A Tropical. All monthly average temperatures are above 64.4°F (18°C).
BS Dry. Semidesert.
BW Dry. Desert.
C Mild mid-latitude. Average temperature of coldest month is between 26.6°F (-3°C) and 64.4°F (18°C).
D Severe mid-latitude. Coldest month average is below 26.6°F (0°C); warmest month average is above 50°F (10°C).
ET Tundra. Warmest month average is below 50°F (10°C), but above freezing.

CLIMATES

Tropical — Af
Dry — BSh, BWh, BWk
Mild mid-latitude (warm) — Cfa, Cfb, Cfc, Csa/Csb
Severe mid-latitude (snow) — Dfa, Dfb, Dfc
Tundra (ice) — ET
Highlands — H

Third Letter:
a Warmest month average temperature is above 71.6°F (22°C).
b Warmest month average temperature is below 71.6°F (22°C). At least 4 months have averages over 50°F (10°C).
c Less than four months with average temperature over 50°F (10°C).
f Constantly moist; rainfall all through the year.
h Hot and dry: average annual temperature is over 64.4°F (18°C).
k Cold and dry; average annual temperature is under 64.4°F (18°C).
s Dry season in summer.
w Dry season in winter.

Weather and Climate

Weather and climate in Kentucky can best be described as variable. Located within the southeastern interior portion of North America, Kentucky has a climate described as humid subtropical (indicating that all monthly average temperatures are above freezing, with some months averaging above 50°F) that is generally similar to other areas of the world located near the southeastern sides of continents. The lack of major physical barriers that could impede air mass movement and the low elevation and relief in central North America are climatically significant. Low elevations to the south of Kentucky permit influxes of warm, moist tropical air from the Gulf of Mexico and tropical Atlantic Ocean. To Kentucky's north and northwest, a similar lack of substantial mountain barriers affords little protection from advancing cold polar air.

Located within the prevailing westerly wind belt, the state is regularly influenced by the migratory cyclonic and anticyclonic weather systems associated with the jet stream. During winter and spring, when the jet stream is closest to the equator, these passing systems are the dominant features on the daily weather map. The low pressure systems, which can include both cold and warm fronts, are responsible for virtually all of the spring and winter precipitation and the cloudy, wet, cool days so typical of the region in the winter months. Occasionally during winter, an air mass associated with a high-pressure system (an anticyclone) plunges unusually far southward out of Canada, providing extreme cold but clear and very dry conditions. In summer and fall the jet stream retreats northward and the southeastern United States is dominated by the semipermanent high-pressure system centered over the western subtropical Atlantic Ocean (commonly known as the Bermuda High). Clockwise airflow around this anticyclonic system insures that summers are hot and humid as air flows up the Mississippi and Ohio Valleys from the south. Spells of hot, humid weather are occasionally broken by a strong migratory low pressure system with attendant cold front. Passage of the cold front usually decreases both the temperature and the humidity for a day or two, but coincident with the frontal passage comes the potential for severe thunderstorms with lightning, heavy rains, and possibly tornadoes.

Temperature patterns across the state are typical for a continental interior, midlatitude location; minima usually occur in January and maxima in July. Mean daily minimum temperatures in January range from the low to mid-20s Fahrenheit; maximum temperatures are typically in the high 30s to low 40s F. In July, mean daily minima range from the low to high 60s with maxima in the high 80s to low 90s. Highest temperatures typically occur in the extreme west, with lowest temperatures found along the upland Ohio and West Virginia borders. The length of the growing season—so important to agriculture—follows a similar pattern, extending from under six months in eastern Kentucky to nearly seven months in the Mississippi Embayment Region. Extreme temperatures below -20°F and above 100°F are recorded on rare occasions.

Precipitation across the state usually accounts for a surplus over evaporation and transpiration in all months. Much of the state falls within a precipitation regime that is unique among continental interiors. Climatic data for similar locations in other continental interiors exhibit annual distributions of precipitation that are in phase with the temperature fluctuations during summer months (i.e., the precipitation maximum occurs in the same month as the temperature maximum). Kentucky's western half, however, exhibits a March precipitation maximum. The state's eastern portion shows a more typical July maximum, although the climatic record also exhibits a strong secondary peak in March. Across the state October is usually the driest month. Kentucky does receive some snowfall during typical winters—usually totaling about twelve inches, with slight variations across the state—but most of the precipitation during this season falls in the form of rain, drizzle, or sleet.

CLIMATES SIMILAR TO KENTUCKY'S

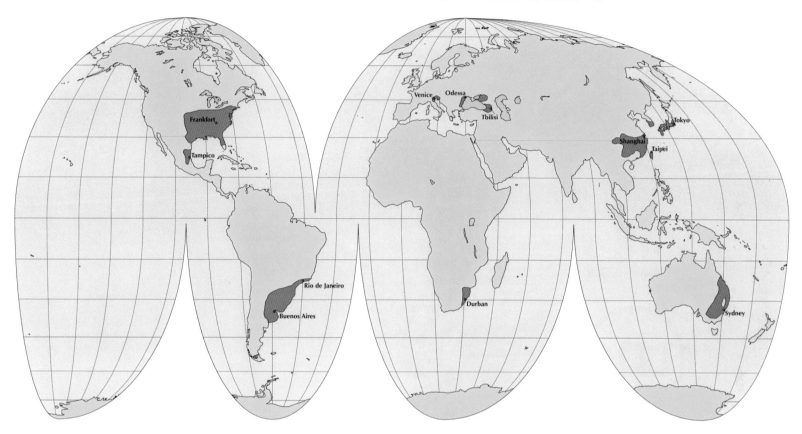

Climate is based on Köppen-Geiger system of climate classification.

AVERAGE ANNUAL TEMPERATURE

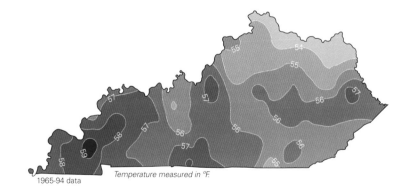

1965-94 data   *Temperature measured in °F.*

AVERAGE GROWING SEASON

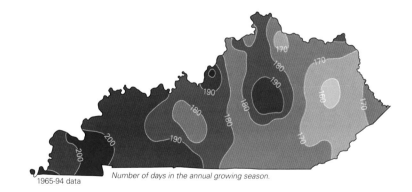

1965-94 data   *Number of days in the annual growing season.*

AVERAGE MAXIMUM JANUARY TEMPERATURE

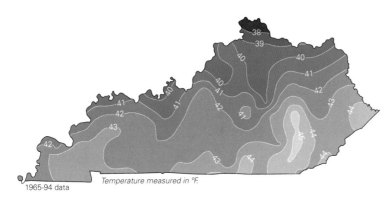

1965-94 data   *Temperature measured in °F.*

MEAN MONTHLY JANUARY SUNSHINE

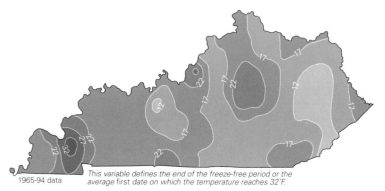

TOTAL HOURS

60
80
100
120
140
160
180
200
220
240

AVERAGE FIRST FALL 32°F DAYS AFTER SEPTEMBER 30

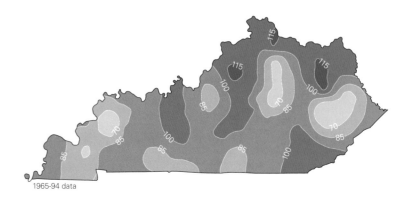

1965-94 data   *This variable defines the end of the freeze-free period or the average first date on which the temperature reaches 32°F.*

AVERAGE MINIMUM JANUARY TEMPERATURE

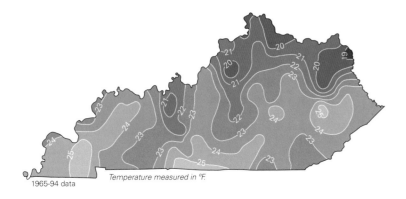

1965-94 data   *Temperature measured in °F.*

AVERAGE NUMBER OF DAYS WITH TEMPERATURE OF 32°F OR LOWER

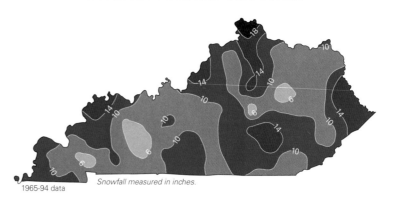

1965-94 data

Ice forms on farm ponds near Seatonville in Jefferson County.

AVERAGE ANNUAL SNOWFALL

1965-94 data   *Snowfall measured in inches.*

AVERAGE MAXIMUM JULY TEMPERATURE

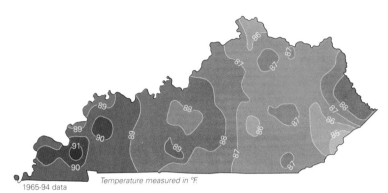

1965-94 data   *Temperature measured in °F.*

AVERAGE MINIMUM JULY TEMPERATURE

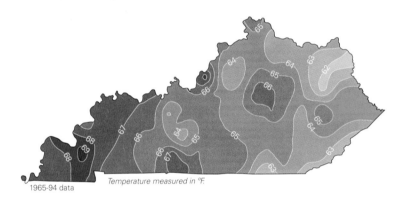

1965-94 data   *Temperature measured in °F.*

MEAN MONTHLY JULY SUNSHINE

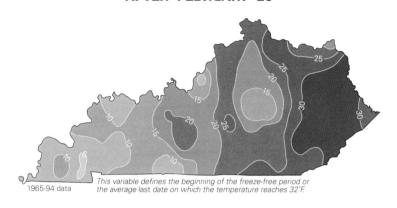

TOTAL HOURS

240
260
280
300
320
340
360

AVERAGE LAST SPRING 32°F DAYS AFTER FEBRUARY 28

1965-94 data   *This variable defines the beginning of the freeze-free period or the average last date on which the temperature reaches 32°F.*

AVERAGE NUMBER OF DAYS WITH TEMPERATURE OF 90°F OR HIGHER

1965-94 data

AVERAGE ANNUAL HEATING DEGREE DAYS

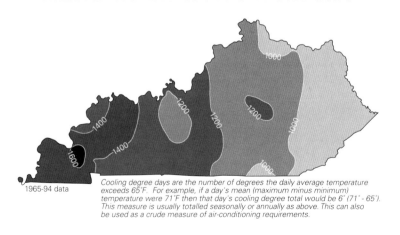

1965-94 data   *Heating degree days are the number of degrees the daily average temperature falls below 65°F. For example, if a day's mean (maximum minus minimum) temperature were 59°F then that day's heating degree total would be 6° (65° - 59°). This measure is usually totalled seasonally or annually as above. This can also be used as a crude measure of heating requirements.*

AVERAGE ANNUAL COOLING DEGREE DAYS

1965-94 data   *Cooling degree days are the number of degrees the daily average temperature exceeds 65°F. For example, if a day's mean (maximum minus minimum) temperature were 71°F then that day's cooling degree total would be 6° (71° - 65°). This measure is usually totalled seasonally or annually as above. This can also be used as a crude measure of air-conditioning requirements.*

AVERAGE ANNUAL PRECIPITATION

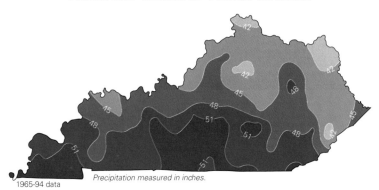

1965-94 data *Precipitation measured in inches.*

MEAN ANNUAL PRECIPITATION

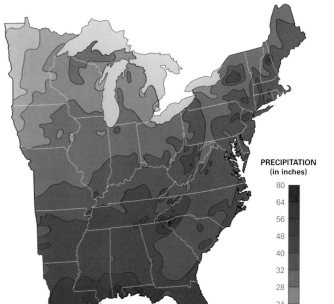

PRECIPITATION
(in inches)

80
64
56
48
40
32
28
24
20
16

AVERAGE MARCH PRECIPITATION

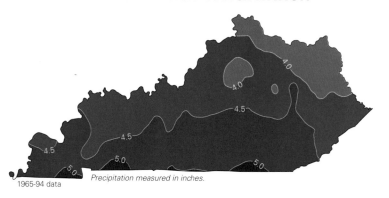

1965-94 data *Precipitation measured in inches.*

AVERAGE NUMBER OF PRECIPITATION DAYS

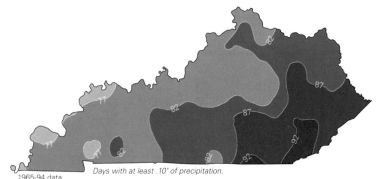

1965-94 data *Days with at least .10" of precipitation.*

AVERAGE OCTOBER PRECIPITATION

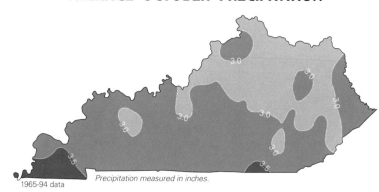

1965-94 data *Precipitation measured in inches.*

AVERAGE ANNUAL PRECIPITATION AND TEMPERATURE FOR KENTUCKY, 1895-1995

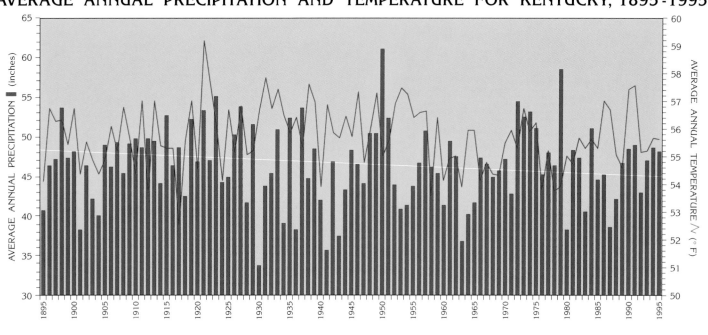

The Ohio River in flood near Louisville.

CLIMOGRAPHS FOR SELECTED CITIES

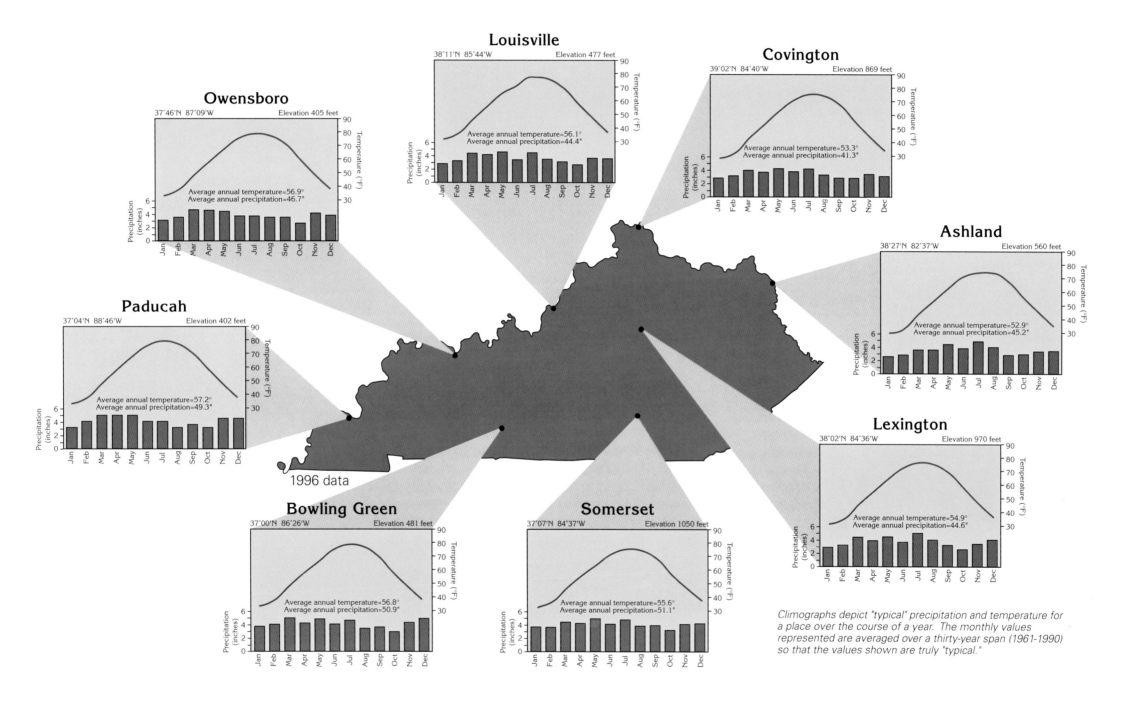

Owensboro
37°46'N 87°09'W — Elevation 405 feet
Average annual temperature=56.9°
Average annual precipitation=46.7"
Precipitation (inches) / Temperature (°F)
Jan Feb Mar Apr May Jun Jul Aug Sep Oct Nov Dec

Louisville
38°11'N 85°44'W — Elevation 477 feet
Average annual temperature=56.1°
Average annual precipitation=44.4"
Precipitation (inches) / Temperature (°F)
Jan Feb Mar Apr May Jun Jul Aug Sep Oct Nov Dec

Covington
39°02'N 84°40'W — Elevation 869 feet
Average annual temperature=53.3°
Average annual precipitation=41.3"
Precipitation (inches) / Temperature (°F)
Jan Feb Mar Apr May Jun Jul Aug Sep Oct Nov Dec

Ashland
38°27'N 82°37'W — Elevation 560 feet
Average annual temperature=52.9°
Average annual precipitation=45.2"
Precipitation (inches) / Temperature (°F)
Jan Feb Mar Apr May Jun Jul Aug Sep Oct Nov Dec

Paducah
37°04'N 88°46'W — Elevation 402 feet
Average annual temperature=57.2°
Average annual precipitation=49.3"
Precipitation (inches) / Temperature (°F)
Jan Feb Mar Apr May Jun Jul Aug Sep Oct Nov Dec

1996 data

Bowling Green
37°00'N 86°26'W — Elevation 481 feet
Average annual temperature=56.8°
Average annual precipitation=50.9"
Precipitation (inches) / Temperature (°F)
Jan Feb Mar Apr May Jun Jul Aug Sep Oct Nov Dec

Somerset
37°07'N 84°37'W — Elevation 1050 feet
Average annual temperature=55.6°
Average annual precipitation=51.1"
Precipitation (inches) / Temperature (°F)
Jan Feb Mar Apr May Jun Jul Aug Sep Oct Nov Dec

Lexington
38°02'N 84°36'W — Elevation 970 feet
Average annual temperature=54.9°
Average annual precipitation=44.6"
Precipitation (inches) / Temperature (°F)
Jan Feb Mar Apr May Jun Jul Aug Sep Oct Nov Dec

Climographs depict "typical" precipitation and temperature for a place over the course of a year. The monthly values represented are averaged over a thirty-year span (1961-1990) so that the values shown are truly "typical."

Air Quality

The quality of the air we breathe is directly related to the amounts of pollutants released into the atmosphere, and to weather and climate conditions. While many pollutants in the atmosphere are directly attributable to human activities, climatic patterns, specific weather events, and other natural factors such as volcanic eruptions can either increase or decrease the threat to humans and the environment. For example, the persistent subtropical high pressure cell that dominates our weather during the summer months tends to enhance the formation and increase the concentration of smog near the surface. Conversely, strong winds can disperse—although not eliminate—pollutants, and precipitation acts as a natural cleansing mechanism by removing pollutants from the air; unfortunately, the resulting rain or snow can then become contaminated. Other aspects of the natural environment can also act to minimize the impact of certain pollutants. The limestones found throughout much of Kentucky, for example, can act as an effective buffer against the impact of runoff from acid rain entering the state's streams and lakes.

ACID RAIN

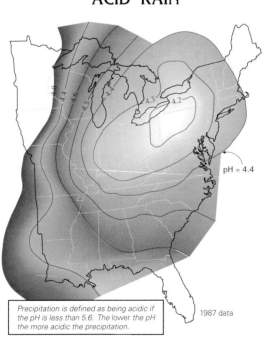

Precipitation is defined as being acidic if the pH is less than 5.6. The lower the pH the more acidic the precipitation.

1987 data

AIR QUALITY CONTROL REGIONS

1996 data

TOXIC AIR EMISSIONS
(FROM REPORTING INDUSTRIES)

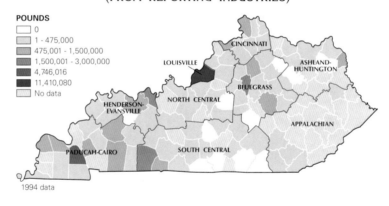

POUNDS
- 0
- 1 - 475,000
- 475,001 - 1,500,000
- 1,500,001 - 3,000,000
- 4,746,016
- 11,410,080
- No data

1994 data

GREENHOUSE GAS EMISSIONS

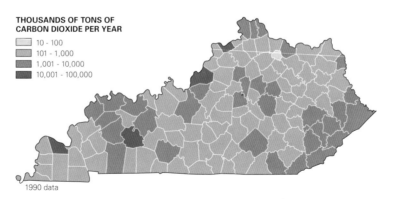

THOUSANDS OF TONS OF
CARBON DIOXIDE PER YEAR
- 10 - 100
- 101 - 1,000
- 1,001 - 10,000
- 10,001 - 100,000

1990 data

AIR CONCENTRATION OF OZONE

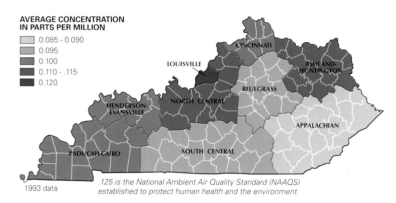

AVERAGE CONCENTRATION
IN PARTS PER MILLION
- 0.085 - 0.090
- 0.095
- 0.100
- 0.110 - .115
- 0.120

1993 data

.125 is the National Ambient Air Quality Standard (NAAQS) established to protect human health and the environment.

OZONE AIR POLLUTION PROBLEMS

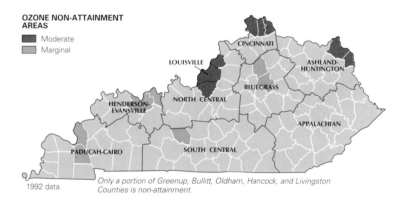

OZONE NON-ATTAINMENT
AREAS
- Moderate
- Marginal

1992 data

Only a portion of Greenup, Bullitt, Oldham, Hancock, and Livingston Counties is non-attainment.

OZONE STANDARD VIOLATIONS

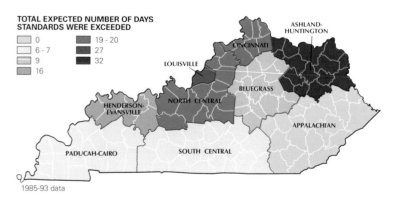

TOTAL EXPECTED NUMBER OF DAYS
STANDARDS WERE EXCEEDED
- 0
- 6 - 7
- 9
- 16
- 19 - 20
- 27
- 32

1985-93 data

Since passage of the federal Clean Air Act in 1963 (and subsequent revisions to that act, most notably in 1970), air quality in the state has improved. In particular, some long-term, large-scale issues associated with atmospheric pollution and human environmental impact were further addressed by the 1990 amendments to the Clean Air Act. Chief among the concerns addressed by these most recent revisions are global warming, depletion of the stratosphere's protective ozone layer, and acid rain. All of these problems are potentially significant for Kentucky's residents and environment.

Although significant environmental problems remain to be addressed in Kentucky, the state's air quality has improved significantly since passage of the original Clean Air Act, a fact reflected in the information gathered at more than fifty data collection stations. These data, combined with additional data gathered at other weather stations, have provided the information necessary to determine the quality of the air in each of the state's nine air quality control regions. Most regions now comply with standards set by the Environmental Protection Agency (EPA) for air content of nitrogen dioxide, lead, carbon monoxide, and particulate matter. But sulfur dioxide, which contributes to the formation of smog and acid rain, and ground-level ozone, which is harmful to plants and animals, continue to exceed air-quality standards in some locations. Typically, the state's urban areas experience greater pollution levels, especially with regard to ozone, than do rural areas. In 1994, 30 percent of the state's population lived in eight metropolitan counties classified as "moderate non-attainment"—that is, in areas where pollution standards for surface ozone were most frequently violated.

AIR CONCENTRATIONS OF SULFUR DIOXIDE

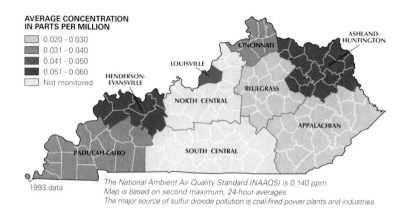

AVERAGE CONCENTRATION IN PARTS PER MILLION
- 0.020 - 0.030
- 0.031 - 0.040
- 0.041 - 0.050
- 0.051 - 0.060
- Not monitored

1993 data

The National Ambient Air Quality Standard (NAAQS) is 0.140 ppm. Map is based on second maximum, 24-hour averages. The major source of sulfur dioxide pollution is coal-fired power plants and industries.

SULFUR DIOXIDE EMISSIONS
FROM COAL-FIRED POWER PLANTS

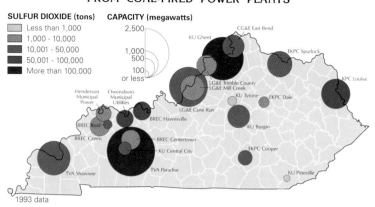

SULFUR DIOXIDE (tons)
- Less than 1,000
- 1,000 - 10,000
- 10,001 - 50,000
- 50,001 - 100,000
- More than 100,000

CAPACITY (megawatts)
- 2,500
- 1,000
- 500
- 100 or less

1993 data

AIR CONCENTRATIONS OF PARTICULATES

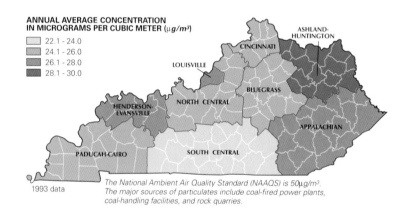

ANNUAL AVERAGE CONCENTRATION IN MICROGRAMS PER CUBIC METER ($\mu g/m^3$)
- 22.1 - 24.0
- 24.1 - 26.0
- 26.1 - 28.0
- 28.1 - 30.0

1993 data

The National Ambient Air Quality Standard (NAAQS) is 50$\mu g/m^3$. The major sources of particulates include coal-fired power plants, coal-handling facilities, and rock quarries.

AIR CONCENTRATIONS OF TOTAL SUSPENDED PARTICULATES

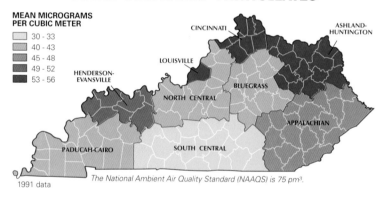

MEAN MICROGRAMS PER CUBIC METER
- 30 - 33
- 40 - 43
- 45 - 48
- 49 - 52
- 53 - 56

1991 data

The National Ambient Air Quality Standard (NAAQS) is 75 pm^3.

AIR CONCENTRATIONS OF CARBON MONOXIDE

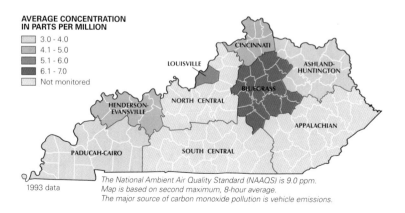

AVERAGE CONCENTRATION IN PARTS PER MILLION
- 3.0 - 4.0
- 4.1 - 5.0
- 5.1 - 6.0
- 6.1 - 7.0
- Not monitored

1993 data

The National Ambient Air Quality Standard (NAAQS) is 9.0 ppm. Map is based on second maximum, 8-hour average. The major source of carbon monoxide pollution is vehicle emissions.

AIR CONCENTRATIONS OF NITROGEN DIOXIDE

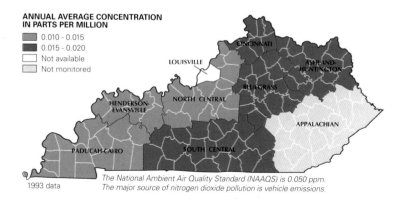

ANNUAL AVERAGE CONCENTRATION IN PARTS PER MILLION
- 0.010 - 0.015
- 0.015 - 0.020
- Not available
- Not monitored

1993 data

The National Ambient Air Quality Standard (NAAQS) is 0.050 ppm. The major source of nitrogen dioxide pollution is vehicle emissions.

TOTAL RELEASES/TRANSFERS OF TOXIC CHEMICALS

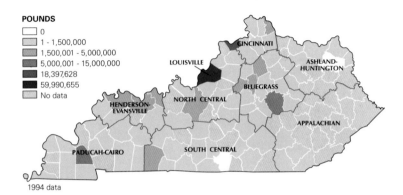

POUNDS
- 0
- 1 - 1,500,000
- 1,500,001 - 5,000,000
- 5,000,001 - 15,000,000
- 18,397,628
- 59,990,655
- No data

1994 data

Environmental Hazards

Environmental hazards in Kentucky can be grouped into three general classes. Geologic hazards such as earthquakes, landslides, and sinkhole formation constitute the first group; atmospheric hazards, including tornadoes, lightning, ice storms, and flooding from heavy rainfall, form the second group; and the third group are those hazards that are human-induced, including landfills, waste, and the release of toxic chemicals into the environment. While the first two groups are considered "natural" hazards, their impact is often greatly magnified as a consequence of human modifications to the environment. For example, it is well-documented that urbanization or surface mining of coal can increase both the frequency and the magnitude of flooding in nearby streams as a consequence of changes in land surface characteristics.

Possibly the single greatest potential natural hazard is earthquakes associated with the New Madrid Fault Zone. This

The Tennessee Valley Authority has installed pollution control equipment on the smokestacks of their Paradise steam plant in Muhlenberg County to reduce sulfur emissions.

The northern portion of the New Madrid Fault Zone is in close proximity to the confluence of the Mississippi and Ohio Rivers. The fault is buried beneath a thick mantle of alluvium and is not visible. This image is an enhanced Landsat TM dataset (bands 4,5,2, dated July 15, 1991). The cities of Cairo, Illinois (above the confluence of the rivers), and Wickliffe, Kentucky (south of the confluence, on the east bank of the Mississippi), are visible as blue light. Note the difference in sediment content between the rivers (medium versus dark blue). Forest is rendered as orange to dark red. Most of the cropland (shades of blue) was under stress as a result of drought conditions during the summer of 1991.

NEW MADRID SEISMIC ZONE

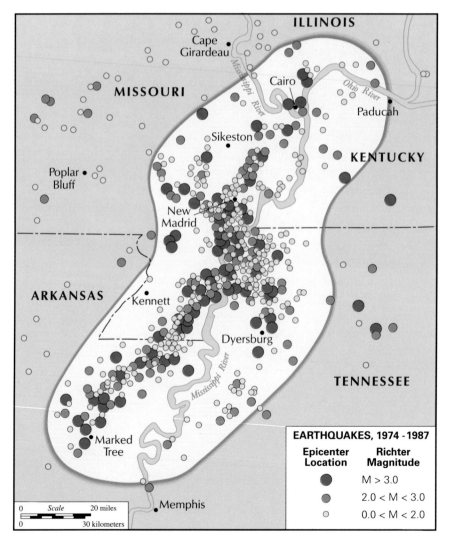

EARTHQUAKES, 1974 - 1987

Epicenter Location	Richter Magnitude
●	M > 3.0
●	2.0 < M < 3.0
○	0.0 < M < 2.0

MAXIMUM SEISMIC INTENSITIES

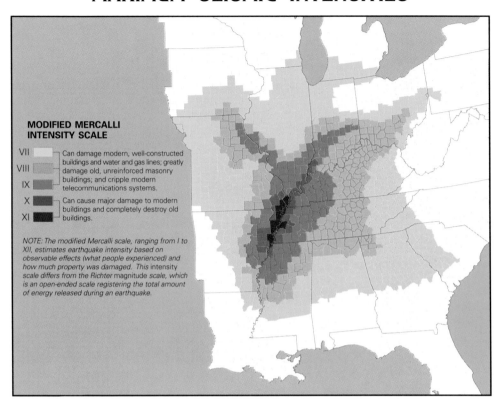

MODIFIED MERCALLI INTENSITY SCALE

VII — Can damage modern, well-constructed buildings and water and gas lines; greatly
VIII — damage old, unreinforced masonry buildings; and cripple modern
IX — telecommunications systems.

X — Can cause major damage to modern buildings and completely destroy old
XI — buildings.

NOTE: The modified Mercalli scale, ranging from I to XII, estimates earthquake intensity based on observable effects (what people experienced) and how much property was damaged. This intensity scale differs from the Richter magnitude scale, which is an open-ended scale registering the total amount of energy released during an earthquake.

SEISMIC RISK

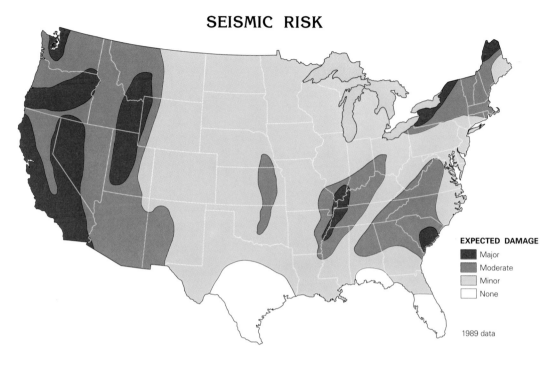

EXPECTED DAMAGE

- Major
- Moderate
- Minor
- None

1989 data

The Mississippi River circumscribes the New Madrid Bend, the western tip of Fulton County. This area is central to the New Madrid Seismic Zone.

fault zone extends southward along the Mississippi River basin from near the confluence of the Ohio and Mississippi Rivers to the northwestern tip of Mississippi. Portions of six states, including Kentucky, lie adjacent to the fault zone. Four of the largest earthquakes ever recorded in the continental U.S. occurred in this area during the winter of 1811-12. These earthquakes have been estimated at magnitudes ranging from 8.3 to 8.7 on the Richter scale, and their effects were felt from the Rocky Mountains on the west to the Atlantic Ocean on the east. Reelfoot Lake in western Tennessee was created by these earthquakes, and reports indicate that the Mississippi River reversed its course for a brief period. Unlike California, where the San Andreas fault system is extensively monitored and earthquakes are frequent, the New Madrid Fault Zone is more problematic to monitor because the faults are deeply buried and very few moderate or large quakes have been recorded in historic time. Consequently, our understanding of the forces at work in the New Madrid area is not well devel-

oped. Geoscientists have concluded that the more brittle crust in interior North America allows earthquake shock waves to travel farther and faster than do those occurring along the western continental margin. An earthquake of the magnitude of those in 1811-1812 would produce disastrous consequences today. Several major cities (Birmingham, Memphis, St. Louis, Little Rock, Nashville, and Louisville) are within 250 miles of the primary fault area and as such would experience considerable damage were a major earthquake to occur. Seismologists have indicated that there is a 90 percent chance that an earthquake of magnitude 6.0 or higher will occur by the year 2050. An earthquake of this magnitude may result in some or all of the following impacts: landslides along steep slopes and most streams, damage to buildings and bridges, liquefaction (which increases the potential for damage to buildings) of alluvial soils prevalent in the Mississippi Embayment Region, disruption of major petroleum transmission pipelines and other utilities, and loss of human life.

RADON LEVEL REGIONS

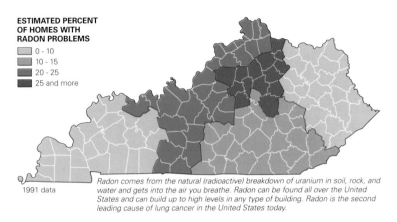

ESTIMATED PERCENT
OF HOMES WITH
RADON PROBLEMS

- 0 - 10
- 10 - 15
- 20 - 25
- 25 and more

1991 data

Radon comes from the natural (radioactive) breakdown of uranium in soil, rock, and water and gets into the air you breathe. Radon can be found all over the United States and can build up to high levels in any type of building. Radon is the second leading cause of lung cancer in the United States today.

HIGHEST RECORDED RADON LEVELS IN HOMES

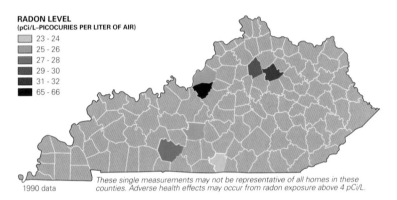

RADON LEVEL
(pCi/L—PICOCURIES PER LITER OF AIR)

- 23 - 24
- 25 - 26
- 27 - 28
- 29 - 30
- 31 - 32
- 65 - 66

1990 data

These single measurements may not be representative of all homes in these counties. Adverse health effects may occur from radon exposure above 4 pCi/L.

Another environmental hazard associated with Kentucky's geology is radon gas, which is derived from the decay of uranium minerals occurring naturally in rock and soil. By itself, radon gas is not harmful, but it quickly decays into ionized particles and other by-products that may become concentrated in enclosed spaces such as basements. These particles combine with other aerosols and lodge in the lungs when inhaled. Further radioactive decay of these by-products then exposes lung tissue to abnormal amounts of harmful radiation; where concentrations are high the threat of lung cancer exists.

Atmospheric hazards include tornadoes, thunderstorms, high winds, heavy precipitation and subsequent flooding, ice storms, and lightning. Of these, tornadoes are the most spectacular and have been the most devastating in Kentucky. These storms, and the thunderstorms with which they are associated, have been observed during all times of the year, although April is the month of most frequent occurrence. Intense heating during afternoon and evening hours is condu-

TORNADO TRACKS
1950-1986

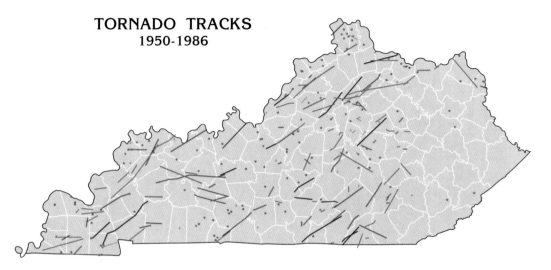

F0: 40-72 m.p.h. winds. Damage is light and might include damage to tree branches, chimneys and billboards. Shallow-rooted trees may be pushed over.

F1: 73-112 m.p.h. winds. Damage is moderate; mobile homes may be pushed off foundations and moving autos pushed off the road.

F2: 113-157 m.p.h. winds. Damage is considerable. Roofs can be torn off houses, mobile homes demolished, and large trees uprooted.

FUJITA INTENSITY SCALE VALUES

—— F0
—— F1
—— F2
—— F3
—— F4
—— F5

F3: 158-206 m.p.h. winds. Damage is severe. Even well constructed houses may be torn apart, trees uprooted, and cars lifted off the ground.

F4: 207-260 m.p.h. winds. Damage is devastating. Houses can be leveled and cars thrown; objects become deadly missiles.

F5: 261-318 m.p.h. winds. Damage is incredible. Structures are lifted off foundations and carried away; cars become missiles. Less than 2% of all tornadoes reach an intensity of this magnitude.

cive to the development of thunderstorms, lightning, and tornadoes. Atmospheric conditions are most favorable during the spring and summer months, when cold fronts bring warm, humid tropical air together with cool, dry continental air—a near-perfect recipe for the formation of severe weather over the states between the Rockies and the Appalachians. Guided by prevailing westerly winds, tornadoes typically move in a southwest-to-northeast direction, although there have been many observed exceptions to this general rule. Rotational wind speeds associated with tornadoes have been estimated to exceed 300 miles per hour in some cases, a force that causes incredible damage. Tornado incidence varies significantly across Kentucky. The western two-thirds of the state borders a region sometimes referred to as "tornado alley," which extends from the Texas and Oklahoma panhandle area northeastward into southern Michigan. The rugged topography of Kentucky's eastern portion upsets the balance of forces necessary to produce a tornado, and thus very few such storms occur in eastern Kentucky and the Appalachians. Strong winds that accompany thunderstorms are sometimes mistaken for tornadoes. These downbursts, as they are called, are strong downward-moving currents of cold air that spread laterally when they reach the surface. In many

CONFIRMED TORNADOES
1950 - 1993

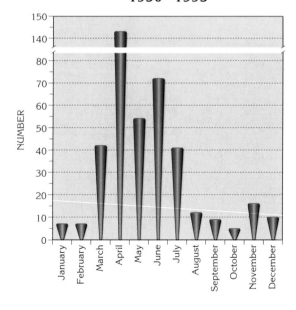

A tornado caused extensive damage to a Louisville suburb in Bullitt County in May 1996.

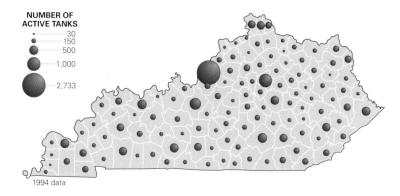

UNDERGROUND STORAGE TANKS

**NUMBER OF
ACTIVE TANKS**

30
150
500
1,000
2,733

1994 data

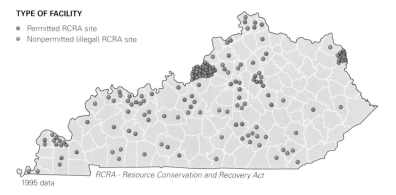

HAZARDOUS WASTE FACILITIES

TYPE OF FACILITY

● Permitted RCRA site
● Nonpermitted (illegal) RCRA site

RCRA - Resource Conservation and Recovery Act

1995 data

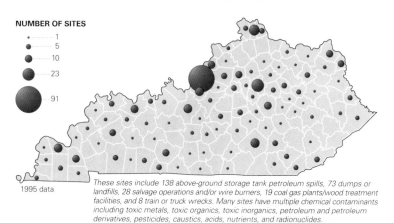

SUPERFUND SITES

NUMBER OF SITES

1
5
10
23
91

1995 data

These sites include 138 above-ground storage tank petroleum spills, 73 dumps or landfills, 28 salvage operations and/or wire burners, 19 coal gas plants/wood treatment facilities, and 8 train or truck wrecks. Many sites have multiple chemical contaminants including toxic metals, toxic organics, toxic inorganics, petroleum and petroleum derivatives, pesticides, caustics, acids, nutrients, and radionuclides.

cases the associated damage is confined to a small area and is thus often assumed to have been caused by a tornado.

Heavy rainfall also poses a frequent and major environmental threat to Kentucky. Heavy rains from a summer thunderstorm may produce localized flooding in small streams or may inundate crops. More widespread flooding, often resulting in property damage, is usually the result of extremely heavy rainfall associated with the passage of a weather front or, on rare occasions, with the remnants of a hurricane that makes its way inland from the Gulf of Mexico or the Atlantic Ocean. Most of Kentucky experiences at least some limited flooding during the late winter and early spring months. At this time the ground is at or near saturation, or sometimes frozen, and heavy rainfall can result from the frequent midlatitude cyclones or low-pressure systems that move across the region. Thus, much of the precipitation that falls is transported directly and rapidly into streams rather than being absorbed by the ground, thereby maximizing the potential for flooding.

Ice is also an environmental hazard for Kentuckians. In the summer, ice in the form of hail produced by severe thunderstorms can inflict considerable crop and property damage. In rare cases hailstones can reach two to three inches in diameter and cause extensive crop loss and serious damage to automo-

biles and dwellings. In the winter months ice storms may result in downed utility lines, automobile accidents that often cause human injury or death, interruption of services and business, and significant property damage.

Although usually not as dramatic as earthquakes or tornadoes, human-induced hazards often pose a greater threat to the environment and to human health and long-term well-being. Kentucky produces nearly 4 million tons of waste each year; much of this waste is disposed of properly and poses little or no threat to the environment. Some waste products, especially those from the industrial sector, have the potential to directly affect human health and environmental quality if not properly disposed of. Open dumps, illegal landfills, and spills of toxic or hazardous waste can directly impact groundwater and stream water quality or contaminate the soil. Underground storage tanks —of which some 40,000 have been registered in Kentucky since 1986—represent another potential source of water and soil pollution. Many of these ultimately must be replaced as leaks develop and as better holding facilities become available. As a method of monitoring and controlling hazardous waste, an active Superfund Site list is maintained that includes waste sites that must meet both federal- and state-led remediation efforts. As of March 1995, this list included 436 sites within the state.

NATURAL REGIONS

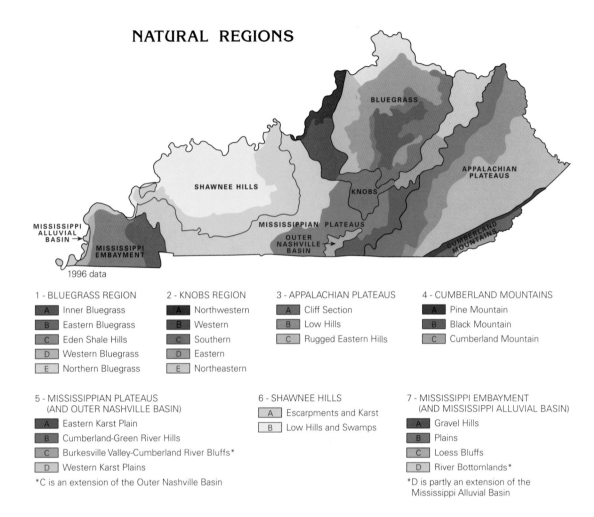

1996 data

1 - BLUEGRASS REGION
- A Inner Bluegrass
- B Eastern Bluegrass
- C Eden Shale Hills
- D Western Bluegrass
- E Northern Bluegrass

2 - KNOBS REGION
- A Northwestern
- B Western
- C Southern
- D Eastern
- E Northeastern

3 - APPALACHIAN PLATEAUS
- A Cliff Section
- B Low Hills
- C Rugged Eastern Hills

4 - CUMBERLAND MOUNTAINS
- A Pine Mountain
- B Black Mountain
- C Cumberland Mountain

5 - MISSISSIPPIAN PLATEAUS
(AND OUTER NASHVILLE BASIN)
- A Eastern Karst Plain
- B Cumberland-Green River Hills
- C Burkesville Valley-Cumberland River Bluffs*
- D Western Karst Plains

*C is an extension of the Outer Nashville Basin

6 - SHAWNEE HILLS
- A Escarpments and Karst
- B Low Hills and Swamps

7 - MISSISSIPPI EMBAYMENT
(AND MISSISSIPPI ALLUVIAL BASIN)
- A Gravel Hills
- B Plains
- C Loess Bluffs
- D River Bottomlands*

*D is partly an extension of the
Mississippi Alluvial Basin

Canada geese (Branta canadensis) at the Ballard Wildlife Management Area in Ballard County. Populations have been increasing over the past decade because of improved conservation practices.

Wildlife

Kentucky, because of its geographic location, climate, and abundant waterways, is one of the nation's premier areas of biodiversity. It is at a crossroads that has facilitated the migration and introduction of new species of both flora and fauna. Focusing on the state's ecological diversity, a number of agencies, most notably the Nature Conservancy of Kentucky and the Kentucky State Nature Preserves Commission, have divided the state into seven ecological or natural regions, a representation that identifies the regions according to their geology, pedology (soils), and biology.

Of the 700 species of freshwater fishes found in the rivers and lakes of the United States and Canada, 242, or more than one-third of the total are found in Kentucky's waters. Along with at least 3,000 species of vascular plants, the state also has 340 species of birds, 75 mammal species, about 100 species of amphibians and reptiles, and a like number of mussels.

Of the larger animal species, several are relative newcomers to the state and several others have been reintroduced after being absent, or nearly so, for decades. Coyotes are an example of the former type, while black bears, ruffed grouse, wild turkeys, ospreys, and river otters are examples of the latter. The first record of a coyote in the state was in 1949, when one was shot in Woodford County. That specimen, as well as reported sightings of two more animals in Fayette and Montgomery Counties in 1950, were all thought to have been part of a group of seven that had been brought to a Franklin County farm and accidentally released. As a result of forest clearing and habitat changes in the eastern United States, coyotes have expanded eastward from their historic western range. By the 1960s coyotes had become well-established in the western portion of the state. By the mid-1970s coyotes

Purple coneflowers, prairie coneflowers, black-eyed Susans, and bergamot carpet a restored tall-grass prairie in Boyle County. Kentucky had approximately three million acres of grasslands at the time of European settlement.

had been reported in every county in the western third of Kentucky. By 1990 every county in the state was known to have coyotes present, although the coyote population was lower in the eastern mountain counties.

The distribution of ruffed grouse, an important game bird in eastern Kentucky, is expanding westward as restoration sites have been located as far west as Marshall County. Between 1984 and 1992 ruffed grouse captured at locations in the eastern part of the state were set free at twelve restoration sites in northern, central, and western Kentucky.

Black bears, which were historically found in the wooded eastern part of the state, had disappeared by the early 1900s as a result of hunting, deforestation, and loss of habitat. As forests have returned to eastern Kentucky,

DIVERSITY OF SELECTED ANIMALS

FRESHWATER FISH

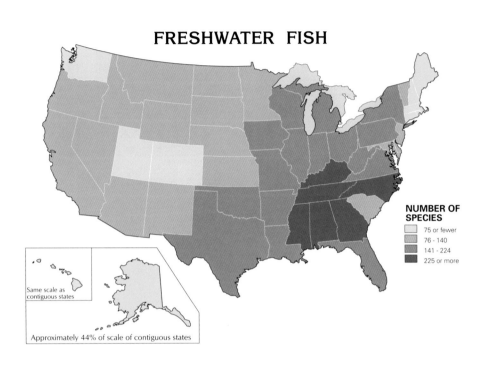

NUMBER OF SPECIES
- 75 or fewer
- 76 - 140
- 141 - 224
- 225 or more

Same scale as contiguous states

Approximately 44% of scale of contiguous states

REPTILES

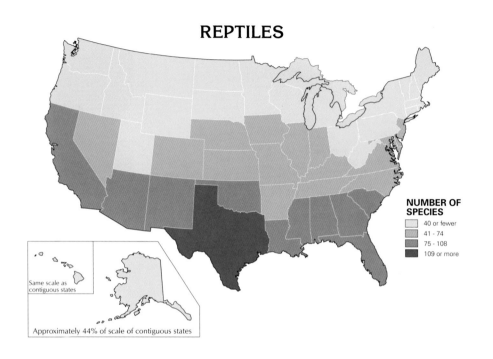

NUMBER OF SPECIES
- 40 or fewer
- 41 - 74
- 75 - 108
- 109 or more

Same scale as contiguous states

Approximately 44% of scale of contiguous states

BIRDS

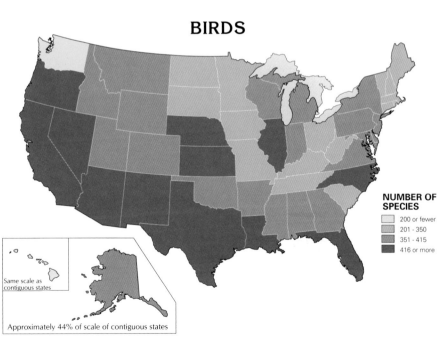

NUMBER OF SPECIES
- 200 or fewer
- 201 - 350
- 351 - 415
- 416 or more

Same scale as contiguous states

Approximately 44% of scale of contiguous states

MAMMALS

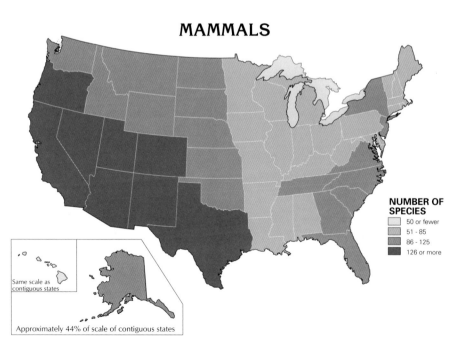

NUMBER OF SPECIES
- 50 or fewer
- 51 - 85
- 86 - 125
- 126 or more

Same scale as contiguous states

Approximately 44% of scale of contiguous states

DIFFUSION OF COYOTES
(*Canis latrans*)

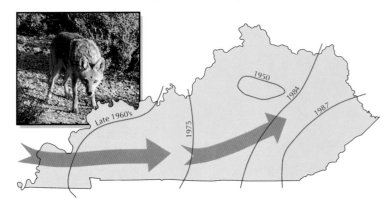

POISONOUS SNAKES is at top right.

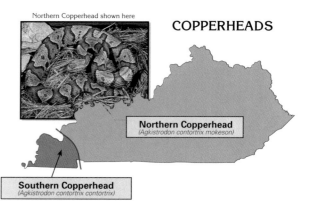

POISONOUS SNAKES

Northern Copperhead shown here

COPPERHEADS

Northern Copperhead
(*Agkistrodon contortrix mokeson*)

Southern Copperhead
(*Agkistrodon contortrix contortrix*)

Copper-belly water snakes like this one are found in the western part of the state.

A wild turkey, Rowan County.

DISTRIBUTION OF RUFFED GROUSE
(*Bonasa umbellus*)

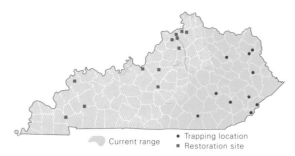

Current range • Trapping location ■ Restoration site

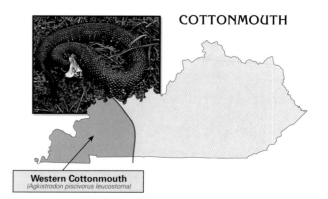

COTTONMOUTH

Western Cottonmouth
(*Agkistrodon piscivorus leucostoma*)

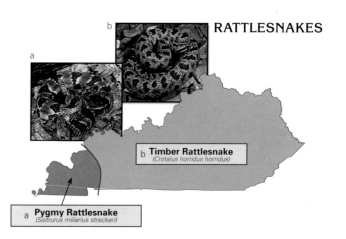

RATTLESNAKES

b **Timber Rattlesnake**
(*Crotalus horridus horridus*)

a **Pygmy Rattlesnake**
(*Sistrurus miliarius streckeri*)

so too have bears come back to the mountains from enclaves in West Virginia and Virginia. There are plans to release black bears along the Big South Fork, a sparsely-settled area near the Kentucky-Tennessee border that has been designated as a national recreation area.

By the late 1800s most of the wild turkeys had disappeared from the state, and in the mid-1940s the only known population was in the Land Between the Lakes area. Beginning in 1978, wild turkey restoration was initiated in the state, and through 1993 more than 5,200 birds were restored all across Kentucky, resulting in a population increase from 2,000 to 40,000 birds. Wild turkeys are found today in every one of the state's counties.

Bobcats are another example of a species on the rebound. Sightings of the small feline have been made across the state, and two areas—Land Between the Lakes and a group of thirty-five eastern Kentucky counties—all have populations sufficiently sizable to allow hunting and trapping.

Four of the nation's seventeen species of poisonous snakes are found in Kentucky. Timber rattlesnakes (*Crotalus horridus)* and copperheads (*Agkistrodon contortrix*) are found throughout the state. The westernmost counties are also home to Kentucky's two other species: the pygmy rattlesnake (*Sistrurus miliarius*) and cottonmouth (*Agkistrodon piscivorous*). One other snake, the nonpoisonous northern copper-bellied water snake, is an example of another animal whose habitat has been disrupted by humans—in this case coal mining in western Kentucky—and is being considered for addition to the federal threatened species list.

As of early 1996, thirty-seven species of flora and fauna found in Kentucky were on the U.S. Department of Fish and Wildlife list of endangered or threatened species. These include sixteen species of mussels, five species of birds (the bald eagle, red-cockaded wood-

pecker, American peregrine falcon, least tern, and piping plover), four fish species (pallid sturgeon, blackside dace, relict darter, and Palezone shiner), one crustacean (Kentucky cave shrimp), three species of mammals (all bats), and eight species of plants (including Price's potato-bean, running buffalo clover, and Short's and white-haired goldenrod). Those animal species that are extinct or have been extirpated from the state include the Eastern cougar, gray wolf, red wolf, elk, golden eagle, Peregrine falcon (although attempts are in progress to restore this species, including the release of several in the city of Lexington in 1993), several species of fishes of the darter family (*Etheostoma*), and the American bison. By 1820 the last wild bison, once numerous in the state, were gone from Kentucky. In the period before settlement, paths made by bison and elk herds were used as trails by the Indians. As early as 1750 Dr. Thomas Walker reported seeing buffalo grazing along the Red River and other areas, but by the time of statehood in 1792, their numbers were already in steep decline. Today captive bison are kept at a number of public parks and

SELECTED ENDANGERED OR THREATENED SPECIES - 1996

FANSHELL
(Cyprogenia stegaria)

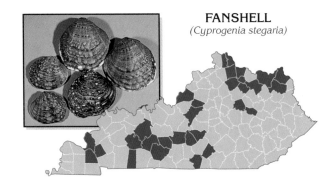

LITTLE-WING PEARLYMUSSEL
(Pegias fabula)

VIRGINIA BIG-EARED BAT
(Plecotus townsendii virginianus)

RED-COCKADED WOODPECKER
(Picoides borealis)

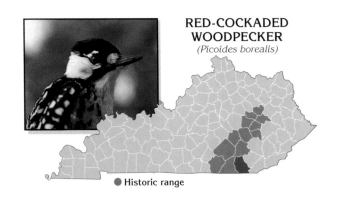

● Historic range

BLACKSIDE DACE
(Phoxinus cumberlandensis)

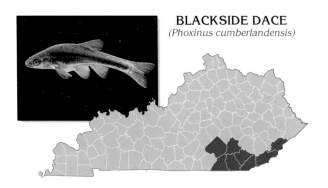

INDIANA MYOTIS (Bat)
(Myotis sodalis)

BALD EAGLE
(Haliaeetus leucocephalus)

● (Breeding Distribution)

VIRGINIA SPIRAEA
(Spiraea virginiana)

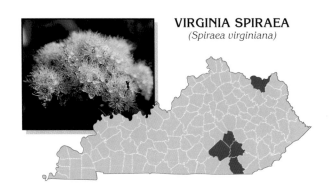

SHORT'S GOLDENROD
(Solidago shortii)

A monarch butterfly on a Missouri golden-rod in McCracken County's West Kentucky Wildlife Management Area. The monarch is one of 143 species of butterflies found in the Commonwealth.

KENTUCKY'S PLANT AND ANIMAL SPECIES

Species	Vascular Plants	Fishes	Mussels	Amphibians and Reptiles	Birds	Mammals
Number in Kentucky	3,000	242	103	105	340	75
Number Federally Threatened/Endangered	8	5[1]	16	0	5	3
Number Presumed Extinct/ Extirpated in Kentucky	5	7	19	1	7	5
Number of State Concern	295	64	35	28	45	16
Number Considered Rare	10%	28%	34%	27%	13%	21%

[1] Includes Kentucky cave shrimp (*Palaemonias ganteri*)   1993 data

A sandstone arch in a Laurel County section of the Daniel Boone National Forest shelters a flowering catchfly, Silene rotundifolia.

Pale purple coneflowers (Echinacea pallida) *greet the dawn at Larue County's Thompson Creek glade.*

private farms. Elk were reintroduced into the Land Between the Lakes National Recreation Area in 1996, and both bison and elk can be observed from an automobile while driving through the 750-acre Elk and Bison Prairie at Land Between the Lakes.

Not all of the newer plant and animal species in the state have been welcome additions. Species are introduced or reintroduced into an ecosystem either purposefully (such as ruffed grouse, osprey, and others noted above) or inadvertently. The two most notorious introductions into Kentucky have occurred accidentally. Zebra mussels (*Dreissena polymorpha*), thought to have arrived in the United States in the mid-1980s as "stowaways" on ships from their native eastern Europe, reached Kentucky's major rivers by the late 1980s. These small mollusks have black or brown and cream-colored stripes, live in colonies, and attach themselves to surfaces such as boat hulls, the shells of other mollusks, turtle shells, and industrial infrastructure that extends into the water. Economic losses and environmental damage caused by the zebra mussel are mounting as they clog pipes, water pumps, and motors, damage boat engines, kill native shellfish (and eventually in Kentucky, endangered species of mussels), and generally alter lake ecology. The other unwelcome import is the plant kudzu (*Pueraria lobata*). When first introduced in 1876 from Japan the vine was used as an ornamental; in the twentieth century it has been used for other purposes, including soil stabilization and control of soil erosion (along highways, for example) before being officially classified as a weed by the U.S. Department of Agriculture in 1970. The vine has spread very rapidly throughout the southeastern portion of the United States, including nearly all of Kentucky, where it has become an eyesore to most people, completely covering other vegetation, barns, and other structures. From an ecosystem perspective, the rapid growth of kudzu potentially threatens the survival of many native plants, especially in southeastern Kentucky.

The physical environment of the Commonwealth can be described as interesting, varied, beautiful. It includes, as we have seen, the dissected "mountains" of the eastern part of the state, the rolling topography of the Bluegrass, the caves and associated karst landscapes of the Pennyroyal, and the low-lying wetlands of western Kentucky. In many ways, the state's physical environment serves as a basic foundation that supports human activity. As Chapter 10 will attest, a diverse topography, together with the central location emphasized in Chapter 1 and the numerous rivers, streams, and lakes and reservoirs, have made Kentucky a destination for tourists from all over the eastern United States, as well as from other, more distant origins. These same reasons help to account for the diverse flora and fauna that reside in the Commonwealth. As Chapter 7 makes clear, the state's geologic history and humid subtropical climate have been ideal for the development of its two primary natural resources: coal and hardwood forests. And certainly the fertile soils in central and western Kentucky have played a critical role in forming the agricultural landscapes that are highlighted in chapter 8. In short, an understanding of the physical environment is crucial if we are to comprehend and nurture the rich, diverse, and continuously evolving human landscapes that are found in Kentucky today.

Historical and Cultural Landscapes

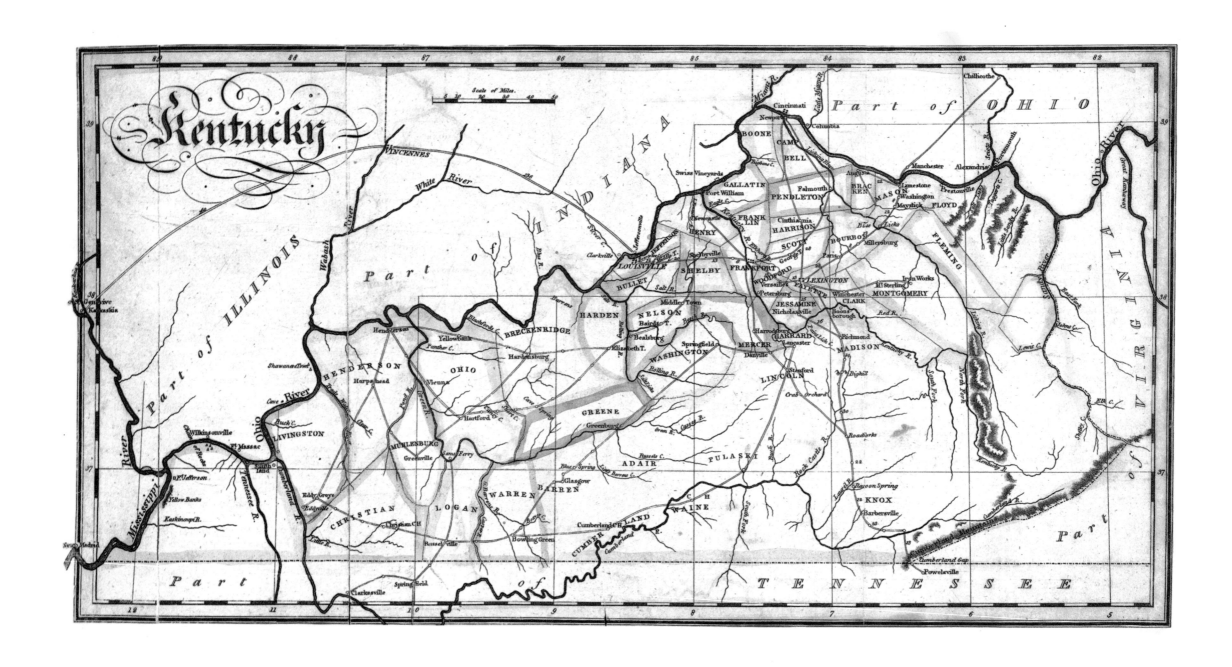

Previous page: *Maysville, a port on the Ohio River in Mason County,*
was the origin point of the Maysville-Lexington Turnpike.

CHAPTER THREE

Historical and Cultural Landscapes

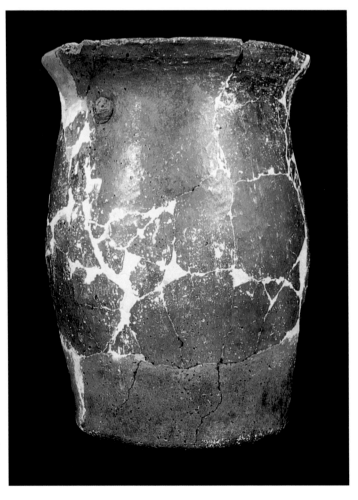

A plain jar made by the Adena people during the Woodland Period.

Native American Occupance

For more than ten millennia before the arrival of explorers and settlers of European ancestry, native peoples lived on Kentucky's lands, hunting game, fishing streams and rivers, and farming fertile bottomlands. As of 1994, researchers had identified 17,738 archaeological sites across the state that evinced traces of prehistoric settlements. Site maps in part reflect those places where archaeological surveys have been conducted, but these maps also record distinctive associations between prehistoric settlements and certain preferred environments.

The Paleoindian period (11,000 B.C. to 9,000 B.C.) bridged the waning centuries of the Pleistocene Ice Age into the present period, the Holocene. Paleoindian peoples were migratory hunters and foragers who sought bison, mammoths, mastodons, and other animals. This period is the least understood of the four prehistoric eras in Kentucky: the population density was probably quite low, and few Paleoindian sites have survived. Postglacial climatic conditions radically altered the character of environmental zones, affecting distributions of flora and fauna; for example, the migration patterns of birds and large animals changed. The number of sites identified in Christian County and adjoining areas suggests that Paleoindian occupancy was concentrated in the western Pennyroyal.

During the eight-thousand-year Archaic period (9,000 B.C. to 1,000 B.C.), American Indians began to restrict their movements to specific territories as they learned how to use a broad range of plant and animal species. The Archaic peoples, however, remained preagricultural. Through intensive fieldwork, archaeologists have identified many Archaic culture sites across Kentucky, especially in the Western Kentucky Coal Field's Green River basin and in the Cumberland Plateau. In western Kentucky, Archaic peoples frequented riverside locations, where they caught fish and shellfish and hunted deer and smaller game animals.

KNOWN ARCHAEOLOGICAL SITES, 1994

PALEOINDIAN PERIOD
11,000 B.C. - 9,000 B.C.

ARCHAIC PERIOD
9,000 B.C. - 1,000 B.C.

NUMBER OF SITES

- No sites
- 1 to 19
- 20 to 39
- 40 to 89
- 90 to 200

WOODLAND PERIOD
1,000 B.C. - A.D. 900

LATE PREHISTORIC PERIOD
A.D. 900 - A.D. 1750

WOODLAND PERIOD CULTURE AREAS

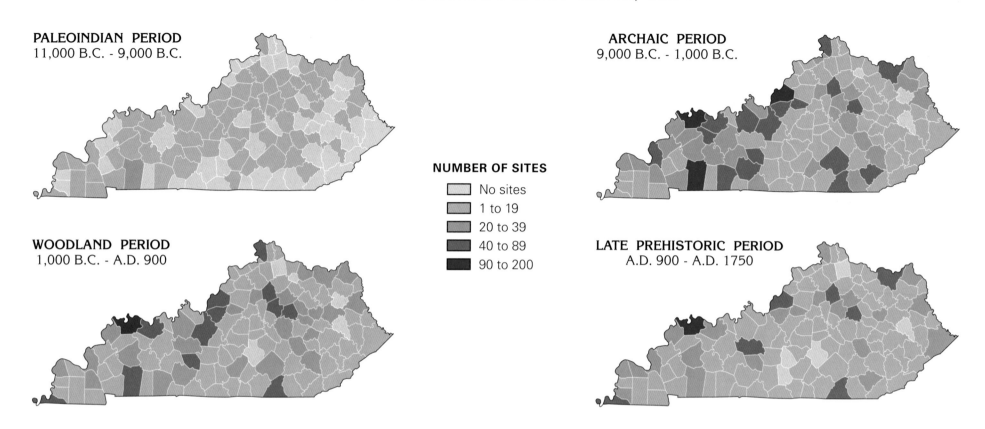

L. Michigan

HAVANA
HOPEWELLIAN

Illinois River

Kaskaskia River

Wabash River

OHIO HOPEWELLIAN

Serpent Mound

Ohio River

CRAB ORCHARD

Crab Orchard

HOPEWELLIAN

Mississippi River

Adena Park

ARMSTRONG
HOPEWELLIAN

Ohio River

Kanawha River

Cumberland River

Culture Regions
- Hopewellian complexes
- Adena-Hopewell heartland
- Other areas of burial mound construction (700 B.C. - A.D. 900)

Burial Mound Sites
- ■ Adena (700 B.C. - 100 B.C.)
- ■ Ohio Hopewellian (100 B.C. - A.D. 900)
- ▲ Crab Orchard Hopewellian (100 B.C. - A.D. 900)

An Adena site embraced by a circular moat on the bank of North Elkhorn Creek in Fayette County.

KENTUCKY BEFORE BOONE

LATE PREHISTORIC PERIOD
A.D. 1000-1700

WOODLAND PERIOD
1000 B.C.-A.D. 1000

ARCHAIC PERIOD
7000-1000 B.C.

PALEOINDIAN PERIOD
10000-7000 B.C.

Read from bottom to top, this diagram summarizes archaeological research findings by visually correlating domestic life scenes, the use of environments and resources, habitation types, and the evolution of tools, points, and pottery over a 12,000-year period.

Cultivation of local plants during the late Archaic period presaged the Woodland period (1,000 B.C. to A.D. 900), when elementary agriculture began to provide consistent foodstuffs to supplement resources obtained from hunting and foraging. Woodland peoples in the Ohio and Mississippi River Valleys grew plants such as sunflowers, squash and gourds, and native tobacco. Woodland sites are more evenly distributed across Kentucky than those of the Archaic period. Two distinct cultural traditions demarcate the Kentucky Woodland era: the mound-building Adena culture of central and northeastern Kentucky, and the non-mound-building Baumer and Crab Orchard cultures of western Kentucky. Kentucky's Adena people were closely associated with groups in southern Ohio and the middle Ohio River Valley, and such affinities placed their tradition within a much larger Ohio Valley Adena Culture region. Large mounds and other extensive earthworks dotted the Adena era countryside. Moats surrounded some mounds, such as the one at Adena Park, north of present-day Lexington. The presence of mounds as well as artifacts worked from Michigan copper, Illinois galena, North Carolina mica, Gulf Coast shell, and other exotic materials suggests that the Adena peoples were linked to other culture groups across the breadth of North America via extensive trade networks. The riverine cultures of western Kentucky, the Baumer and Crab Orchard groups—named for two important archaeological sites in southern Illinois—were closely related to Woodland groups elsewhere in the lower Ohio Valley, southern Illinois, and the middle Mississippi Valley.

By A.D. 900 the Late Prehistoric period was under way, and Kentucky was again occupied by representatives of two distinct cultural traditions: the temple mound-building Mississippian Culture in the west and the non-mound–building Fort Ancient Culture in the northeast. The diffusion of maize and beans northward from Mexico allowed Mississippian peoples to add important sources of nutrients to their diets. This development re-

duced the emphasis upon hunting but engendered a need to develop farming techniques and tools. Mississippians built large settlements along the banks of the Mississippi, the Tennessee, and the lower Ohio Rivers. They favored sites where fertile river floodplains could support centers based on intensive horticulture. Their stockade-enclosed villages coalesced around flat-topped temple mounds. In a general way, the agricultural Fort Ancient peoples were the ancestors of historic groups such as the Miami, the Delaware, and the Shawnee. Similarly, the Mississippian cultures were related to the historic Chickasaw and the Choctaw. Some Late Prehistoric sites in southeastern Kentucky may be related to the ancestral Cherokee.

The extent of native peoples' tribal areas at the time of initial European contact is not well known, and consequently maps showing the distributions of these areas must be regarded with care. Because most Indian groups encountered by seventeenth-century colonists were those living near coastal European settlements, records of coastal tribal areas are reasonably accurate. Knowledge of interior nations, on the other hand, those whose territories lay beyond the day-to-day contact with coastal settlements, is very generalized. Kentucky and the surrounding countryside had been occupied by many different native groups since the Paleoindian period; nevertheless, out-migration had already resulted in relative depopulation by the time whites arrived to record the presence of the native peoples. Cherokee people had once occupied a portion of the southeastern Cumberland Plateau, south of the Cumberland River, and the Chickasaw held the land that lay between the Tennessee and Mississippi Rivers. Shawnee had occupied an extensive area of what would become central Ohio, and their forays into Kentucky's Bluegrass in the 1770s threatened the security of white frontier settlements.

The trans-Appalachian Indian lands were crisscrossed by a network of bison, or "buffalo,"

LATE PREHISTORIC PERIOD CULTURE AREAS

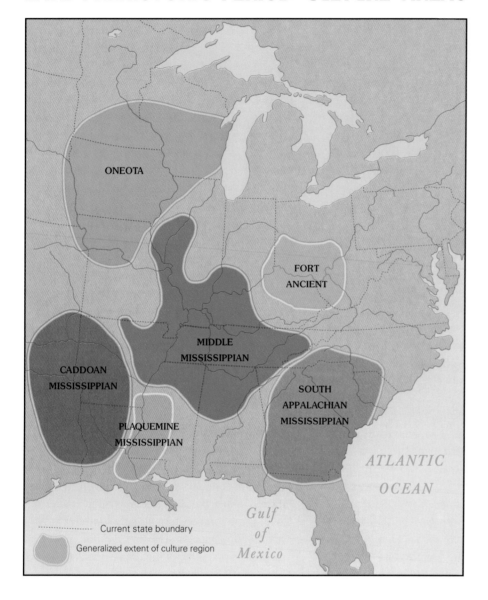

ONEOTA

FORT ANCIENT

MIDDLE MISSISSIPPIAN

CADDOAN MISSISSIPPIAN

SOUTH APPALACHIAN MISSISSIPPIAN

PLAQUEMINE MISSISSIPPIAN

ATLANTIC OCEAN

Gulf of Mexico

- - - - Current state boundary

Generalized extent of culture region

TRIBAL TERRITORIES AT TIME OF EUROPEAN CONTACT

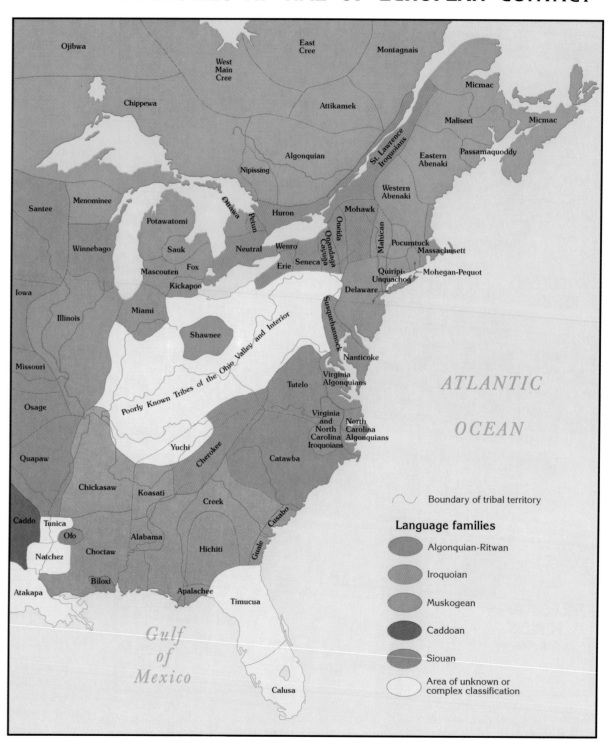

Boundary of tribal territory

Language families

Algonquian-Ritwan

Iroquoian

Muskogean

Caddoan

Siouan

Area of unknown or complex classification

PIONEER AND INDIAN TRAILS

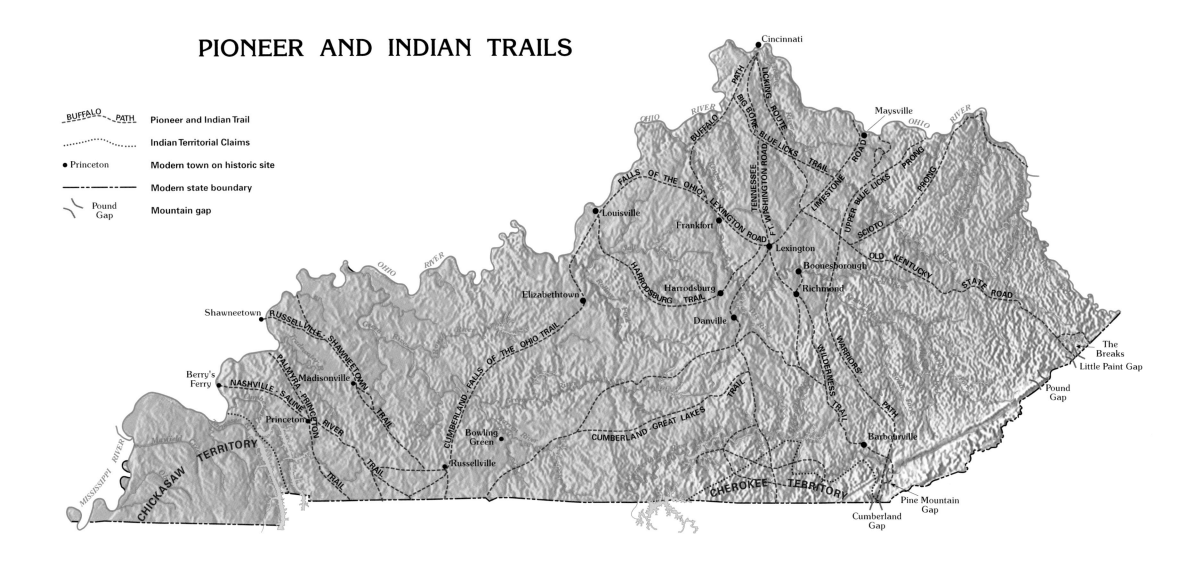

traces, overland trails, and stream and river routes. Some trails followed ancient bison paths that might have focused on salt licks; others linked up with distant routes that carried Indian traders far to the north or south of Kentucky. During the 1750s scouts such as Christopher Gist and Thomas Walker explored the Cumberland Mountains and traced routes into the Kentucky country via the Ohio River on the north and via the Indian trail along the Cumberland Gap and Cumberland River–Pine Mountain Gap route in the southeast. In 1774 Pennsylvanian James Harrod established a settlement that would bear his name south of the Kentucky River. The fort built on the site functioned as a

link between the Falls of the Ohio and Lexington via the Harrodsburg Trail. Other explorers and hunters, including Daniel Boone, carried reports back to residents in the East of expansive fertile lands beyond the Cumberlands.

By the 1780s, when white settlers began arriving in Kentucky in substantial numbers to claim Virginia military land warrants or to purchase land, few Indian settlements remained to contend the advance. Formerly, a quarter-million Cherokee in the southeastern states had hunted in Kentucky; by about 1650, however, diseases introduced by European explorers had reduced the Cherokee population to one-tenth of its former size. Treaties

with land companies forced the Cherokee to abandon claim to Kentucky hunting lands by the mid-1770s. Shawnee may have occupied villages near the Ohio River in northeastern Kentucky; from this strategic location their traders could reach the Great Lakes, the Atlantic, or the Gulf Coast. In the 1670s the Shawnee were driven from the area by the Iroquois, although they returned almost a century later, only to find that treaties and white land speculators had removed any claim they had to Kentucky lands. In Kentucky's far west, the Chickasaw had avoided encounters with French fur traders and treaties with land speculation companies, and by the 1780s they still retained the territory between

TRIBAL LAND CLAIMS
circa 1787

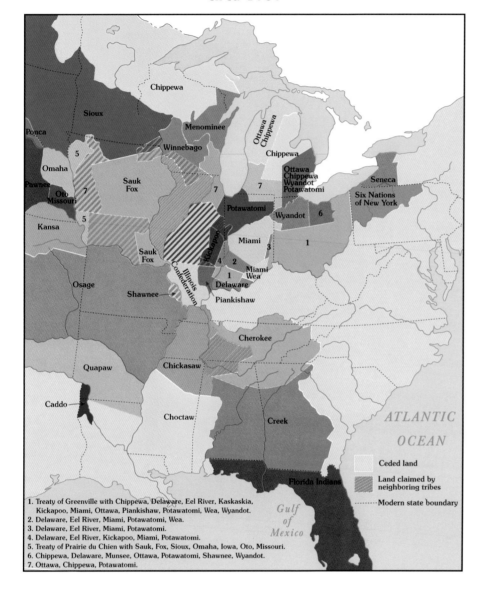

1. Treaty of Greenville with Chippewa, Delaware, Eel River, Kaskaskia, Kickapoo, Miami, Ottawa, Piankishaw, Potawatomi, Wea, Wyandot.
2. Delaware, Eel River, Miami, Potawatomi, Wea.
3. Delaware, Eel River, Miami, Potawatomi.
4. Delaware, Eel River, Kickapoo, Miami, Potawatomi.
5. Treaty of Prairie du Chien with Sauk, Fox, Sioux, Omaha, Iowa, Oto, Missouri.
6. Chippewa, Delaware, Munsee, Ottawa, Potawatomi, Shawnee, Wyandot.
7. Ottawa, Chippewa, Potawatomi.

TRAVEL ROUTES OF THE CHEROKEE REMOVAL
(TRAIL OF TEARS -1838)

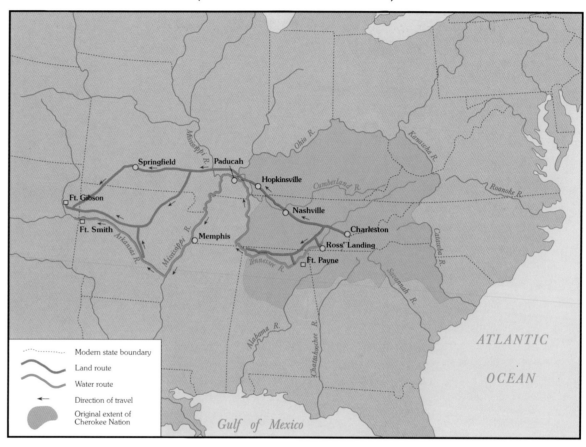

the Tennessee and Mississippi Rivers. In 1818 Kentucky Governor Isaac Shelby and General Andrew Jackson negotiated the purchase of approximately eight thousand square miles of Chickasaw land for three hundred thousand dollars; just over two thousand square miles lay in Kentucky, and this territory became known as the Jackson Purchase. The land proved to have rich loess soils and enormous potential as a transportation hub.

Two decades after the acquisition of the Jackson Purchase, Kentucky land was traversed a last time by some of the native peoples who had hunted its woodlands and farmed its soils for generations. In 1838 President Martin Van Buren ordered that the U.S. Army remove the Cherokee from their Appalachian mountain homeland in Tennessee, North Carolina, Alabama, and

Georgia. Under General Winfield Scott, about nineteen thousand Cherokee were evicted from their homes and confined in stockades. They were then sent to designated Indian lands in what would later be called Oklahoma. One group traveled via steamboats and barges on the Tennessee River, passing through western Kentucky relatively quickly to reach Indian Territory by the summer of 1838. A much larger contingent traveled overland along three major routes. Those who took the Northern Route traversed the western Kentucky Pennyroyal by way of Hopkinsville and Mantle Rock, crossing the Ohio River at Berry's Ferry into southern Illinois. So many Cherokee died along the way that the route became known as the Trail of Tears. A park in Hopkinsville commemorates the event.

Trail of Tears Park, Hopkinsville.

"A New Map of North America," 1794.

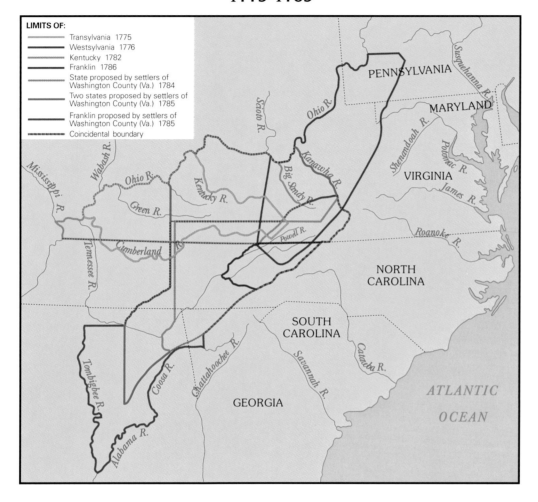

PROPOSED STATES
WEST OF THE ALLEGHENY MOUNTAINS
1775-1785

LIMITS OF:
- Transylvania 1775
- Westsylvania 1776
- Kentucky 1782
- Franklin 1786
- State proposed by settlers of Washington County (Va.) 1784
- Two states proposed by settlers of Washington County (Va.) 1785
- Franklin proposed by settlers of Washington County (Va.) 1785
- Coincidental boundary

Early Settlement Patterns of European Americans

Despite the prior interests of Native Americans, Europeans and European Americans indicated their own claim to lands that would become Kentucky on the earliest imperial maps of eastern North America. Grants to the Virginia Company by James I in 1603 and to Robert Heath by Charles I in 1629 included parts or all of the future state, although the lands were located far inland from the seventeenth-century settlement frontier. American independence from Great Britain prompted a speculative land rush beyond the crest of the Appalachian Mountains. This rush embroiled the new federal government, the original states, private land development companies, and settlers, as boundary lines and jurisdictions were argued and formalized.

New York maintained a claim on Kentucky that it did not cede until 1782. In 1778 and 1781 Virginia set aside the land around the Green River to reward soldiers for their military service. Schemes such as the Transylvania Company included parts of Kentucky in proposals for new colonies or states west of the Appalachians. Further, Thomas Jefferson divided the future commonwealth among two of the fourteen new states proposed in the Ordinance of 1784. The present state was designated as part of Virginia's Fincastle County, formed in 1772. In 1776 a group meeting in Harrodsburg elected George Rogers Clark and John Gabriel Jones to petition the Virginia legislature to designate the western portion of Fincastle County as Kentucky County. The legislature abolished Fincastle County, replacing it with Montgomery, Washington, and Kentucky Counties. Except for the future Jackson Purchase, Kentucky County was recognizable as the modern state. Eighteenth-century Kentucky was part of Virginia's western territory. Isolated from the Virginia government in Richmond, early residents wanted to establish an independent state and a government to which they had ready access to assure representation of their economic and social interests. The state was delimited by boundaries that followed a major latitudinal parallel that separated Kentucky from Tennessee to the south, and the Ohio River, which formed a natural boundary with the Northwest Territories to the north. The eastern boundary followed the Big Sandy River and the crest of Cumberland Mountain in large measure. The ultimate western boundary was the Mississippi River.

Meanwhile, individual explorers and settlers gradually increased their knowledge of the frontier region. These explorers included men such as Thomas Walker, who worked for the Loyal Land Company in 1750, and Christopher Gist, who worked for the Ohio Company in 1750–52. The most famous explorer was Daniel Boone, whose association with Kentucky began with his travels along the Warriors' Path in 1767. Boone also cut a road into the territory for the Transylvania Company in 1775, and he eventually played a role in fostering the settlement of Boonesborough, a village that grew from Boone's Fort on the Kentucky River. Knowledge gleaned by these and

COUNTY EVOLUTION

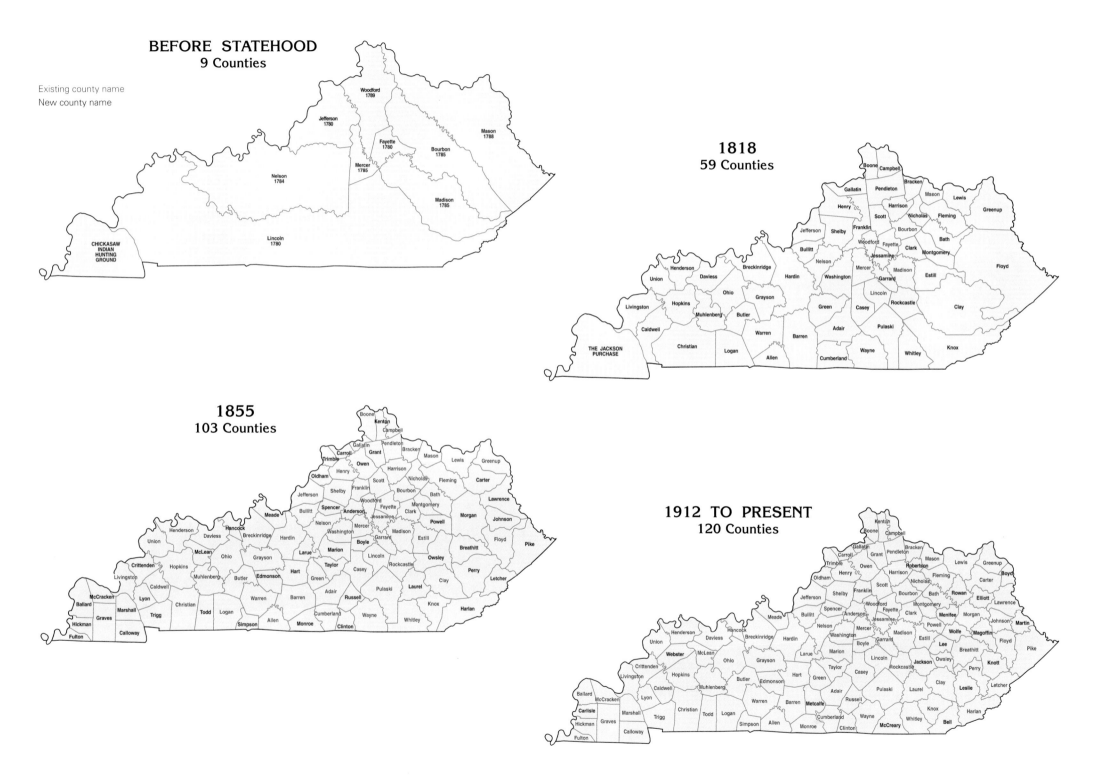

BEFORE STATEHOOD
9 Counties

Existing county name
New county name

Woodford
1789

Jefferson
1780

Mason
1788

Fayette
1780

Bourbon
1785

Mercer
1785

Nelson
1784

Madison
1785

Lincoln
1780

CHICKASAW
INDIAN
HUNTING
GROUND

1818
59 Counties

Boone
Campbell
Bracken
Gallatin
Pendleton
Mason
Lewis
Henry
Harrison
Scott
Nicholas
Fleming
Greenup
Jefferson
Shelby
Franklin
Bourbon
Woodford
Fayette
Bath
Clark
Montgomery
Henderson
Breckinridge
Bullitt
Nelson
Jessamine
Madison
Estill
Floyd
Union
Daviess
Hardin
Washington
Mercer
Garrard
Ohio
Grayson
Lincoln
Rockcastle
Clay
Livingston
Hopkins
Butler
Green
Casey
Caldwell
Warren
Adair
Pulaski
Christian
Barren
Wayne
Whitley
Knox
Logan
Allen
Cumberland

THE JACKSON
PURCHASE

1855
103 Counties

Boone
Kenton
Campbell
Gallatin
Pendleton
Carroll
Grant
Bracken
Trimble
Owen
Mason
Lewis
Greenup
Oldham
Henry
Harrison
Nicholas
Fleming
Carter
Jefferson
Shelby
Franklin
Scott
Bourbon
Bath
Spencer
Anderson
Woodford
Fayette
Montgomery
Lawrence
Henderson
Meade
Bullitt
Jessamine
Clark
Powell
Morgan
Johnson
Daviess
Breckinridge
Nelson
Mercer
Madison
Estill
Union
Hancock
Hardin
Washington
Boyle
Garrard
Floyd
Pike
McLean
Ohio
Larue
Marion
Lincoln
Owsley
Breathitt
Crittenden
Hopkins
Grayson
Taylor
Casey
Rockcastle
Perry
Livingston
Muhlenberg
Butler
Hart
Green
Clay
Letcher
Caldwell
Edmonson
Adair
Russell
Laurel
Ballard
McCracken
Lyon
Warren
Barren
Pulaski
Knox
Harlan
Graves
Marshall
Christian
Todd
Logan
Simpson
Allen
Monroe
Cumberland
Wayne
Whitley
Hickman
Calloway
Fulton

1912 TO PRESENT
120 Counties

Kenton
Boone
Campbell
Gallatin
Grant
Pendleton
Bracken
Carroll
Mason
Lewis
Greenup
Trimble
Owen
Harrison
Robertson
Fleming
Boyd
Oldham
Henry
Scott
Nicholas
Rowan
Carter
Elliott
Franklin
Bourbon
Bath
Montgomery
Jefferson
Shelby
Woodford
Clark
Menifee
Morgan
Lawrence
Spencer
Anderson
Fayette
Powell
Wolfe
Johnson
Martin
Jessamine
Magoffin
Meade
Bullitt
Nelson
Mercer
Madison
Estill
Lee
Floyd
Pike
Washington
Garrard
Breathitt
Knott
Henderson
Hancock
Breckinridge
Hardin
Boyle
Lincoln
Jackson
Owsley
Perry
Union
Daviess
Larue
Marion
Rockcastle
Leslie
Letcher
Webster
McLean
Ohio
Grayson
Taylor
Casey
Clay
Crittenden
Hart
Green
Pulaski
Laurel
Livingston
Hopkins
Butler
Edmonson
Adair
Russell
Harlan
Ballard
McCracken
Lyon
Caldwell
Muhlenberg
Warren
Barren
Metcalfe
Knox
Bell
Carlisle
Marshall
Christian
Todd
Logan
Simpson
Allen
Cumberland
Wayne
Whitley
Hickman
Graves
Trigg
Monroe
Clinton
McCreary
Fulton
Calloway

SETTLEMENT PATTERNS, 1790-1910

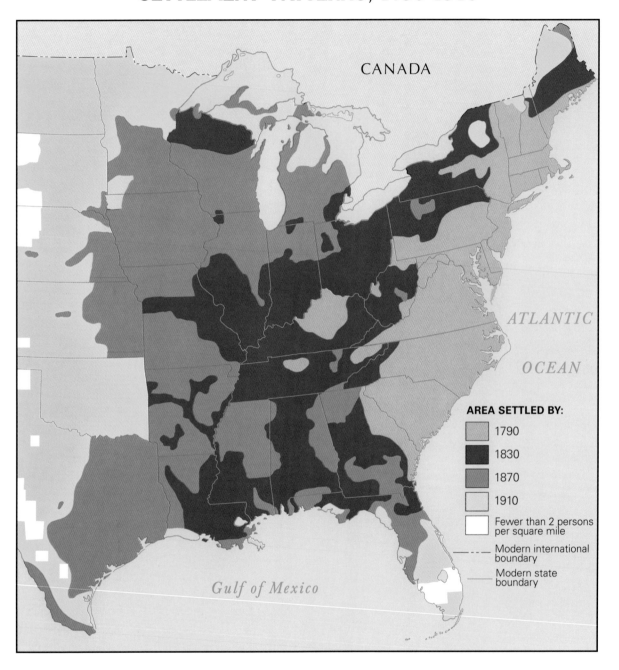

CANADA

ATLANTIC

OCEAN

AREA SETTLED BY:

1790

1830

1870

1910

Fewer than 2 persons per square mile

Modern international boundary

Modern state boundary

Gulf of Mexico

other early European American explorers was applied in the preparation of contemporary maps of the region, facilitating the area's inclusion in American plans for imperial westward expansion.Active settlement of the Kentucky Bluegrass region quickly followed the American Revolution, and migrants from the East arrived in increasing numbers via the Wilderness Road, which ran from Cumberland Gap and Pine Mountain Gap to Lexington; the Midland Trail, from Ashland to Lexington and Louisville; and the Limestone Road, from Limestone (Maysville) on the Ohio River to Lexington. Many of these routes formed the backbone of the Commonwealth's first transportation network, and much of this network still exists. For example, the Limestone Road became U.S. Highway 68, and sections of the Wilderness Road extending north and west from Cumberland Gap and Pine Mountain Gap toward Boonesborough became U.S. 25 and U.S. 25E.

Although statehood was secured in 1792 and routes into the new state were increasingly frequented by migrating settlers, Kentucky remained a remote frontier. Displaced Native Americans continued to attack white settlements. In April 1793, for example, Native Americans captured Morgan's Station east of Mount Sterling and carried away nineteen women and children. Nevertheless, larger settlements such as Lexington and Louisville were relatively secure and were effectively linked to eastern coastal cities by way of overland trails and Ohio River flatboats. Lexington, though not on a navigable stream, received regular shipments of goods from eastern ports by way of the Ohio River and the Limestone Road. Residents could shop at any of several dry goods stores and could purchase a surprisingly broad range of commodities.

In 1792 a Lexington newspaper announced that John Moylan, whose store stood next door to the Buffalo Tavern, had received a fresh shipment of dry goods and groceries. Moylan's ledger for the year lists purchases by community residents. Accounted in pounds, shillings, and pence, the ledger reveals the diversity of commodities then available in Lexington:

Household items: knives and forks, iron spoon, pewter hardware, glassware, looking glass, coffee pot, frying pan, Dutch oven, thread on bobbins, silk thread, needles, knitting needles, thimble, scissors, linen, indigo, starch, rosin, bed sheets, blankets, carpet, beeswax, paste board, Castile soap, shaving soap, candles, spectacles, stationary [sic], spelling book.

Construction materials and tools: nails, brads, hinges, locks and latches, door knobs, putty, plank, axe, roofers' adze, coopers' adze, draw knife, spade, hand vise.

Clothing and yard goods: boots, ladies shoes, leather shoes, coarse shoes, shoe buckles, silk handkerchiefs, broad cloth, white cloth, muslin, blue coating material, calico, silk material, ribbon, buttons, wool hats, castor [beaver] hats, worsted hosiery.

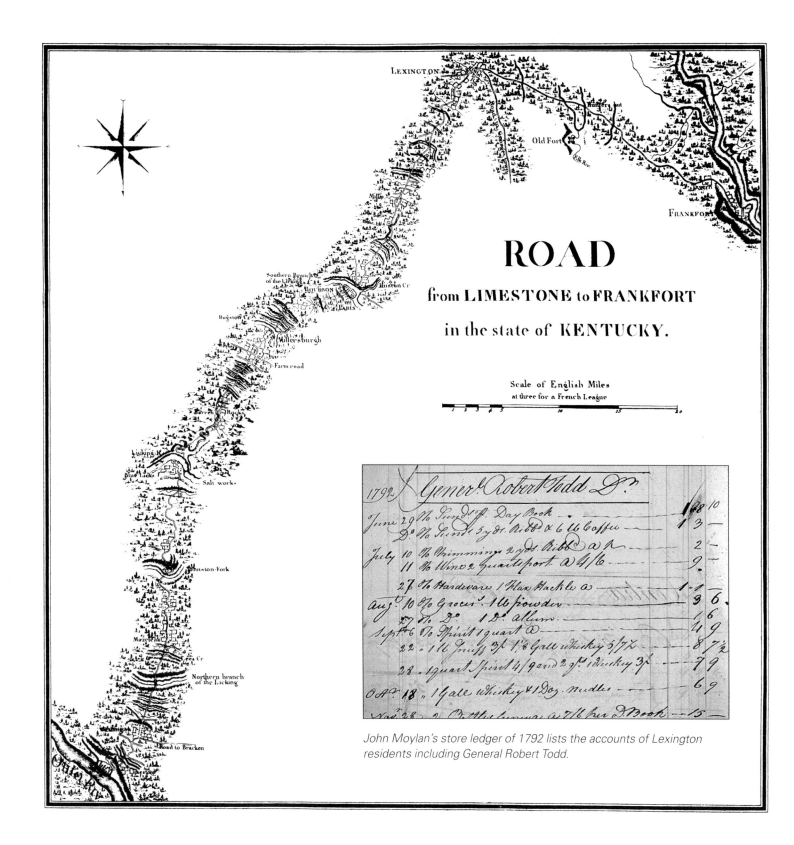

ROAD

from **LIMESTONE** to **FRANKFORT**

in the state of **KENTUCKY**.

Scale of English Miles
at three for a French League

John Moylan's store ledger of 1792 lists the accounts of Lexington residents including General Robert Todd.

Food stuffs and spirits: wine, Lisbon wine, port, sherry, gin, whiskey, butter, cheese, salt, mustard in bottles, cayenne pepper, cinnamon, cloves, nutmeg, mace, Hyson tea, Louchong tea, raisins, sugar, loaf sugar, vinegar, bacon.

Other goods: snuff, rifles, flints, gun locks, powder and shot, lead, whip, riding whip, bridle, girth buckles, twine and fish hooks, surveyors' chain, heavy duty chain, brimstone, saddle cloth, saddle bags, oats.

Kentucky's historic settlement by eastern Americans took place within the turbulent context of war between the American colonists and their mother country, Great Britain. The contentious political climate meant that attempts to settle Kentucky would involve even more adversity than was common elsewhere on the advancing frontier. Principal among these hardships was the potential for raids carried out by native groups such as the Shawnee. The settlers had to be prepared to defend themselves at any moment.

The necessity of building a defensible structure that could also be used as a residence yielded an unusual type of structure and site, a place abandoned as soon as the political situation improved. Early residents referred to such fortified houses or groups of houses as "stations." Settlers established more than 150 stations in the Inner Bluegrass Region, where the state's earliest postrevolutionary settlement took place. Upland settings with permanent freshwater springs were preferred for station sites. A trail or trace frequently ran near the site, and stations often served as landmarks along the nascent road network, providing protection and shelter for traveler and new settler alike. Defensive precautions varied widely. Station form varied from a lone barricaded cabin to a group of cabins protected within a stockade that housed stables and a blockhouse. Some sites had no stockade and had only one or more cabins that could be barricaded; others comprised several log cabins arranged in a square or rectangle, with log picketing or stockading filling in the gaps between the walls of adjacent structures. Strode's Station near present-day Winchester, for example, was described as having fourteen or fifteen cabins on each of two long sides of a rectangle and eight cabins on the shorter ends. Blockhouses stood at three of the corners. This site was a smaller version of the much larger sites, called forts, built at Lexington, Harrodsburg, Boonesborough, and Louisville.

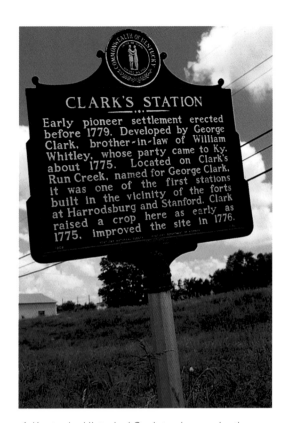

CLARK'S STATION

Early pioneer settlement erected before 1779. Developed by George Clark, brother-in-law of William Whitley, whose party came to Ky. about 1775. Located on Clark's Run Creek, named for George Clark, it was one of the first stations built in the vicinity of the forts at Harrodsburg and Stanford. Clark raised a crop here as early as 1775, improved the site in 1776.

A Kentucky Historical Society sign marks the site of George Clark's Station near Danville in Boyle County.

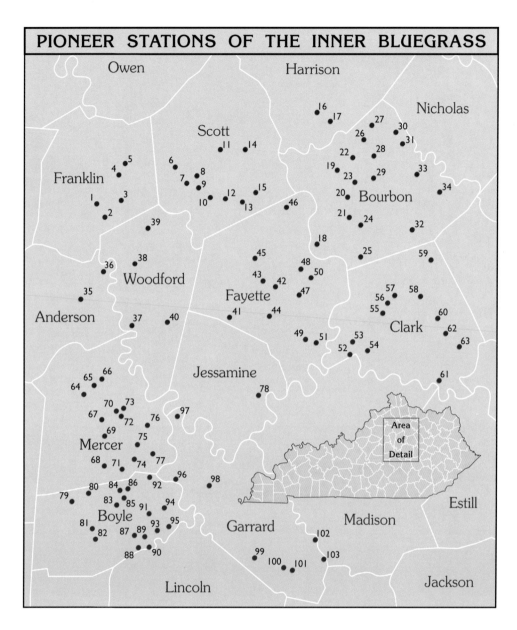

PIONEER STATIONS OF THE INNER BLUEGRASS

PIONEER STATIONS

1. William Haydon
2. James Arnold
3. John Major
4. William Goar
5. Henry Innes
6. Anthony Lindsay
7. John Scott
8. Thomas Herndon
9. William Campbell
10. Robert Johnson
11. Ash Emison
12. John McClelland
13. Matthew and Francis Flournoy
14. Stephen Archer
15. Laban Shipp
16. Benjamin Harrison
17. Isaac Ruddell
18. John Grant
19. William McGee
20. William McConnell
21. William Thomas
22. John Cooper
23. Robert Clarke
24. Dr. Henry Clay
25. Samuel Curtwright
26. John Kiser
27. Samuel McMillan
28. John Martin
29. Peter and James Houston
30. William Miller
31. John Miller
32. Swinney
33. James Sodowski
34. Henry Wilson
35. Jacob Coffman
36. Samuel Hutton
37. General Charles Scott
38. Jesse Graddy
39. George Blackburn
40. John Craig
41. Levi Todd
42. Francis McConnell
43. William McConnell
44. John Todd
45. James Masterson
46. William Grant
47. John Craig
48. William Bryant
49. Daniel Boone
50. John Rogers
51. John Boofman
52. John Holder
53. David McGee
54. William Bush
55. Stephen Boyle
56. John Strode
57. John Constant
58. William Bramblett
59. John Donaldson
60. Edmund Raglund
61. Dunaway
62. William Scholl
63. Frazier
64. James McAfee
65. John Meaux
66. John Bunton
67. Isaac Hite
68. Henry Wilson
69. William McAfee
70. David Williams
71. Gabriel Madison
72. Hugh McGary
73. Thomas Denton
74. Azor Rees
75. Lewis Rose
76. John Gordon
77. John Bowman
78. Jacob Hunter
79. John Harberson
80. Silas Harlan
81. Irvine
82. Robert Caldwell
83. William Field
84. James Harrod
85. John Cowan
86. Low Dutch
87. James Brown
88. William Warren
89. James Wilson
90. Edward Worthington
91. John Crow
92. William McBride
93. George Clark
94. Stephen Fisher
95. John Dougherty
96. Samuel and George Scott
97. William Grant
98. James Smith
99. Lewis Craig
100. Zophar Carpenter
101. John Kennedy
102. William Miller
103. Humphrey Best

Even though stations were usually occupied for short periods, they played a vital supporting role in Kentucky's settlement and proved to be an effective solution to the need for relatively safe housing. They served as places of defense and sanctuary, as residences, as way stations, and as nodes on an emerging regional communication network.

Probably the best known of Kentucky's early forts is Fort Boonesborough, which was founded in 1775 on a broad Kentucky River floodplain, a topographic rarity on that river, which otherwise features a narrow entrenchment within nearly vertical limestone palisades. The site was intended to be the headquarters for Richard Henderson's Transylvania colony. Henderson's surveyor laid out sixty-six two-acre town lots that were assigned to individuals, and Henderson himself drew up a plan for a fort that was to incorporate four corner blockhouses with eight cabins and a central gate in each long side and five cabins along each short side. Fort construction required several years, and by 1778 several cabins, some blockhouses, and the wall picketing were in place. The Virginia legislature nullified Henderson's interests in the settlement in 1778, and the following year a charter was granted to a group of settlers for a town to be laid out at the site. A decade later Boonesborough became a tobacco inspection and shipping point, and a tobacco warehouse was erected. Early Boonesborough was never a large town, and in 1810 the U.S.

BOONESBOROUGH, KENTUCKY
circa 1790

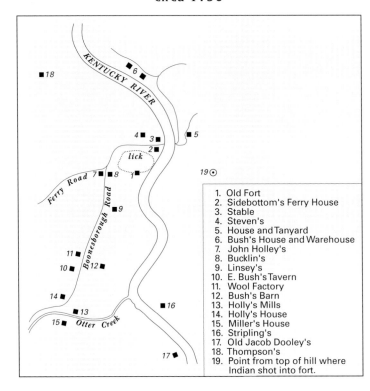

1. Old Fort
2. Sidebottom's Ferry House
3. Stable
4. Steven's
5. House and Tanyard
6. Bush's House and Warehouse
7. John Holley's
8. Bucklin's
9. Linsey's
10. E. Bush's Tavern
11. Wool Factory
12. Bush's Barn
13. Holly's Mills
14. Holly's House
15. Miller's House
16. Stripling's
17. Old Jacob Dooley's
18. Thompson's
19. Point from top of hill where Indian shot into fort.

BOONESBOROUGH TOWN PLAN, 1809

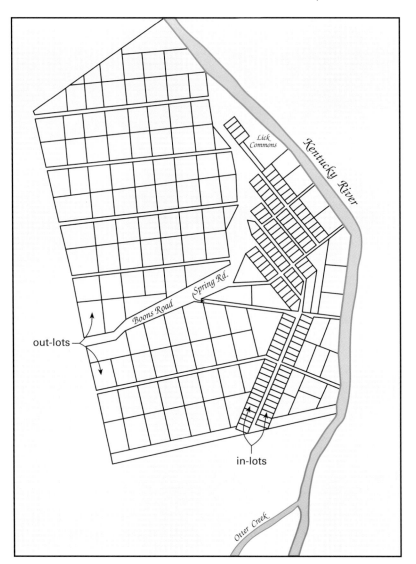

Fort Boonesborough State Park, Madison County. This modern fort is a reproduction built above the original fort's flood plain site. Employees dress in period costume and demonstrate craft work for visitors.

census listed only eight households made up of sixty-eight people. An ambitious town plan was filed in the Madison County clerk's office that same year, but few houses were built and occupied by townspeople. Most of the town lots would be consolidated and farmed. In 1963 almost sixty acres of the original Boonesborough site was deeded to the Kentucky Department of Parks. A swimming beach, campground, and picnic area now occupy the land, and an idealized replica of a fort was built as a tourist attraction on a hill overlooking the site.

Constructing defensible stations and forts required that settlers consider a topographic vantage point, a reliable supply of fresh water, and access to cross-country trails. Establishment of a town was another matter. Were towns simply to evolve, adding streets and building clusters wherever someone decided to put them? Almost from the beginning of effective European American settlement, the commercial Kentucky frontier was served by towns such as Harrodsburg, Lexington, Louisville, Lancaster, and Maysville (Limestone). These new urban places took advantage of both site and situation. Lexington, for instance, was located in the middle of a rich agricultural region; Paris, along a busy turnpike road; and Louisville, at the Falls of the Ohio, where the presence of dangerous rapids ensured that rational boat pilots would put ashore and portage.

The forms of Kentucky towns were not new inventions. Rather, early residents often embraced a formal plan derived from previous experience or copied from a town or city "back east." Many surveyors of early Kentucky towns laid out streets in a grid. Several towns, Harrodsburg and Lexington among them, allotted land to residents according to a two-tiered system of in-lots and out-lots. A settler could obtain an in-lot of about one-half acre as a residence site and an out-lot of four to six acres at the settlement's perimeter. On this out-lot the settler could establish a pasture, a small farm, or a garden. Virginia settlements had used this type of land subdivision as early as the late sixteenth century. The Virginia legislature created Kentucky County in 1776; Harrodsburg would be the first county seat. The Virginia Land Act of 1779 specified that the standard size for future towns would be 640 acres, or one square mile.

EARLY TOWN PLAN OF HARRODSBURG
circa 1786-1787

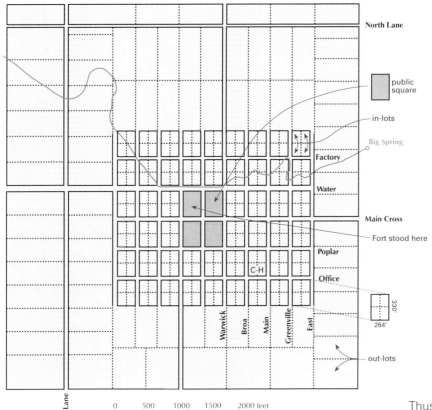

North Lane

public square

in-lots

Big Spring

Factory

Water

Main Cross

Fort stood here

Poplar

Office

out-lots

Warwick · Broa · Main · Greenville · East

C-H

West Lane

330'

264'

0   500   1000   1500   2000 feet

Approximate Scale

Above, *Paris was established on the south bank of Stoner Creek along the Maysville–Lexington Turnpike. The old route passes the couthouse and coincides with Main Street. Paralleled by streets one block to either side, this town form is similar to National Road towns in Pennsylvania and Ohio.* Right, *Paris, 1877.*

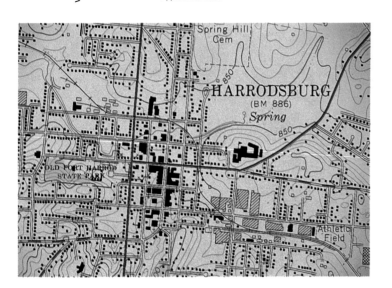

Harrodsburg, 1952.

Thus Harrodsburg, Lexington, and other early town sites followed the in-lot, out-lot form; the lots were oriented in a grid and were delimited by streets. The present-day map of Harrodsburg exhibits many modifications to the original plat, yet the basic mile-square street grid remains intact. Other town plans also evolved. Paris, the seat of Bourbon County, for example, was established in 1789, and its plan closely resembles a Pennsylvania road town, although it also appears to have included the Virginia in-lot and out-lot system. Main Street is the town's spine and is coincident with the Maysville-to-Lexington turnpike. The town core extends for ten blocks parallel to the turnpike and is four blocks wide.

In many Kentucky counties the most important town was the county seat. This was the place of county governance and included the courthouse as well as offices of attorneys and surveyors. It also functioned as a service center for surrounding farms. Surveyors often platted county seat towns by following models of grid-patterned towns in the eastern states, especially Pennsylvania. William Penn had planned

Philadelphia as a grid-patterned city in 1682. The Philadelphia plan placed an open square in the center of a uniform street grid, and this square offered an ideal place to erect an important public building. Surveyors used similar town layouts in the counties west of Philadelphia, where the courthouse was erected on a central square, the focus of through-streets. One of these towns was Lancaster, Pennsylvania. In 1797 people from Lancaster moved to Garrard County, Kentucky, and founded Lancaster as the county seat. The new Kentucky town was platted in the same style as its Pennsylvania progenitor, and residents erected a courthouse in the square in 1799.

During the eighteenth century, East Coast colonies expanded westward, and European settlers extended their frontier across the Appalachians, often into lands being "developed" by eastern or European investors. Land company promoters often surveyed land and platted town sites in advance of settlement and then attempted to recruit prospective residents. Some promotional towns in Kentucky were settled, but others, such as Lystra, were never more than plans on paper. A group of speculators in London, England, bought fifteen thousand acres of land along the Salt River's Rolling Fork and planned Lystra as a grid-patterned town, complete with central square, lots, and street names, in 1794. Buyers never materialized; neither did the town. Out of such frontier ferment Lexington arose as the first important town of the trans-Appalachian west, although by 1820 Louisville had overtaken it as the state's primary urban center.

Behind the explorations, legal divisions, and town foundings were the needs and desires of the people who increasingly looked to the trans-Appalachian west for commercial and agricultural opportunity, especially after the American Revolution. Following the early explorations and reports, a handful of frontier settlements existed by 1780. By 1790 parts of what would soon become the Commonwealth of Kentucky could be called "well settled"—especially northern Kentucky across from Cincinnati, the Bluegrass Region between Lexington and Louisville, and the fertile lands connecting the Bluegrass with the Nashville Basin in Tennessee. Although Kentucky may have been isolated from the East Coast states in its earliest years of statehood, that situation was temporary. River and overland connections to the rest of the country were crucial to the state's commercial success. Kentucky's integration into the fabric of the new American nation is illustrated by the state's position in a national system of post roads that existed early in the nineteenth century. Kentucky's farmers, merchants, and, soon, industrialists quickly adapted to this most extensive political, economic, and social framework.

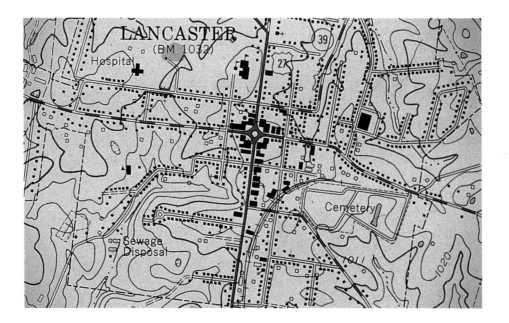

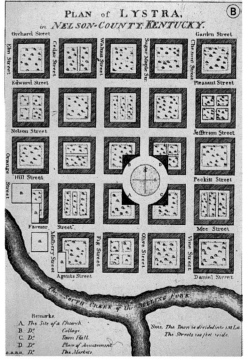

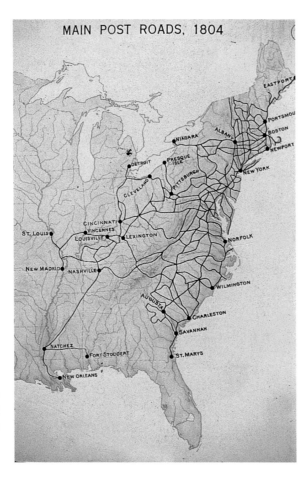

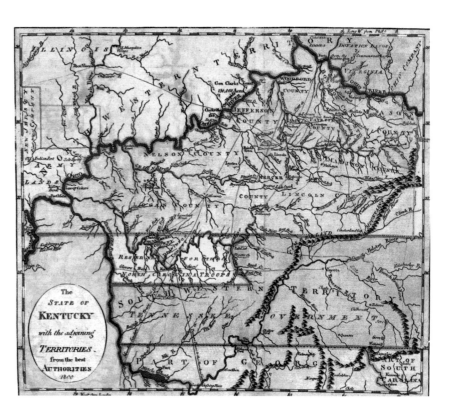

Top left, *Lancaster, 1952.* Top right, *Plan of Lystra, 1795.* Lower left, *main post roads, 1804.* Lower right, *Kentucky with Adjoining Territories, 1800.*

NATURAL VEGETATION PRIOR TO STATEHOOD

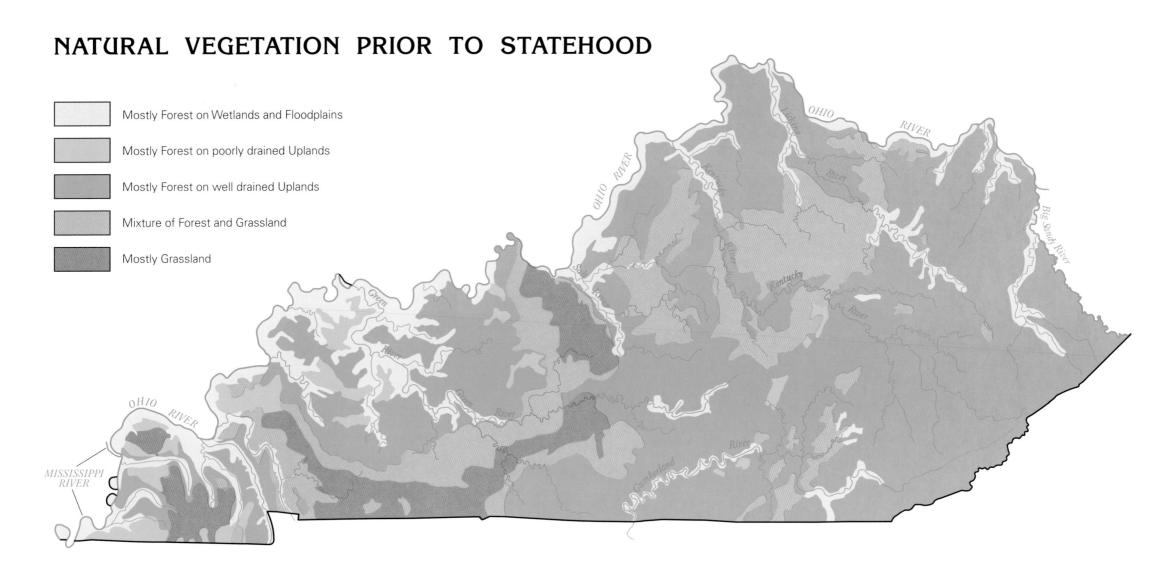

Mostly Forest on Wetlands and Floodplains

Mostly Forest on poorly drained Uplands

Mostly Forest on well drained Uplands

Mixture of Forest and Grassland

Mostly Grassland

Occupying the Land

Political Organization.   As a political unit, the Commonwealth of Kentucky has a governmental jurisdiction delimited by its eighteenth-century boundaries. Within that territory the state is responsible for administering a variety of programs and services that it provides for its residents. The state's administrative geography also includes nested regional and local governmental units. The state's territory is divided into 120 counties that are charged by the constitution with specific responsibilities. Cities and towns also have their own governmental jurisdictions. Counties are important political units, in part because they were created by local or regional interest groups. Between 1792 and 1912 new coun-

ties split from established ones because of population growth, economic rivalries, or family interests. New boundaries were delimited and demarcated. The state had 4 counties in 1784: Jefferson, Fayette, Lincoln, and Nelson. By 1800 there were 42. By the time Kentucky achieved statehood in 1792, it had been subdivided into 9 counties with the addition of Bourbon, Madison, Mercer, Woodford, and Mason. Between 1801 and 1820, surveyors laid out an additional 26; between 1821 and 1840, 22 were added, followed by another 20 in the next two decades. Knott and Carlisle became counties between 1881 and 1900, and the legislature approved the last county, McCreary, in 1912. University of Kentucky historian Robert Ireland has called these local

units "little kingdoms," a sobriquet reflecting the influential role that powerful individuals or families have played in local politics. Today, Kentucky has the third largest number of counties in the United States, after Texas with 245 and Georgia with 159.

Farms and Plantations.   According to Virginia land law, settlers could legally occupy Kentucky land by obtaining a warrant, filing an acreage entry with a county surveyor, getting the land surveyed, and recording the new plat with the appropriate land office. The land office would then issue to the settler a patent to

SHINGLED LAND CLAIMS
SHELBY COUNTY

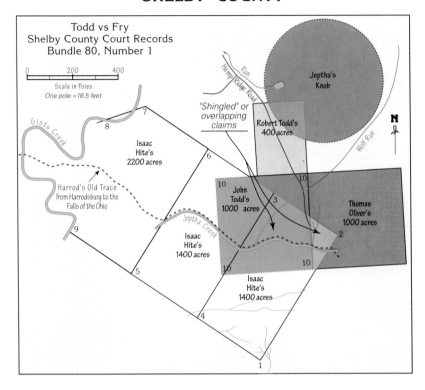

Todd vs Fry
Shelby County Court Records
Bundle 80, Number 1

0   200   400
Scale in Poles
One pole = 16.5 feet

Gists Creek

Harrod's Old Trace
from Harrodsburg to the
Falls of the Ohio

Hemp Ridge Road

Run

Jeptha's Knob

Wolf Run

"Shingled" or
overlapping
claims

Robert Todd's
400 acres

Isaac
Hite's
2200 acres

John
Todd's
1000 acres

Jeptha Creek

Isaac
Hite's
1400 acres

Thomas
Oliver's
1000 acres

Isaac
Hite's
1400 acres

N

SLAVES HELD, 1860

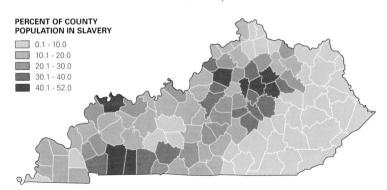

PERCENT OF COUNTY
POPULATION IN SLAVERY

- 0.1 - 10.0
- 10.1 - 20.0
- 20.1 - 30.0
- 30.1 - 40.0
- 40.1 - 52.0

TOBACCO INSPECTION POINTS

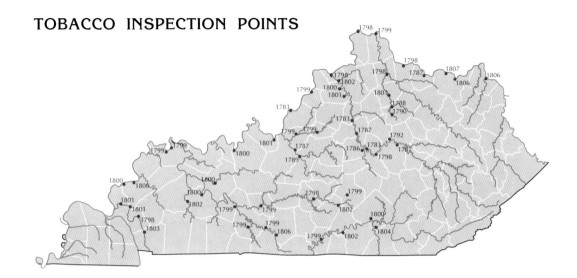

the land parcel. Kentucky lands were surveyed, as lands within the original colonies had been, according to metes and bounds, a practice that described land plats by reference to prominent natural features, buildings, roads, and existing survey lines. The boundaries of land claims might follow stream courses or ridge lines or lines surveyed with seemingly little concern for cardinal directions. The corners of a property might be marked by prominent trees or rock outcrops. Virginia militia and Continental army veterans were eligible for free land warrants in Kentucky, as were other settlers who cleared and improved land or who brought immigrants into the new territory. Warrants could also be purchased. In part because no accurate, comprehensive survey had been made of the Kentucky lands before settlement, and in part because English or Virginia officials granted warrants that covered more land than actually existed, Kentucky land claims were confused and inaccurate. Frequently settlers found that land they thought they owned was actually part of an older claim held by someone else. As a result, thousands of overlapping or "shingled" land claims were patented, and the state's courts were inundated with land claim cases.

Slaves and Slavery. Black slaves accompanied the first European American explorers and settlers to Kentucky. Slaves were used primarily as agricultural workers, espe-

cially in hemp and tobacco production. Thus slave concentrations were highest in areas with the most productive land, especially the central Bluegrass and the southwestern Pennyroyal. Although slaves never exceeded 24 percent of the total state population, some counties actually had majority slave populations at times. Slaves were also used to a lesser extent in mining and manufacturing.

Tobacco Ports, Farm Production, and Marketing. Some of Kentucky's first settlers carried tobacco culture with them from Virginia. The old tobaccos were dark, heavy, and low-growing and were raised primarily in the fertile bottomlands of western Kentucky, where production was centered before the Civil War. To assure quality and value, tobacco required careful curing and storage. Once air- or fire-curing had been completed on the farm, the crop was often packed into thousand-pound hogsheads for transport to manufacturers or for export to New Orleans. Unwieldy hogsheads could not be readily transported over primitive nineteenth-century roads, so tobacco shipping ports grew up along the state's navigable rivers, at first along the Kentucky River adjacent to the first postrevolutionary settlements, and then in western Kentucky. Shipping ports also became tobacco inspection stations, a legacy of a Virginia law of 1783 that mandated public warehouses for tobacco inspection, storage, and taxation. During the first decade

Warehouses on the Kentucky River and Its Tributaries
Established by Special Acts of the Kentucky Legislature
Between 1792 and 1810 for the Storage and Inspection of Tobacco

December 20, 1792:

Cleveland's	Fayette	Cleveland's Landing
Stafford's	Fayette	Stafford's Landing
Holder's	Clark	Mouth of Howard's Creek
Bush's	Clark	Opposite Boonesborough

February 10, 1798:

Hogan's	Fayette	Mouth of Hickman Creek, Jessamine County
Hickman's	Garrard	Land of James Hogan, opposite mouth of Hickman Creek
Curd's	Mercer	Mouth of Dix River
Harrod's Landing	Mercer	Land of Walter Beall, town of Warwick
Boone's	Madison	Boonesborough
Biggerstaff's		Land of Samuel Biggerstaff
Scott's	Woodford	Land of Charles Scott, mouth of Craig's Creek
Frankfort	Franklin	Daniel Weisiger's warehouse built on "warehouse lot"
Samuel Johnston's	Fayette	At ferry below mouth of Hickman Creek

December 13, 1798:

Port William	Gallatin	Mouth of river
Silver Creek	Madison	Land of Green Clay, mouth of creek
Quantico	Garrard	Land of William Davis, mouth of Sugar Creek

December 19, 1798:

Hind's	Madison	Land of William Mayo, Jr.

December 12, 1799:

Froman's	Woodford	Land of Jacob Froman, mouth of Brushy Run
Hart's	Clark	Land of Nathaniel Hart, mouth of Four-mile Creek
Stone's	Madison	Land of Green Clay, Stone's Ferry
South Frankfort	Franklin	
Jack's Creek	Madison	Land of Green Clay between Elk Branck and Jack's Creek
Swinney's	Montgomery	Land of Swinney and Collins, on Red River (tributary)

December 13, 1800:

Warwick	Madison	Land of Robert Clarke, below mouth of Four-mile Creek at Clarke's Ferry
Drennon's Creek	Henry	Land of Hite and Hogg, mouth of Creek

December 19, 1801:

Prestonville	Gallatin	Mouth of river, land of John Smith and Francis Preston
Sullenger's	Henry	Land of Robert Sullenger at crossing of the Newcastle-Big Bone Lick Road
White's	Madison	Land of James White, mouth of Middle Fork of Station Camp Creek, Estill County (tributary)

December 16, 1802:

Gullion's	Gallatin	Land of Jeremiah Gullion, opposite the mouth of Eagle Creek

December 26, 1805:

Howard's	Clark	Land of John Howard, mouth of upper Howard's Creek
Drowning Creek	Madison	Mosby's land at mouth of Creek
Red River	Clark	Near Powell County at Clarke's and Smith's Iron Works (tributary)

November 26, 1806:

Benson	Franklin	Land of Christopher Greenup, west side of river below Benson Creek
M'Coun's	Mercer	Land of Samuel M'Coun, at ferry, near Clear Creek
Rough's Run	Woodford	Land of Jeremiah Buckley, mouth of Rough's Run Creek

December 2, 1806:

Newton's	Jessamine	Land of Newton Curd, a half mile below the mouth of Dix River

January 1, 1808:

Wilkin's	Woodford	Land of Thomas Turpin, north side of river near Delaney's Ferry

January 25, 1808:

South Frankfort	Franklin	Land of John Smart

February 3, 1808:

Tate's Creek	Madison	Land of William M'Bean, one fourth of a mile below the mouth of Tate's Creek
Haydon's	Madison	Land of Richard Haydon, mouth of Muddy Creek

February 23, 1808:

Hieronimous's	Clark	Land of Henry Hieronimous
Saunder's	Gallatin	Land of Nathaniel Saunders, on Eagle Creek (tributary)

TOBACCO PRODUCED, 1860

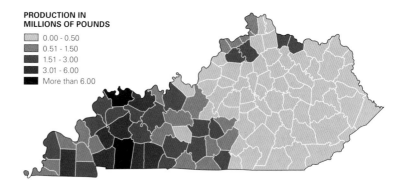

PRODUCTION IN
MILLIONS OF POUNDS

- 0.00 - 0.50
- 0.51 - 1.50
- 1.51 - 3.00
- 3.01 - 6.00
- More than 6.00

TOBACCO PRODUCED, 1900

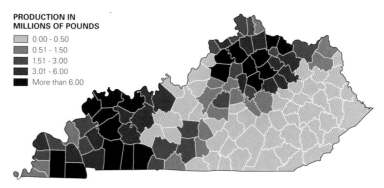

PRODUCTION IN
MILLIONS OF POUNDS

- 0.00 - 0.50
- 0.51 - 1.50
- 1.51 - 3.00
- 3.01 - 6.00
- More than 6.00

after the law's enactment, riverside warehouses were authorized just below the Falls of the Ohio at Shippingport; at Leestown (no warehouse was built there); at the mouth of Hickman Creek on the Kentucky River; at the mouth of the Dix River in Mercer County; at Harrod's Landing; at the mouth of Craig Creek in Fayette County; at Boonesborough; at John Collier's land in Madison County; at the mouth of Stone Creek at Steel's Landing in Fayette County; and at Jack's Creek, also in Fayette County.

The state inspection system, especially for commodities destined for foreign ports, was continued by the new Kentucky government. A legislative act of 1792 established warehouses and provided for governor-appointed inspectors. By 1807 there were more than fifty tobacco inspection stations in Kentucky. The state's extensive inspection and shipping warehouse system emphasized the importance of commercial agricultural exchange to Kentucky settlers, the transfer of agricultural practices from Virginia, and the importance of

WHEAT PRODUCED, 1860

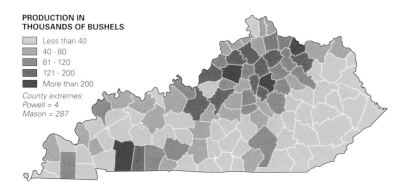

PRODUCTION IN
THOUSANDS OF BUSHELS

- Less than 40
- 40 - 80
- 81 - 120
- 121 - 200
- More than 200

County extremes:
Powell = 4
Mason = 287

WHEAT PRODUCED, 1900

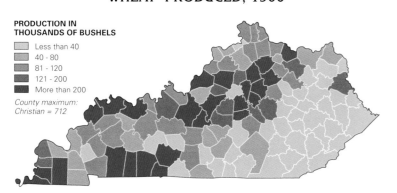

PRODUCTION IN
THOUSANDS OF BUSHELS

- Less than 40
- 40 - 80
- 81 - 120
- 121 - 200
- More than 200

County maximum:
Christian = 712

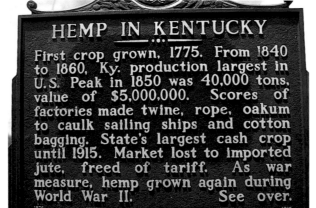

CLARK COUNTY HEMP

One of the ten Bluegrass counties which produced over 90 per cent of the entire country's yield in late 1880s. Production increased from 155 tons in 1869 to over 1,000 tons in 1889, valued at about $125 per ton. In 1942, Winchester selected as site of one of 42 cordage plants built throughout country to offset fiber shortage during war. See over.

HEMP IN KENTUCKY

First crop grown, 1775. From 1840 to 1860, Ky. production largest in U.S. Peak in 1850 was 40,000 tons, value of $5,000,000. Scores of factories made twine, rope, oakum to caulk sailing ships and cotton bagging. State's largest cash crop until 1915. Market lost to imported jute, freed of tariff. As war measure, hemp grown again during World War II. See over.

HEMP PRODUCED, 1860

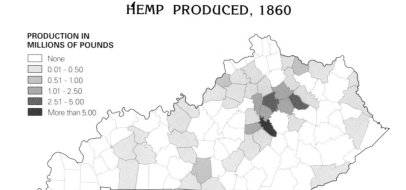

PRODUCTION IN
MILLIONS OF POUNDS

- None
- 0.01 - 0.50
- 0.51 - 1.00
- 1.01 - 2.50
- 2.51 - 5.00
- More than 5.00

HEMP PRODUCED, 1900

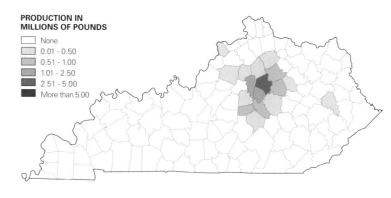

PRODUCTION IN
MILLIONS OF POUNDS

- None
- 0.01 - 0.50
- 0.51 - 1.00
- 1.01 - 2.50
- 2.51 - 5.00
- More than 5.00

---

waterways for early transportation connections. Some inspection stations, such as Frankfort and Warwick, became towns in their own right. But the tobacco inspection ports would not form the backbone of the state's urban system as such riverine settlements had in the East Coast states. The difficulty of access to the Kentucky River because of its steep banks, the siting of county seats, and the importance of overland routes from the Ohio Valley and the East to places like Lexington ensured a different development of the state's urban areas.

Dark tobaccos, genetic descendants of those eighteenth-century tobaccos, are still grown in the state's western counties, in air-cured and fire-cured varieties. White burley tobacco was first grown in Brown County, Ohio, in 1864, and after the Civil War its rapid adoption by Bluegrass farmers brought an eastward shift in tobacco production. Air-cured burley was especially mild compared with the dark varieties that had been used primarily for

manufacturing chewing products, and after the Civil War burley was increasingly used in cigarettes. As demand for burley increased, production spread from the Bluegrass to almost every county of the state.

Wheat and Hemp. Wheat was a frontier staple that evolved into one of Kentucky's primary commercial crops. By 1860 production mirrored the state's most fertile lands and its highest population densities: the greater Bluegrass, the western Pennyroyal, and the western Ohio River counties were all principal producing areas. By 1900 wheat growing had intensified within these areas, abetted by mechanization and improvements in transportation.

The British brought the hemp plant to America as a source of fiber for stout cordage and cloth. The first crop was grown near Danville in 1775, and the Bluegrass Region soon became a

major hemp producer. Rope walks and bagging factories supplied Kentucky hemp products to the plantation South before the Civil War. Hemp production was a labor-intensive process, and the commodity could be especially profitable if produced by slaves. Consequently, a close areal association between hemp and slaves developed. After the Civil War the demand for Kentucky hemp declined steadily, and production retreated into the core producing areas of the Inner Bluegrass.

THE CIVIL WAR AND KENTUCKY

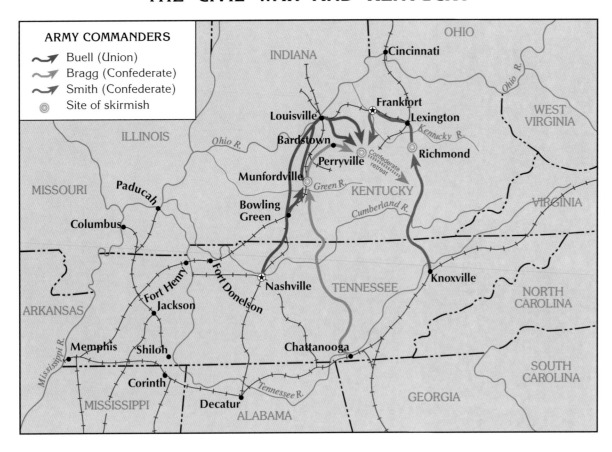

Civil War history buffs in period dress and equipage reenact the Battle of Perryville each summer.

The Civil War and Kentucky

In the decade prior to the opening of the Civil War in 1861, Kentucky's economy was heavily oriented toward trade with the South. When viewed from the Deep South slave states, Kentucky must have seemed a nearly limitless cornucopia of foodstuffs, mules and horses, and slaves. By simply following the regional river systems—the Ohio and its two tributaries (the Tennessee and the Cumberland), and the Mississippi—Kentucky traders could deliver commodities via economical water transportation directly to Memphis, New Orleans, Nashville, or Knoxville, the major cities of the mid- and western South. North–south rail lines augmented the rivers, providing overland connections between these cities and Kentucky's Ohio River ports at Louisville and Paducah. Considered strategically, Kentucky's geographical position would prove the South's undoing during the war, despite its supposedly neutral status as a border state. Union General U.S. Grant was free to position troops and war material along the Ohio River's 450-mile north bank, and when Confederate soldiers took Kentucky's Mississippi River port at Columbus in September 1861, General Grant initiated a campaign that advanced naval and ground forces south across western Kentucky following the same river and rail trade corridors that had supplied the South with commodities.

Confederate defensive positions on the Tennessee River at Fort Henry and Fort Donelson on the Cumberland fell to Union forces in February 1862, thereby opening the route along the Tennessee all the way to northern Alabama and precipitating further Confederate losses in major battles at Shiloh, Tennessee, and Corinth, Mississippi. Grant's forces also followed the Cumberland River to claim Nashville, the capital of Tennessee and a major supply and manufacturing center. With the Union Army controlling the Tennessee and Cumberland Rivers, as well as the west Tennessee railroads, the Confederate defensive position that had been established in Bowling Green, Kentucky, was compromised and had to be abandoned.

In retaliation, Southern armies attempted a substantive incursion northward into Kentucky, beginning in August 1862, to disrupt transportation and elicit support from the state's Southern sympathizers. Confederate General Edmund Kirby-Smith marched an army north from Knoxville to engage Union troops successfully at Richmond, Kentucky, whereupon he marched into Lexington to await General Braxton Bragg who was moving north from Chattanooga toward Munfordville, an important Louisville and Nashville Railroad bridging point across the Green River. Bragg's forces obtained the surrender of the bridge's Union defenders in September, and then, instead of heading to Louisville, marched northeast toward Bardstown and then on to Perryville, where they encountered the Union troops of General Don Carlos Buell. The Perryville battle opened on October 8, 1862, and resulted in more than 7,700 killed or wounded in both armies. The combined Confederate forces lost this engagement and retreated into Tennessee. Perryville marked the last important Southern incursion into Kentucky, although Confederate cavalry led by John Hunt Morgan continued harassing raids within the state as late as 1864.

Master stone mason Richard Tufnell reviews the reconstruction of a nineteenth-century Irish-built rock fence in Bourbon County. Tufnell, a member of the Drystone Conservancy, is reviving the art of drystone wall construction through a program of instruction and certification modeled on similar efforts in Great Britain.

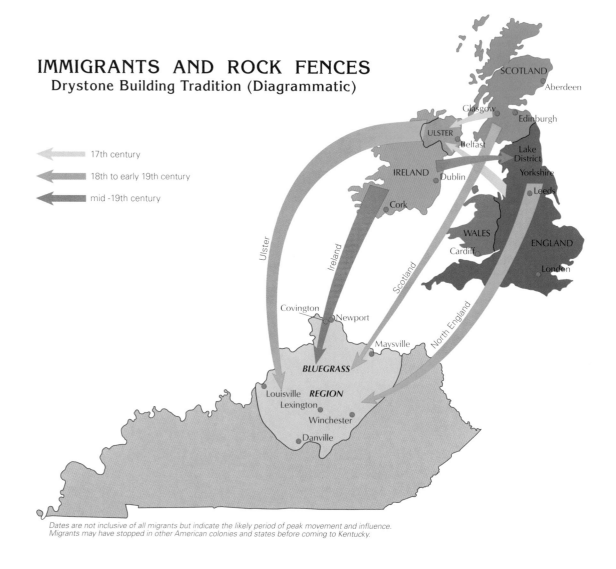

IMMIGRANTS AND ROCK FENCES
Drystone Building Tradition (Diagrammatic)

17th century

18th to early 19th century

mid -19th century

Dates are not inclusive of all migrants but indicate the likely period of peak movement and influence. Migrants may have stopped in other American colonies and states before coming to Kentucky.

Immigration

During the eighteenth and early nineteenth centuries, the immigrants of European heritage who moved into Kentucky were primarily of English, Scots, or Scots-Irish descent. Many had not moved directly from their northern European homelands to Kentucky but had lived for a generation or more in the East before venturing west of the Appalachians. European migration to America began to accelerate after about 1820. Germans and Irish, especially, left their countries in increasing numbers. In the middle and late decades of the nineteenth century, immigrants from southern and eastern Europe joined the exodus and came to America. Kentucky, however, attracted few European immi-

grants during this time, with the exception of Germans who settled in Ohio River towns such as Covington, Louisville, and Owensboro, and small numbers of eastern Europeans who were recruited as miners in the Eastern Kentucky Coal Field. Nor did substantial numbers of European immigrants seek permanent residence in other southern states, in part because they were reluctant to try to compete with slave labor. Somewhat larger numbers of Catholic Irish, fleeing the Potato Famine during the 1840s, moved into Kentucky and other southern states. The Irish were able to find jobs as day laborers on farms, as workers on road and railroad construction crews, and as stonemasons. In

the Kentucky Bluegrass Region, for example, farmers hired Irish stonemasons to erect rock fences to enclose fields and farmsteads. Stonemasons from Scotland, England, and Ulster also contributed to the construction of Bluegrass rock fences and other stone structures. Their knowledge of traditional Irish and British construction techniques was passed to slaves, and freedmen later dominated the early twentieth-century rock fence construction trade.

Except for Covington and Louisville, Kentucky did not experience the massive nineteenth-century European immigrations that flooded the Midwest and the nation's northern industrial cit-

FOREIGN-BORN POPULATION, 1880

PERCENT OF COUNTY POPULATION BORN ABROAD

Fewer than 0.1
0.1 - 0.5
0.6 - 1.0
1.1 - 2.5
2.6 - 5.0
More than 5.0

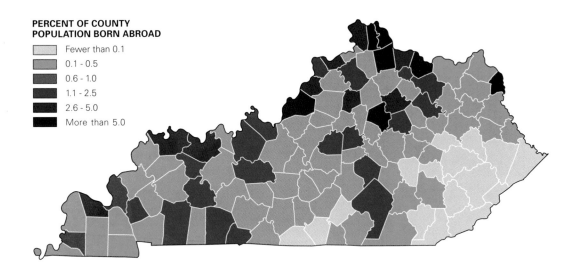

FOREIGN-BORN POPULATION, 1920

PERCENT OF COUNTY POPULATION BORN ABROAD

Fewer than 0.1
0.1 - 0.5
0.6 - 1.0
1.1 - 2.5
2.6 - 5.0
More than 5.0

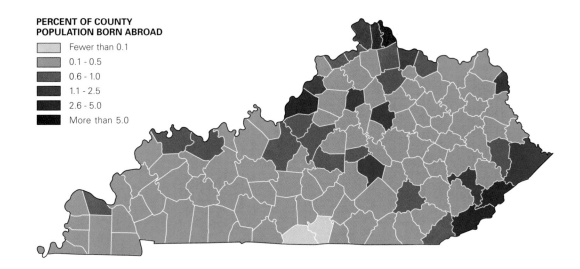

ies. In 1880 and 1920, just over 13 percent of the U.S. population was foreign-born, whereas in Kentucky the comparable figures were 3.7 and 1.3 percent, respectively. The 1880 and 1920 dates represent national immigration watersheds. The first marked the transition from the migration of northern and western Europeans to the migration of eastern and southern Europeans. The latter group dominated U.S. immigration until the federal legislature passed severe restrictions in 1921 and 1924. In both 1880 and 1920 immigrant groups in Kentucky tended to cluster in urban areas, but by 1920 coal company recruiters had been successful in attracting immigrants to live and work in several counties of the Eastern Kentucky Coal Field.

By 1880 Irish and German residents dominated Kentucky's foreign-born population, accounting for more than 80 percent of the foreign-born. The Irish and Germans remained dominant through 1920, when together they comprised 47 percent of the state's foreign-born residents. They had been joined by small numbers of immigrants from Russia (9 percent of the total), Italy (6 percent), England (6 percent), and Switzerland, Hungary, and Poland (a total of 11 percent). No other foreign-born groups were present in significant numbers.

Around the turn of the century, many large coal companies recruited miners from Europe to work in newly opened eastern Kentucky mines. Mules drew loaded coal cars from the working face underground to a railside tipple.

RAILROAD CONSTRUCTION IN EASTERN KENTUCKY

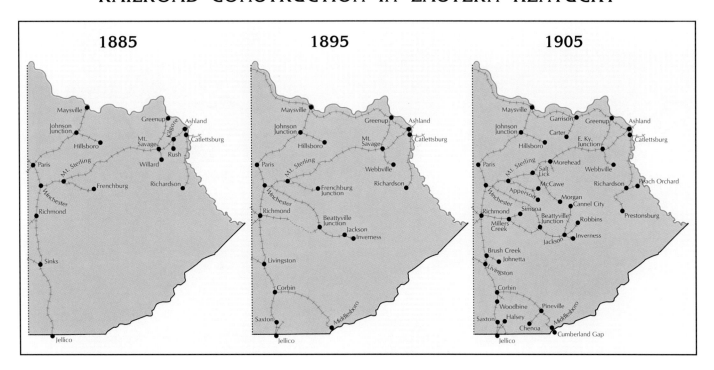

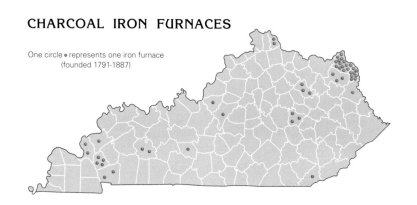

CHARCOAL IRON FURNACES

One circle ● represents one iron furnace (founded 1791-1887)

Resource Exploitation Patterns

Timber. The magnificent virgin hardwood forest that covered much of Kentucky when white settlement began in the eighteenth century was initially cut by people seeking to clear land for farming. Harvesting timber for commercial sale, especially in eastern Kentucky, began as early as 1830. Mountain farmers cut logs from their land and floated them on high spring floodwaters to sawmills along major rivers. Frankfort, on the Kentucky River, and Catlettsburg, a port town near the confluence of the Big Sandy River and the Ohio, were prominent sawmill towns. By 1870 state sawmills produced more than 217 million board feet of lumber annually, ranking the state fifteenth in the nation. Large-scale timber cutting awaited the arrival of the railroads. Timber buyers followed the railroad tracks into eastern Kentucky. State timber production peaked in 1907, when the industry employed thirty thousand people and produced 913 million board feet.

Iron. Kentucky's first charcoal-fired iron furnace was built about 1791, and by 1845 more than twenty-five furnaces were operating within the Commonwealth. Most furnaces were either in the Hanging Rock and Red River districts of eastern Kentucky or along the Green and Cumberland Rivers in the west. The 1840s saw the heyday of iron production in the state and, indeed, in the nation. Iron smelted with charcoal was especially suited to a predominantly agricultural economy; the product was malleable and was easily fashioned into horseshoes, axes, scythes, and sickles by local blacksmiths. Several factors precipitated the decline of charcoal-produced iron in Kentucky by the 1850s, including the depletion of local ores and timber, economic depression, changing product demands driven by an increasingly industrial regional and national economy, the invention of the Bessemer process in steel production, and the shift to anthracite coal and coke as fuel. Although the industry increased

Crafted with dry-laid sandstone, the large double iron furnace at Fitchburg still stands in Estill County.

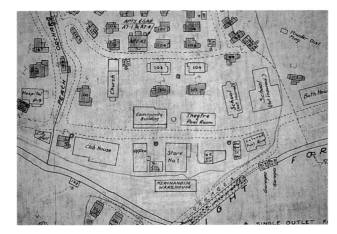

Far left, *Blueprint for Wheelwright town center, 1954.* Left, *Wheelwright, 1954.*

SUPPLEMENT
To Plans
For Manager's House
ELKHORN COAL CORPORATION
Wheelwright, Kentucky
Drawing No. 3A   November 10, 1916

MANAGER'S HOUSE

SUPPLEMENT
To Plans
For Three Room Box House
ELKHORN COAL CORPORATION
Wheelwright, Kentucky
Drawing No. 4E   November 10, 1916

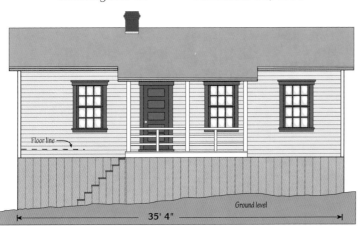

MINER'S HOUSE

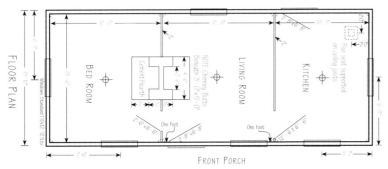

FLOOR PLAN FOR MINER'S HOUSE

production in response to the demands of the Civil War, by 1880 Kentucky's charcoal furnaces had given way to the new steel industry at Ashland, Newport, and Owensboro. Today many remnant charcoal iron furnaces still stand abandoned in the old iron-producing regions.

Coal. Although early mountain farmers with coal outcrops on their land extracted small quantities for home heating purposes, the large-scale exploitation of the state's mammoth bituminous coal deposits had to await the extension of railroads into eastern Kentucky and the Western Kentucky Coal Field. Rail access allowed entrepreneurs to begin developing large mines, employing five hundred miners or more each, by 1910. Before the railroads entered the eastern mountains, roads were few, and most farm families produced few commodities for commercial sale. Because the farming population was scattered and few towns of any size existed outside the county seats, mining companies assembled and controlled their labor pools by building new towns to house miners and by providing the miners with services.

Hundreds of company towns sprang up across eastern Kentucky by 1920. Among the largest and most modern was Wheelwright, named for Jere Wheelwright, president of the Consolidation Coal Company. The town had a linear form, of necessity, because creek bottoms were too narrow to accommodate more than an access road, a railroad track, and a single line of houses. Service buildings, including a company store, schools, a church, a theater, a community building, a clubhouse, a hospital, and an automobile service station, clustered together at an Otter Creek tributary junction. Miners' houses, such as the "Three Room Box House," lined the creek bottoms. Mine managers and executives lived in more spacious houses. Small portals in adjacent hillsides led to underground mines, unseen yet extensive nxetworks of horizontal tunnels driven or "drifted" into the coal seams. A single mine had multiple parallel tunnels. Some tunnels allowed movement of miners and equipment in and out of the mine and provided a route for coal removal. Other tunnels permitted a continual circulation of fresh air to the active mining faces. Once access tunnels had been established, miners opened side tunnels to remove coal, leaving large coal "pillars" behind to support the mine roof. Conveyors brought coal from underground to a tipple or preparation plant for cleaning, sizing, and loading for rail shipment to customers.

CHURCH MEMBERS

PERCENT OF COUNTY
POPULATION
- [ ] Fewer than 40
- [ ] 40 - 60
- [ ] 61 - 75
- [ ] 76 - 90
- [ ] More than 90

State average = 60.4%

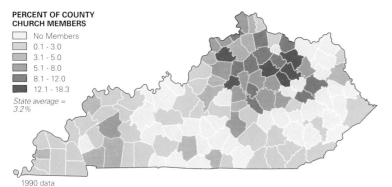

1990 data

CATHOLIC

PERCENT OF COUNTY
CHURCH MEMBERS
- [ ] No Members
- [ ] 0.3 - 5.0
- [ ] 5.1 - 10.0
- [ ] 10.1 - 20.0
- [ ] 20.1 - 40.0
- [ ] 40.1 - 59.0

State average = 74%

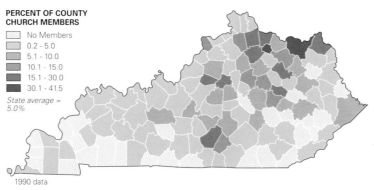

1990 data

CHRISTIAN CHURCH (DISCIPLES OF CHRIST)

PERCENT OF COUNTY
CHURCH MEMBERS
- [ ] No Members
- [ ] 0.1 - 3.0
- [ ] 3.1 - 5.0
- [ ] 5.1 - 8.0
- [ ] 8.1 - 12.0
- [ ] 12.1 - 18.3

State average = 3.2%

1990 data

CHRISTIAN CHURCH AND CHURCH OF CHRIST

PERCENT OF COUNTY
CHURCH MEMBERS
- [ ] No Members
- [ ] 0.2 - 5.0
- [ ] 5.1 - 10.0
- [ ] 10.1 - 15.0
- [ ] 15.1 - 30.0
- [ ] 30.1 - 41.5

State average = 5.0%

1990 data

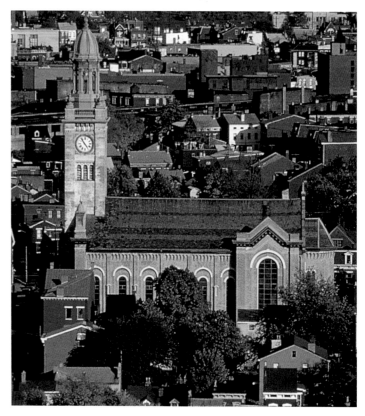

Mother of God Roman Catholic Church, built in 1871, rises above the Covington skyline.

Cultural Attributes

Religion. Religion remains a primary component of culture and identity in the United States, a nation founded at least in part upon the principle of religious freedom. Many of Kentucky's early settlers departed established communities in Virginia to seek greater latitude of religious practice in the Bluegrass. Among them were Lewis Craig and his Baptist congregation, who trekked from Spotsylvania, Virginia, to Gilbert's Creek, near what would later be the town of Lancaster in Garrard County, where they founded a Baptist church in 1781. Presbyterians from North Carolina built the Cane Ridge Meeting House in Bourbon County in 1791, the site of a major religious revival in 1801. The United Society of Believers in Christ's Second Appearing, known as Shakers, established Pleasant Hill, now a Mercer County outdoor museum, in 1806 and a community at South Union, near Bowling Green, in 1807.

In 1990 about 123 million Americans, or approximately 50 percent of the national population, claimed adherence to one of 133 religious organizations. Many others tacitly belong to religious groups that are less formally organized and are not part of a regional or national body; those memberships can only be inferred. In Kentucky, for those religious groups for which information is available, religious adherents totaled 2,227,747 in 1990, or 60 percent of the state's population.

Kentucky lies at the northern edge of the Bible Belt, which extends from Virginia and the Carolinas west to Oklahoma and Texas and is closely associated with the area traditionally known as the South. This area is distinguished by the prevalence of conservative, evangelical Protestant religious groups, many of which believe in a literal interpretation of the Bible. Kentucky's largest formally organized church group is the Southern Baptist Convention, which accounts for 43 percent of Kentucky's church adherents. The Roman Catholic Church is second, with 16 percent of the state's church members, followed by the United Methodist Church (10 percent), the Christian Church and Church of Christ (4 percent), African American Baptist churches (4 percent), and the Church of Christ (3 percent). (In the nineteenth century, a new branch of Protestantism developed that recognized only the authority of the New Testament and the autonomy of local congregations. Known generally as the Restoration Movement, this religious body eventually organized into three branches based on theological differences. The Christian Church [Disciples of Christ] is the most liberal of the three; the second branch, the Christian Church and Churches of Christ, is regarded as theologically moderate; and the third group, the Churches of Christ, is the most conservative, permitting only a cappella church

UNITED METHODIST

PERCENT OF COUNTY
CHURCH MEMBERS

☐ No Members
☐ 0.6 - 10.0
☐ 10.1 - 15.0
☐ 15.1 - 20.0
☐ 20.1 - 30.0
☐ 30.1 - 45.5

*State average =
12.9%*

1990 data

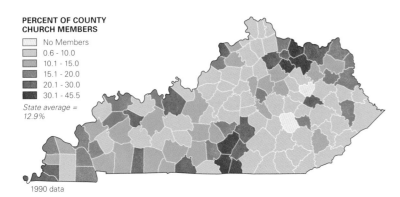

CHURCHES OF CHRIST

PERCENT OF COUNTY
CHURCH MEMBERS

☐ No Members
☐ 0.2 - 3.0
☐ 3.1 - 7.0
☐ 7.1 - 10.0
☐ 10.1 - 18.4
☐ 30.1

*State average =
3.3%*

1990 data

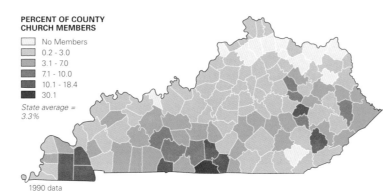

*The 1930s Pleasant Green Baptist Church stands on the
site of the "oldest African-American church west of the
Allegheny Mountains" in Lexington, where slave Peter
Duerett founded the congregation in 1790.*

SOUTHERN BAPTIST CONVENTION

PERCENT OF COUNTY
CHURCH MEMBERS

☐ 5.6 - 30.0
☐ 30.1 - 45.0
☐ 45.1 - 60.0
☐ 60.1 - 75.0
☐ 75.1 - 87.7

*State average =
48.7%*

1990 data

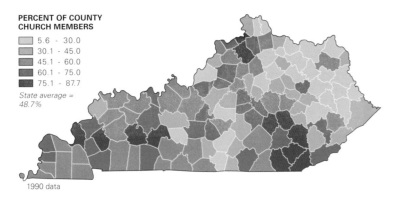

AFRICAN AMERICAN BAPTIST

PERCENT OF COUNTY
CHURCH MEMBERS

☐ No Members
☐ 0.4 - 2.0
☐ 2.1 - 4.0
☐ 4.1 - 8.0
☐ 8.1 - 9.4
☐ 17.9

*State average =
1.6%*

1990 data

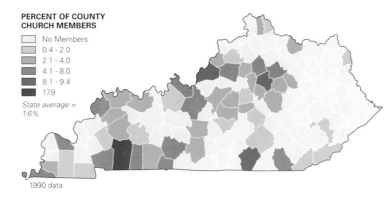

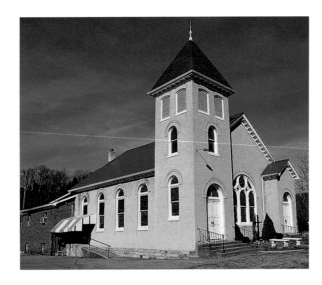

Red House Methodist Church, Madison County.

Cane Ridge Meeting House, Bourbon County.

Burkesville Baptist Church, Cumberland County.

SYNAGOGUES

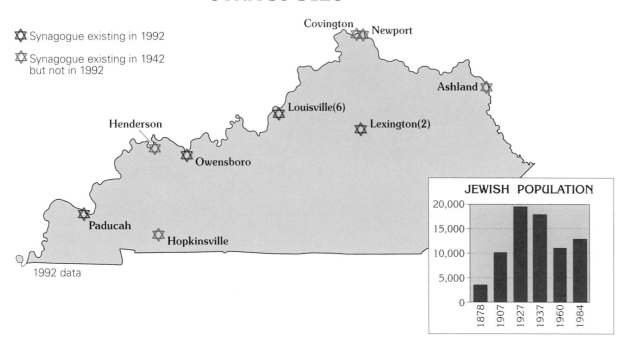

☆ Synagogue existing in 1992

✡ Synagogue existing in 1942
but not in 1992

Covington Newport

Ashland

Louisville(6)

Lexington(2)

Henderson

Owensboro

Paducah

Hopkinsville

1992 data

JEWISH POPULATION

20,000

15,000

10,000

5,000

0

1878 1907 1927 1937 1960 1984

Temple Adath Israel, Lexington.

music and stressing the autonomy of local congregations, for example.)

Patterns of total church membership in Kentucky in part reflect population density. A somewhat more representative picture of the distribution of religious practice compares church membership with the total population. In northern, central, and western Kentucky, relatively high proportions of the population are affiliated with major religious organizations. Because church membership is enumerated at the church address and not at members' home addresses, it is possible for a county to have a church membership figure that is higher than the county's total population. Washington County, for example, draws members to its churches from surrounding counties, and it has a church membership that is 108 percent of the county's population. The state's eastern counties exhibit the lowest official church memberships, with Menifee County reporting only 13 percent. These counties, however, have traditionally been the home of numerous independent churches for which no systematic information is available.

Each major church group represented in the state exhibits a distinctive areal distribution. Members of the Southern Baptist Convention are most prominent across the southern and western portions of the state. Catholics occupy a belt that parallels

the Ohio River, the avenue of movement of nineteenth-century German immigrants, extending from Mason County in the north to Union County in the west. The oldest Catholic settlements tend to retain the highest levels of membership, as in Covington and Newport in northern Kentucky and in Nelson, Marion, and Washington Counties. Here also are important Catholic institutions, such as the Trappist abbey Our Lady of Gethsemane and St. Joseph Proto-Cathedral, both in Nelson County. St. Joseph, consecrated in 1819, was the first Catholic cathedral built in the United States west of the Alleghenies. In Washington County, the St. Rose Church and Proto-Priory was established as the first Catholic school in the state in 1806. Methodism claims membership in two distinct regions: in the northeastern Bluegrass and in a belt extending from the eastern Pennyroyal at Adair, Clinton, and Cumberland Counties west and northward to the Mississippi and Ohio Rivers. The Churches of Christ has its strongest representation in the eastern counties and in the Pennyroyal and Jackson Purchase. The Christian Church (Disciples of Christ) was born in the fervor of the revivals held at the Cane Ridge Meeting House in 1801. Though initially Presbyterian, the Christian Church group split off in 1804 to organize an independent church. Today the Christian Church continues its strongest representation at its his-

toric core, Bourbon County and a large cluster of surrounding Bluegrass counties. African-American Baptist churches are closely associated with those places where the state's black population has traditionally been concentrated: urban areas, the Bluegrass, and the central Pennyroyal. Of special note is Christian County: 24 percent of the population is black, and the county's African-American Baptist churches account for about 18 percent of total county church membership.

Jewish congregations have historically clustered in the state's largest urban centers. Ohio River towns could count seven synagogues in 1892. By 1942 the total number had expanded to fifteen, six of which were in Louisville. In 1992 ten synagogues remained in four cities: Louisville, Lexington, Owensboro, and Paducah. Islamic mosques can also be found in the larger urban centers.

BLACK HAMLETS
IN THE INNER BLUEGRASS REGION

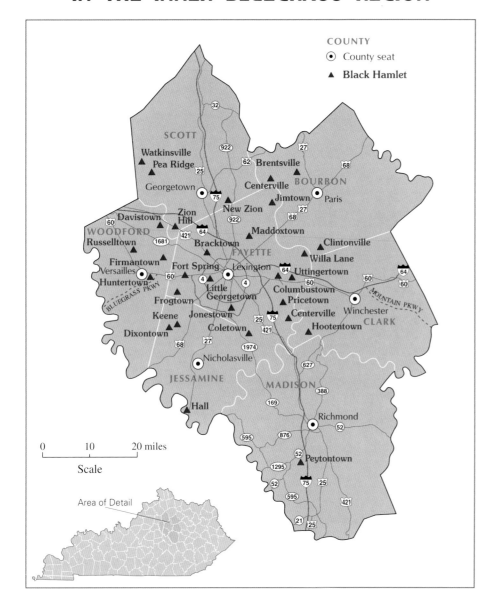

COUNTY
◉ County seat
▲ Black Hamlet

Area of Detail

POSTBELLUM INNER BLUEGRASS FARM
AND BLACK HAMLET

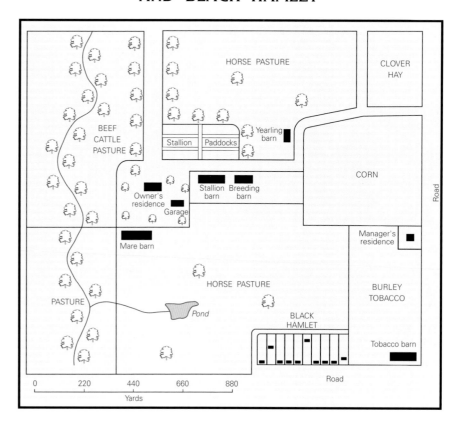

Ethnicity. Post-Civil War black residence patterns in Kentucky have derived primarily from the dominant pattern of agricultural slaveholding in the antebellum period. Although the period 1870-1990 saw little change in the basic distributional pattern of blacks across the state, several processes acted in concert to influence their evolving residential pattern. First, immediate post-Civil War rural reorganization of agricultural production produced a number of black hamlets in the countryside, as former slaves moved off plantations into a new role as agricultural laborers. Second, since the Civil War and to the present day, migration from locations formerly characterized by the intensive use of slave labor has taken two forms: migration from the countryside to cities, such as Louisville, Lexington, and Hopkinsville; and general out-

migration from Kentucky, which has resulted in a consistently decreasing percentage of blacks who are state residents. In 1860, 20.4 percent of the population were blacks; in 1870, 16.7 percent; in 1900, 13.2 percent; in 1930, 8.6 percent; in 1960, 7.1 percent; and in 1990, 7.1 percent. Third, the need for labor in the Eastern Kentucky Coal Field resulted in an increase of the black population in some parts of that region after 1900.

The first U.S. census, taken in 1790, reported that 14 percent of Kentucky's population claimed German ancestry. By 1990 that proportion had increased to 21 percent. While Germans were among the state's frontier settlers, the largest numbers migrated to the state after the failed 1848 German revolution. The present residential pattern of people with German ancestry is similar to that of the past, with the largest

EUROPEAN ANCESTRY

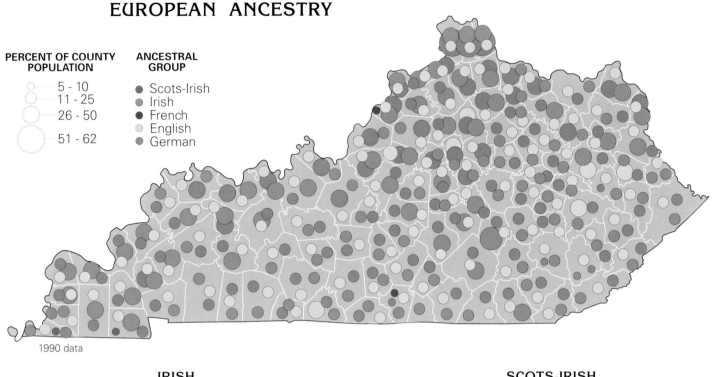

PERCENT OF COUNTY POPULATION
- 5 - 10
- 11 - 25
- 26 - 50
- 51 - 62

ANCESTRAL GROUP
- Scots-Irish
- Irish
- French
- English
- German

1990 data

concentrations in Boone, Kenton, and Campbell Counties and in Jefferson, Daviess, and Hancock Counties.

Irish immigrants, like the Germans and the English, were among the state's earliest residents. The migration of Irish to America, and to Kentucky, increased dramatically during the Potato Famine of the 1840s. The 1990 census recorded about 696,000 people of Irish ancestry in Kentucky, or almost 19 percent of the state's total population. They are more widely distributed than people of German ancestry, although both groups show a concentration along the Ohio River.

Because Kentucky was a county of Virginia, an English colony, one might expect Kentucky's early residents to have been overwhelmingly English. But the state's population had varied ancestries from the earliest years of settlement, and that variety remains evident today. About 19 percent of the state's population in 1990 declared themselves to have English heritage. These people are generally distributed across the state, and in no county does their concentration exceed the levels of the most concentrated Irish or German groups.

Many of the first settlers to occupy the remote valleys of central Appalachia—now West Virginia, eastern Kentucky, and western Virginia—were Protestants from Northern Ireland, often called the Scots-Irish. Today they comprise about 2.9 percent of the state's residents. Ironically, their numbers are relatively sparse in Kentucky's Appalachian counties; they are best represented in the Bluegrass and the Jackson Purchase.

Residents of French ancestry also comprise less than 3 percent of the state's total population. Outside Cumberland County, where they account for about 9 percent of the population, their distribution most closely resembles that of the state's residents of German ancestry.

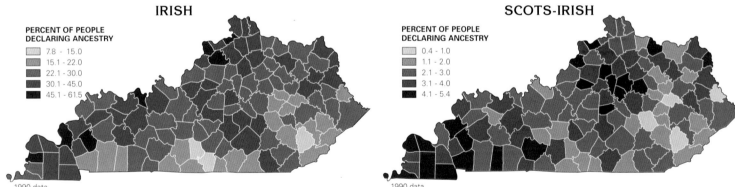

IRISH

PERCENT OF PEOPLE DECLARING ANCESTRY
- 7.8 - 15.0
- 15.1 - 22.0
- 22.1 - 30.0
- 30.1 - 45.0
- 45.1 - 61.5

1990 data

SCOTS-IRISH

PERCENT OF PEOPLE DECLARING ANCESTRY
- 0.4 - 1.0
- 1.1 - 2.0
- 2.1 - 3.0
- 3.1 - 4.0
- 4.1 - 5.4

1990 data

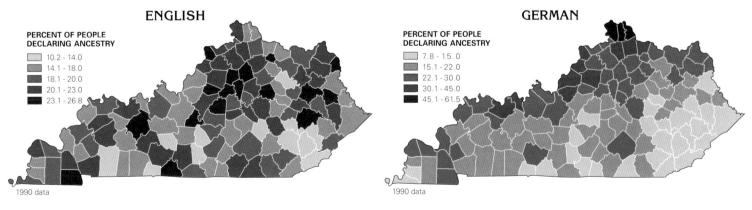

ENGLISH

PERCENT OF PEOPLE DECLARING ANCESTRY
- 10.2 - 14.0
- 14.1 - 18.0
- 18.1 - 20.0
- 20.1 - 23.0
- 23.1 - 26.8

1990 data

GERMAN

PERCENT OF PEOPLE DECLARING ANCESTRY
- 7.8 - 15.0
- 15.1 - 22.0
- 22.1 - 30.0
- 30.1 - 45.0
- 45.1 - 61.5

1990 data

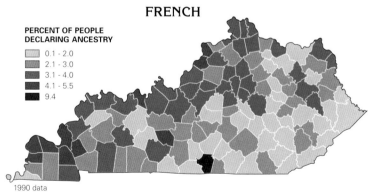

FRENCH

PERCENT OF PEOPLE DECLARING ANCESTRY
- 0.1 - 2.0
- 2.1 - 3.0
- 3.1 - 4.0
- 4.1 - 5.5
- 9.4

1990 data

Population

CHAPTER FOUR

Population

The crowd at a Louisville public swimming pool in mid-summer represents a rough cross-section of a neighborhood's population.

THE PEOPLING OF what is today Kentucky dates back in time to well before the first European explorers established trails through the state and built forts and small communities. Indeed, white settlement simply marks one of many stages in population growth and change that began long before recorded history, with the southeastward migration of Paleoindians along the margins of the great Pleistocene ice sheets. Since that early period about twelve thousand years ago, the abundance of resources necessary to sustain life has strongly influenced the size and distribution of Kentucky's population. Grazing animals attracted hunters in search of food, deposits of chert provided a necessary material for tool making, and rich clays formed the basis of pottery that was used for food preparation and storage and for trade. Vast forests yielded essential materials for construction, and valley bottoms contained soils that helped families establish perennial crops. Eventually, rivers allowed easy travel and provided sources of power, and numerous springs supplied abundant fresh water for early permanent settlements.

Contemporary populations are, of course, quite different from those of long ago. The total number of people is certainly much larger, and it continues to grow. In fact, the number of people *added* to the state's population between 1970 and 1990 is more than double the number of residents of the state in 1800, when the census enumerated 220,955 people. The types of resources that attract people have evolved considerably, to include coal, natural gas, limestone, and the benefits of centralized capital and market influences. The state's inhabitants tend to be increasingly clustered within large cities and their immediate surroundings, with smaller towns waxing and waning in size according to changes in the coal industry, agricultural market prices, and even the price of gasoline. While about 22 percent of Kentuckians lived in urban areas in 1900, this figure had expanded to almost 52 percent by 1990. Furthermore, Kentucky's population is more diverse now than ever before.

A shopping mall in Louisville. Malls now serve as community centers, attracting a cross-section of the population, not only for shopping but also for recreational activity.

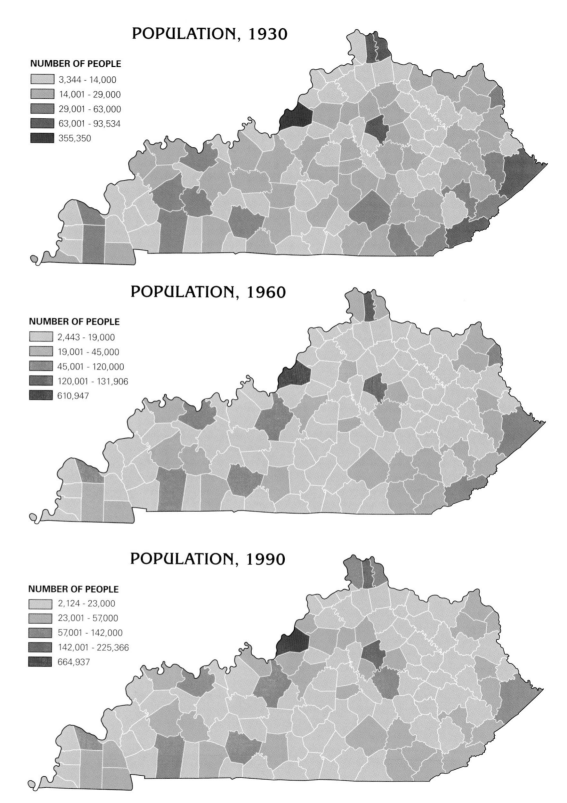

POPULATION, 1930

NUMBER OF PEOPLE

	3,344 - 14,000
	14,001 - 29,000
	29,001 - 63,000
	63,001 - 93,534
	355,350

POPULATION, 1960

NUMBER OF PEOPLE

	2,443 - 19,000
	19,001 - 45,000
	45,001 - 120,000
	120,001 - 131,906
	610,947

POPULATION, 1990

NUMBER OF PEOPLE

	2,124 - 23,000
	23,001 - 57,000
	57,001 - 142,000
	142,001 - 225,366
	664,937

KENTUCKY'S POPULATION GROWTH

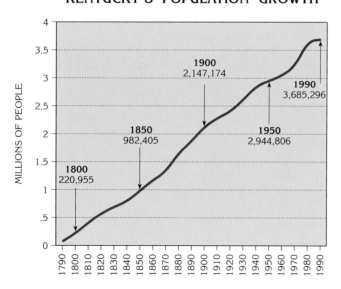

MILLIONS OF PEOPLE

1800
220,955

1850
982,405

1900
2,147,174

1950
2,944,806

1990
3,685,296

POPULATION DENSITY

1930

PERSONS PER SQUARE MILE

	75 or fewer
	76 - 150
	151 - 300
	301 - 600
	More than 600

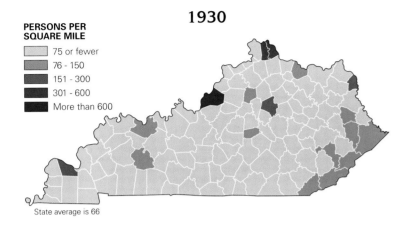

State average is 66

1960

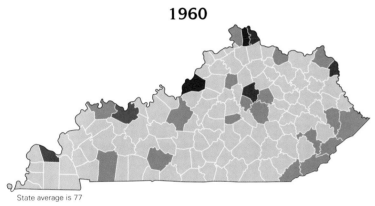

State average is 77

1990

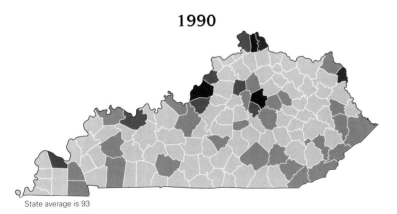

State average is 93

CHANGE IN POPULATION DENSITY

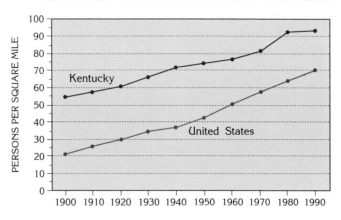

A study of Kentucky's population should, of necessity, examine not only the current characteristics of the population but also the many processes that have caused changes in these characteristics in the past and most certainly will cause changes in the future. In describing the "mix" of the state's population, it is important to recognize such characteristics as age, gender, race, and family. These characteristics are central to understanding local, state, and federal decisions relating to education, health care, housing, and a host of social services. Processes of change include elements of fertility and mortality; the prevalence of one over the other results in natural population increase or decline. Deathrates nationwide have fallen dramatically during the past century, and Kentucky offers no exception. Birthrates, on the other hand, have remained fairly high in the state until recently. Another process that is especially important at the county level is migration, which is vital for understanding the rapid population changes that have occurred throughout Kentucky's history. Early in-migration of settlers from the eastern United States caused explosive growth in certain parts of Kentucky during the 1800s, particularly with the emergence of coal as a principal resource in a rapidly industrializing American society. Migration to cities caused Louisville, Lexington, and Cincinnati–Northern Kentucky to become the major urban centers that exist today, and the search for higher wages and more amenable living environments has drained the populations of many rural counties.

Distribution, Density, and Change

The distribution and density of Kentucky's population, as measured in persons per square mile, have changed considerably during the twentieth century. Most noteworthy is the progressive growth of suburbs and commuter zones around the cities of Louisville and Lexington and the Kentucky counties adjacent to Cincinnati, Ohio. As the central counties of Jefferson and Fayette have become increasingly dense, a spillover of settlement has emerged in surrounding counties. This settlement is supported by improvements in road networks and the extension of utilities—including gas, water, and sewerage—from the cities to outlying areas. Most rural counties have experienced either stability or decline in their population sizes and densities. The far eastern and western counties have been particularly hard hit, with streams of people leaving the area during the Depression and the post–World War II decades for employment in such cities as Cincinnati, Columbus, Dayton, Indianapolis, Detroit, and St. Louis. For example, from 1930 to 1990, the Appalachian counties of Bell and Harlan experienced significant declines in population: Bell County's population fell from 38,747 in 1930 to 31,506 in 1990; Harlan County's, from 64,557 to 36,574. Similarly, the Jackson Purchase counties of Carlisle and Hickman experienced notable declines during this period.

It is also interesting to examine the state's historic population density in comparison to that of the nation.

POPULATION CHANGE

UNITED STATES POPULATION CHANGE

1960-1970

PERCENT CHANGE
- -36.8 to -20.1
- -20.0 to -10.1
- -10.0 to -0.1
- 0.0 to 10.0
- 10.1 to 20.0
- 20.1 to 89.3

County extremes:
Harlan = -36.8%
Bullitt = +39.7%

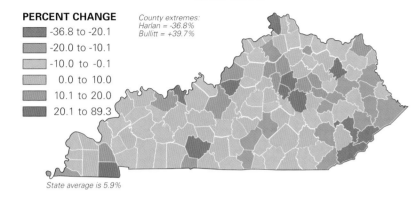

State average is 5.9%

1970-1980

County extremes:
Fulton = -11.9%
Oldham = +89.3%

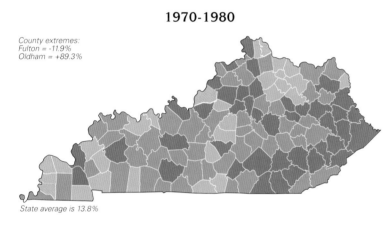

State average is 13.8%

1980-1990

County extremes:
Harlan = -12.8%
Boone = +25.5%

State average is 0.7%

DECENNIAL GROWTH RATES

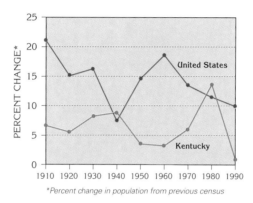

Percent change in population from previous census

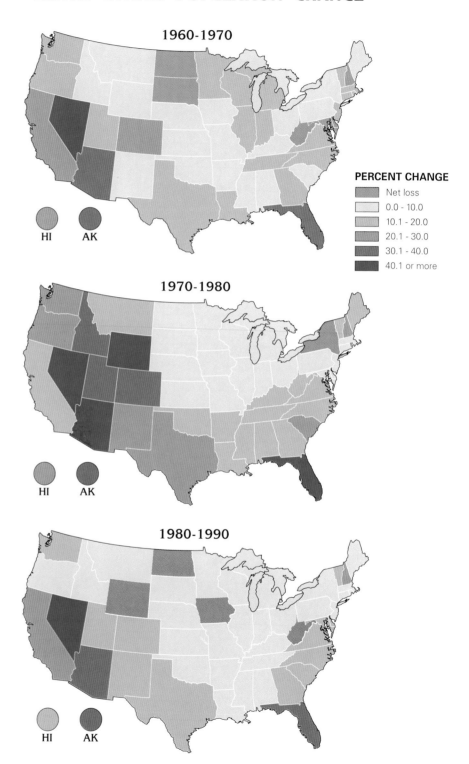

1960-1970

PERCENT CHANGE
- Net loss
- 0.0 - 10.0
- 10.1 - 20.0
- 20.1 - 30.0
- 30.1 - 40.0
- 40.1 or more

HI AK

1970-1980

HI AK

1980-1990

HI AK

During the late 1930s, for example, while the United States experienced a slowing of the growth in population density, Kentucky's continued to increase. This trend is in part a response to the state's role in supplying vital resources for the war products industry. Also instrumental, however, were the migration patterns of native Kentuckians living in other states who often returned home after losing a job during the Depression. Yet during the 1940s, just the opposite was true: Kentucky's population density reached a plateau, and the nation's resumed a vigorous increase. After a period of gradually increasing density within the state, the 1990s again find Kentucky's population density experiencing stability and low increase.

For much of the twentieth century, Kentucky's population has grown more slowly than that of the country as a whole. Indeed, the state's population increased by only 5.9 percent from 1960 to 1970, at less than half the nation's 13.4 percent growth rate. The 1970–80 period, however, witnessed a dramatic reversal in this trend: Kentucky's ten-year growth jumped to 13.8 percent—several percentage points above the national figure—primarily as a result of nationwide movement away from the North toward the southeastern and western states. Favorable coal markets during the 1970s also helped to retain mine workers and their families. The 1980–90 decade saw yet another reversal, with Kentucky's growth falling to the lowest level in a long time, at only about 0.7 percent. Two factors have caused this slowing of Kentucky's population growth: net out-migration of residents and reductions in fertility levels.

The "Mix" of Kentucky's Population

Age and Sex Characteristics. The processes that affect an area's population size also result in changes in population composition. Age structure is one element of this composition that may be strongly influenced by such factors as fertility and migration. Age also has direct effects on society, ranging from school enrollments and the demand for teachers to the number of new housing starts and housing vacancy rates, from the demand for health services that cater to geriatric populations to the sales of products in local stores.

The dependency ratio is one measure that provides a rough but informative first view of Kentucky's age structure. In its basic calculation, this ratio compares the number of both young and old individuals with the number of individuals within the traditional laboring age group of fifteen through sixty-four years. The premise is that the population in the labor force supports the dependent young and old populations, so high ratios indicate populations that are more stressed to provide both financial and social support through taxes or familial connections. In general, the least densely settled rural counties tend to have the highest total dependency ratios in Kentucky. In most cases this is caused by an out-migration of the young, unmarried labor force,

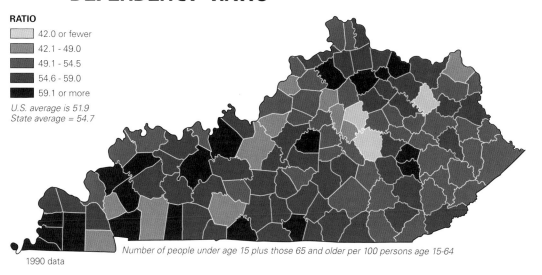

DEPENDENCY RATIO

RATIO

- 42.0 or fewer
- 42.1 - 49.0
- 49.1 - 54.5
- 54.6 - 59.0
- 59.1 or more

U.S. average is 51.9
State average = 54.7

1990 data

Number of people under age 15 plus those 65 and older per 100 persons age 15-64

A Methodist retirement community in Jessamine County.

CHILDHOOD DEPENDENCY RATIO

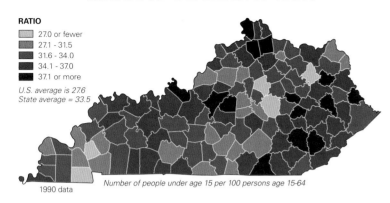

RATIO
- 27.0 or fewer
- 27.1 - 31.5
- 31.6 - 34.0
- 34.1 - 37.0
- 37.1 or more

U.S. average is 27.6
State average = 33.5

1990 data — *Number of people under age 15 per 100 persons age 15-64*

PERSONS YOUNGER THAN FIFTEEN

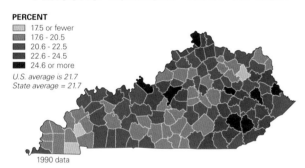

PERCENT
- 17.5 or fewer
- 17.6 - 20.5
- 20.6 - 22.5
- 22.6 - 24.5
- 24.6 or more

U.S. average is 21.7
State average = 21.7

1990 data

PERSONS YOUNGER THAN FIVE

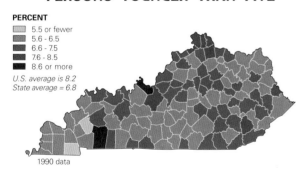

PERCENT
- 5.5 or fewer
- 5.6 - 6.5
- 6.6 - 7.5
- 7.6 - 8.5
- 8.6 or more

U.S. average is 8.2
State average = 6.8

1990 data

leaving behind married couples with children and also elders. Low-ratio counties typically have large colleges or universities (e.g., Fayette, Madison, Rowan, Warren, and Calloway Counties), military bases (Hardin and Christian), or job opportunities for young professionals (Franklin and Woodford). The dependency ratio can be subdivided into two additional measures: a childhood measure, which gauges the relative size of the youngest population; and an elderly measure, which monitors the relative size of the group of people over the age of sixty-five. The basic pattern that emerges in Kentucky from these two measures is one of high young-age dominance in the east and high old-age dominance in the south central and western portions of the state.

A slightly more refined measure of age structure examines the percentage of a population that falls within certain age categories. The maps showing the percentage of the population under five years and under fifteen years of age highlight certain areas of the state where either fertility is high (as in a cluster of southeastern counties) or the number of older people is relatively low (as in far northern counties). Most counties of the state have percentages of persons younger than five that are below the state average of 6.8 percent. Maps showing the percentages of the state population aged sixty-five and over and eighty-five and over again show the dominance of western counties as home to comparatively old populations. Among the group aged sixty-five and over, retirement migration remains a central process for concentrating or diffusing populations. The west, especially the area around the Land Between the Lakes, has become home to a large number of retirees drawn to its recreational opportunities, while the east has experienced extensive retirement out-migration. A focus on the group aged eighty-five and over demonstrates the influence of aging-in-place and identifies the counties that require health care facilities to serve the needs of older Kentuckians. Additionally, fewer of the state's oldest residents live in the remote eastern mountains; a greater proportion live in the Appalachian fringe, where services in such cities as Lexington and Cincinnati are more accessible.

Gender is another characteristic commonly emphasized in population studies. A comparison of the number of males to that of females using "sex ratios" begins to

ELDERLY DEPENDENCY RATIO

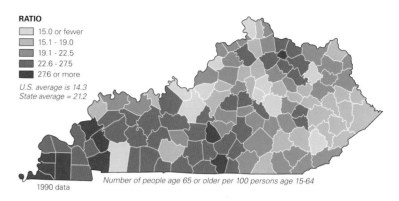

RATIO
- 15.0 or fewer
- 15.1 - 19.0
- 19.1 - 22.5
- 22.6 - 27.5
- 27.6 or more

U.S. average is 14.3
State average = 21.2

1990 data — *Number of people age 65 or older per 100 persons age 15-64*

PERSONS SIXTY-FIVE AND OLDER

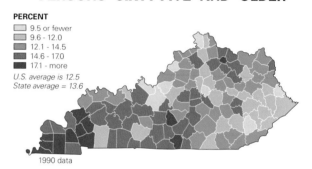

PERCENT
- 9.5 or fewer
- 9.6 - 12.0
- 12.1 - 14.5
- 14.6 - 17.0
- 17.1 - more

U.S. average is 12.5
State average = 13.6

1990 data

PERSONS EIGHTY-FIVE AND OLDER

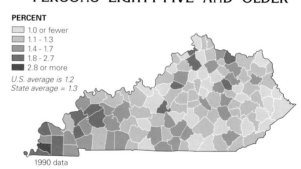

PERCENT
- 1.0 or fewer
- 1.1 - 1.3
- 1.4 - 1.7
- 1.8 - 2.7
- 2.8 or more

U.S. average is 1.2
State average = 1.3

1990 data

SEX RATIOS

AGES 15 TO 39

MALES PER 100 FEMALES
- ☐ 97.0 or fewer
- ☐ 97.1 - 109.0
- ☐ 109.1 - 140.0
- ☐ 140.1 - 180.0
- ■ More than 180.0

U.S. average is 101.1
State average = 100.1

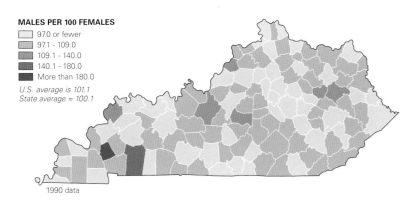

1990 data

AGES 40 TO 64

MALES PER 100 FEMALES
- ☐ 91.0 or fewer
- ☐ 91.1 - 95.0
- ☐ 95.1 - 99.0
- ☐ 99.1 - 105.0
- ■ More than 105.0

U.S. average is 93.8
State average = 96.0

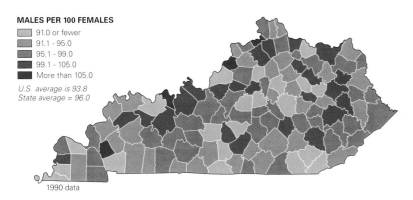

1990 data

AGES 65 AND OLDER

MALES PER 100 FEMALES
- ☐ 62.0 or fewer
- ☐ 62.1 - 67.0
- ☐ 67.1 - 71.0
- ☐ 71.1 - 78.0
- ■ More than 78.0

U.S. average is 67.0
State average = 70.2

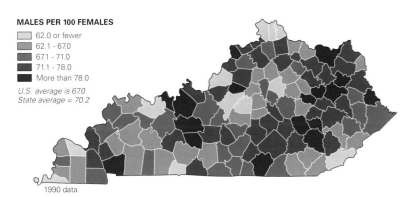

1990 data

HISTORIC AGE-SEX POPULATION PYRAMIDS

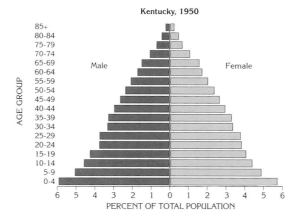

Kentucky, 1950

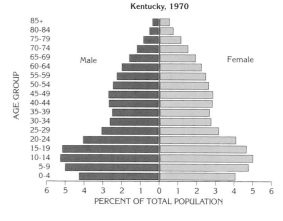

Kentucky, 1970

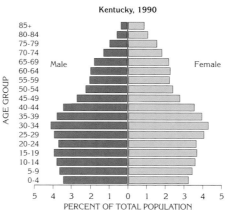

Kentucky, 1990

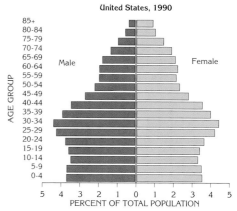

United States, 1990

uncover the variation according to gender in the processes of demographic change. In general, and virtually everywhere, more males are born than females—a ratio of 105 males to 100 females. Males have long had higher levels of mortality across most of the age spectrum. Consequently, more females than males survive to older ages. This broad trend is clearly evident in the Kentucky and U.S. averages of sex ratios: a ratio for males at birth of about 105 drops to around 100 by age forty and continues to fall to approximately 70 by age sixty-five and over. County patterns, however, allude to male dominance in the extraction industries of the state's coalfields and to female dominance in more remote rural counties that young males have left in search of jobs.

Perhaps the most instructive means of simultaneously examining both the age and the sex structure of a population is through the use of population pyramids. Such diagrams clearly illustrate the influence of high fertility, as in the 1950 Kentucky pyramid with its wide base; the aging of the Baby Boom cohort through the teen years that resulted from earlier high fertility, seen in the 1970 pyramid; and the continued aging of this cohort and its declining fertility levels, shown in the 1990 pyramid with a narrowing base. These diagrams also demonstrate that gender imbalance becomes more pronounced as age increases.

At the county level, population pyramids begin to show the variable influence of both historic and contem-

COUNTY DIFFERENCES IN AGE-SEX STRUCTURE

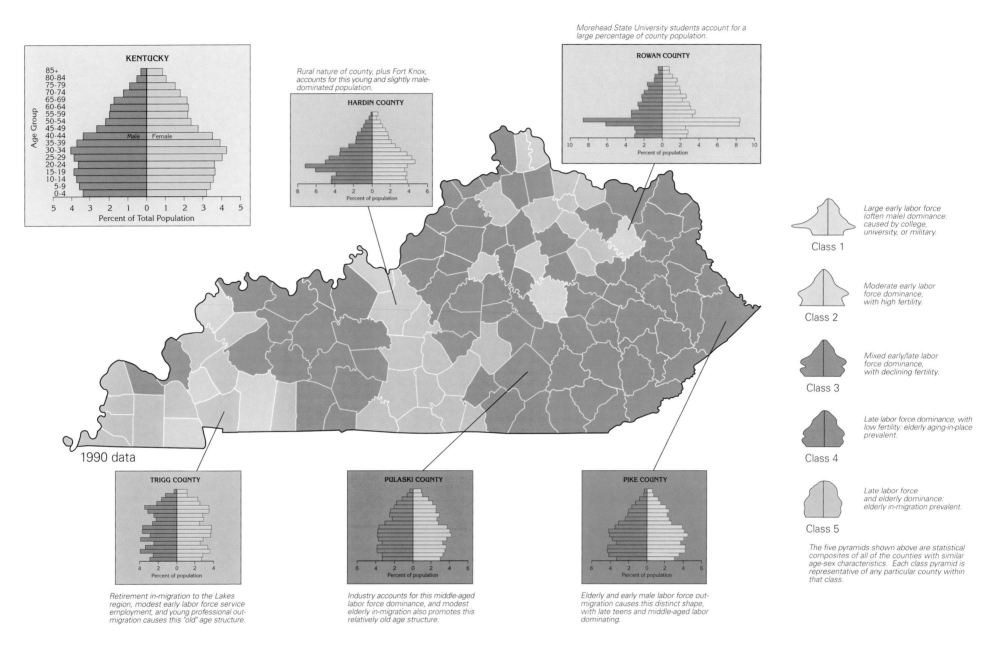

Morehead State University students account for a large percentage of county population.

Rural nature of county, plus Fort Knox, accounts for this young and slightly male-dominated population.

KENTUCKY

Age Group: 85+, 80-84, 75-79, 70-74, 65-69, 60-64, 55-59, 50-54, 45-49, 40-44, 35-39, 30-34, 25-29, 20-24, 15-19, 10-14, 5-9, 0-4

Male / Female

Percent of Total Population

HARDIN COUNTY
Percent of population

ROWAN COUNTY
Percent of population

1990 data

TRIGG COUNTY
Percent of population

Retirement in-migration to the Lakes region, modest early labor force service employment, and young professional out-migration causes this "old" age structure.

PULASKI COUNTY
Percent of population

Industry accounts for this middle-aged labor force dominance, and modest elderly in-migration also promotes this relatively old age structure.

PIKE COUNTY
Percent of population

Elderly and early male labor force out-migration causes this distinct shape, with late teens and middle-aged labor dominating.

Class 1
Large early labor force (often male) dominance: caused by college, university, or military.

Class 2
Moderate early labor force dominance, with high fertility.

Class 3
Mixed early/late labor force dominance, with declining fertility.

Class 4
Late labor force dominance, with low fertility: elderly aging-in-place prevalent.

Class 5
Late labor force and elderly dominance: elderly in-migration prevalent.

The five pyramids shown above are statistical composites of all of the counties with similar age-sex characteristics. Each class pyramid is representative of any particular county within that class.

porary migration patterns. The map showing county differences in age-sex structure was developed by first analyzing key measures of age percentages and sex ratios by age and then clustering counties according to general pyramidal shapes that portray these measures. Class 1 counties, for example, are influenced by a large and transient young adult population, which is drawn to an area most often by opportunities for higher education (with a gender balance) or by military commitments (with male dominance, as in Christian County). Class 2 counties have a young population that is slightly less mobile than the residents of class 1 counties. The case of Hardin County shows how the demographic effects of the military population of Fort Knox are somewhat muted by the influence of the major metropolitan area of nearby Louisville. Progressing through class 5, counties exhibit increasing aging, caused first by declining fertility and aging-in-place (classes 3 and 4)—note the Eastern and Western Coal Fields—and eventually retirement in-migration (class 5), of which Trigg County provides an excellent example.

BLACK POPULATION

1930

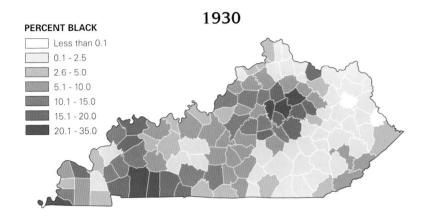

1960

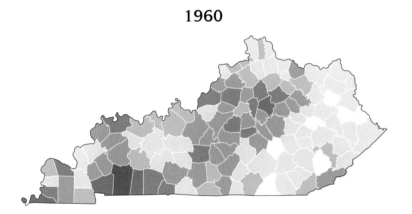

1990

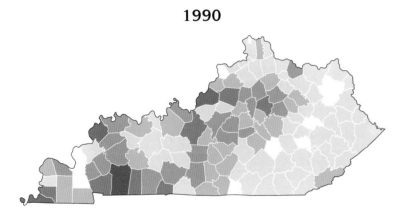

U.S. RACIAL COMPOSITION

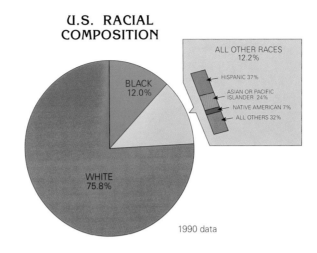

ALL OTHER RACES
12.2%

HISPANIC 37%
ASIAN OR PACIFIC
ISLANDER 24%
NATIVE AMERICAN 7%
ALL OTHERS 32%

BLACK
12.0%

WHITE
75.8%

1990 data

KENTUCKY RACIAL COMPOSITION

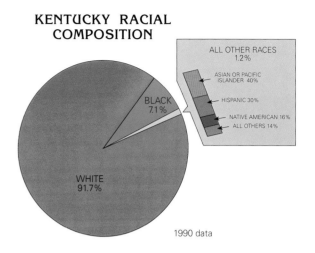

ALL OTHER RACES
1.2%

ASIAN OR PACIFIC
ISLANDER 40%
HISPANIC 30%
NATIVE AMERICAN 16%
ALL OTHERS 14%

BLACK
7.1%

WHITE
91.7%

1990 data

BLACK POPULATION, 1790-1990

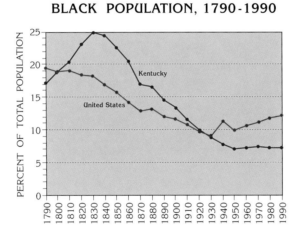

Kentucky

United States

PERCENT OF TOTAL POPULATION

The Louisville Urban League, at 16th Street and Broadway, is one of more than one hundred Urban League organizations across the United States that coordinate a variety of neighborhood development projects.

Race, Ethnicity, and Nativity. Since the early days of the state's settlement, Kentucky's population has become increasingly diverse with respect to race, ethnicity, and place of birth. Native American dominance through the eighteenth century changed rapidly with the influx of European settlers. The African population grew during periods of legal slavery and railroad construction, and many other racial and ethnic groups, including Asian, have more recently been drawn by the educational and employment opportunities of the state's urban areas. Despite this history of change, Kentucky's racial composition remains far more homogeneous than that of the nation as a whole: the white population, proportionally much larger in the state than in the nation, represents the majority of Kentuckians, blacks represent a significantly smaller share, and all other races combined account for a very small fraction of the state's total population.

Kentucky's black population is by far the dominant minority and has played a central part in the state's progress. Maps of the black population's concentration by county show the largest shares in the non-Appalachian western two-thirds

KENTUCKY-BORN RESIDENTS, 1970

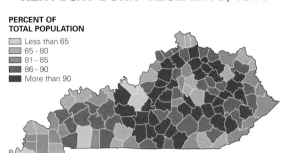

PERCENT OF TOTAL POPULATION
- Less than 65
- 65 - 80
- 81 - 85
- 86 - 90
- More than 90

KENTUCKY-BORN RESIDENTS, 1990

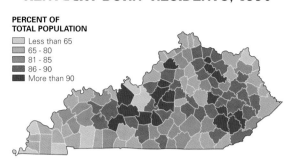

PERCENT OF TOTAL POPULATION
- Less than 65
- 65 - 80
- 81 - 85
- 86 - 90
- More than 90

FOREIGN-BORN RESIDENTS, 1990

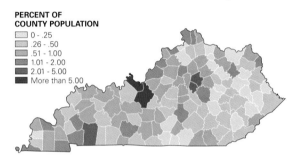

PERCENT OF COUNTY POPULATION
- 0 - .25
- .26 - .50
- .51 - 1.00
- 1.01 - 2.00
- 2.01 - 5.00
- More than 5.00

POPULATION NATIVE TO STATE OF RESIDENCE

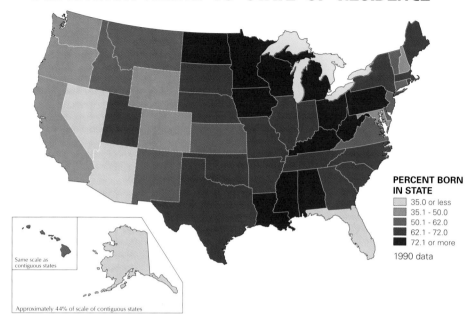

PERCENT BORN IN STATE
- 35.0 or less
- 35.1 - 50.0
- 50.1 - 62.0
- 62.1 - 72.0
- 72.1 or more

1990 data

Same scale as contiguous states

Approximately 44% of scale of contiguous states

of the state, a pattern that has remained fairly constant since at least 1930. Harlan County, along with several neighboring counties, provides an exception to this pattern. Early in the century, more than six thousand blacks lived in Harlan County alone, employed mostly in railroad construction. This population has gradually declined since 1930, and by 1990 only twelve hundred blacks resided in Harlan County. Many rural counties of western Kentucky have experienced similar declines. Indeed, Kentucky's black population today lives predominantly in urban areas, and the state total increased by less than 20 percent in the sixty years between 1930 and 1990, from 224,845 to 262,057.

A final characteristic that illustrates the key influence of demographic processes on population change is nativity, or place of birth. This characteristic is most indicative of migration levels: low nativity percentages indicate areas that have either large numbers of native-born people who move away and do not return, or a relatively immobile birth population with extensive in-migration of nonnatives. Kentucky currently has one of the highest percentages of native-born residents in the nation (77.4 percent in 1990). Only three states—Iowa, Louisiana, and Pennsylvania—have higher nativity rates. In Kentucky's case, a historic tradition of ties to one's land and kin not only has kept out-migration fairly low but has also encouraged natives to return to the state after years of living elsewhere. Central and mostly rural counties within the state are more likely to have high nativity percentages, and virtually all counties have experienced declining nativity during the past twenty years. This is one indication of Kentucky's growing diversity. Although in-migrants remain predominantly Anglo-American, their spatial and cultural experiences are becoming more varied.

NATIVITY OF KENTUCKY RESIDENTS

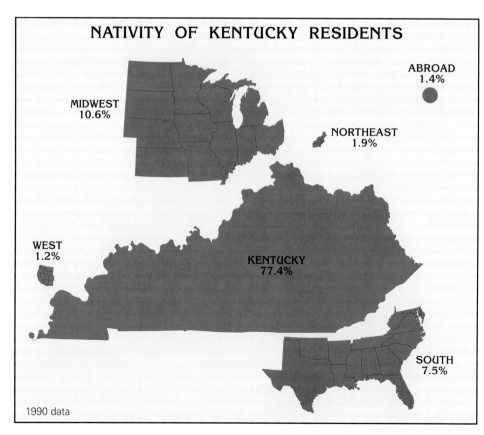

ABROAD 1.4%

MIDWEST 10.6%

NORTHEAST 1.9%

WEST 1.2%

KENTUCKY 77.4%

SOUTH 7.5%

1990 data

Marriage and the Family. Marriage is among the more important transitions experienced along the life course. People marry for many and varied reasons, including companionship and love, the desire for financial stability, and peer and cultural pressure or obligation. Regardless of the circumstances, marriage tends to have far-reaching demographic impacts. Fertility is highest among married women, and an older age at first marriage is strongly related to fertility decline and smaller sizes of completed families. Marriage has also been linked to extended life expectancies, especially among males.

Finally, marriage and family development almost always result in mobility, whether a short-distance move away from the parental home or a long-distance migration to a larger house or a better job. A migration decision made within marriage always involves and influences at least two people.

Levels of marriage, and especially its counterpart, divorce, have undergone striking changes during the past century. In 1930 Kentuckians were more likely to be married than residents of the nation in general. After an overall increase in the percentage of married

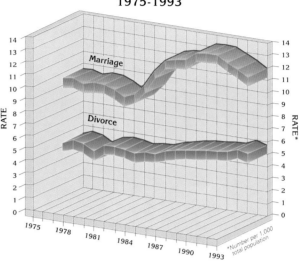

MARRIAGE AND DIVORCE IN KENTUCKY
1975-1993

MARRIAGE RATE

AVERAGE ANNUAL RATE PER
THOUSAND TOTAL POPULATION
- ☐ 4.6 - 10.0
- ☐ 10.1 - 13.0
- ☐ 13.1 - 18.0
- ■ 18.1 - 26.0
- ■ 26.1 - 39.8

U.S. average is 9.2
State average = 12.5

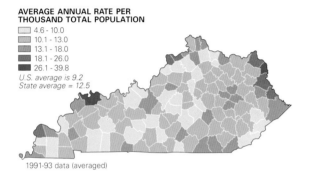

1991-93 data (averaged)

DIVORCE RATE

AVERAGE ANNUAL RATE PER
THOUSAND TOTAL POPULATION
- ☐ 0 - 3.0
- ☐ 3.1 - 6.0
- ☐ 6.1 - 7.0
- ■ 7.1 - 10.0
- ■ 10.1 - 15.5

U.S. average is 4.7
State average = 5.8

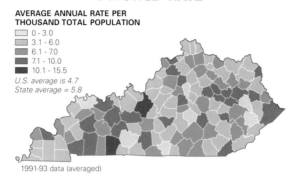

1991-93 data (averaged)

MARRIED MALES

PERCENT OF
MALES AGED 15+
- ☐ Less than 45.0
- ☐ 45.0 - 50.0
- ☐ 50.1 - 55.0
- ■ 55.1 - 60.0
- ■ More than 60.0

State average = 57.8

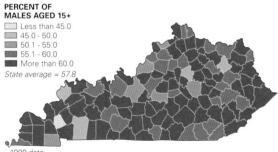

1990 data

DIVORCED MALES

PERCENT OF
MALES AGED 15+
- ☐ Less than 6.0
- ☐ 6.1 - 8.0
- ☐ 8.1 - 9.0
- ■ 9.1 - 10.0
- ■ More than 10.0

State average = 7.1

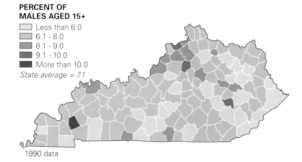

1990 data

MARRIED FEMALES

PERCENT OF
FEMALES AGED 15+
- ☐ Less than 45.0
- ☐ 45.0 - 50.0
- ☐ 50.1 - 55.0
- ■ 55.1 - 60.0
- ■ More than 60.0

State average = 55.1

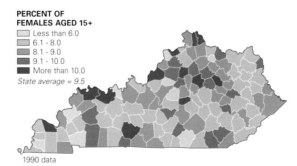

1990 data

DIVORCED FEMALES

PERCENT OF
FEMALES AGED 15+
- ☐ Less than 6.0
- ☐ 6.1 - 8.0
- ☐ 8.1 - 9.0
- ■ 9.1 - 10.0
- ■ More than 10.0

State average = 9.5

1990 data

KENTUCKY AND U.S. MARITAL STATUS COMPARISONS

Married

		1930	1940	1950	1960	1970	1980	1990
Kentucky	Male*	62.8	62.3	66.4	67.7	65.1	65.6	57.8
United States	Male*	60.0	61.2	66.9	68.1	64.7	62.5	59.3
Kentucky	Female*	63.6	63.1	65.1	64.2	61.0	60.7	55.1
United States	Female*	61.1	61.0	65.1	64.9	60.1	57.8	55.0

*Percent of male/female population aged 15 and over who are married.

Divorced

		1930	1940	1950	1960	1970	1980	1990
Kentucky	Male*	1.0	1.2	1.8	2.0	2.6	5.2	7.1
United States	Male*	1.1	1.3	1.9	2.1	2.6	5.3	7.4
Kentucky	Female*	1.2	1.5	2.2	2.7	3.5	6.8	9.5
United States	Female*	1.3	1.7	2.4	2.9	3.8	7.1	9.5

*Percent of male/female population aged 15 and over who are divorced.

HOUSEHOLD SIZE

MEDIAN NUMBER OF PERSONS PER HOUSEHOLD
- 2.10 - 2.26
- 2.27 - 2.37
- 2.38 - 2.47
- 2.48 - 2.64
- 2.65 - 2.86

State average = 2.40

1990 data

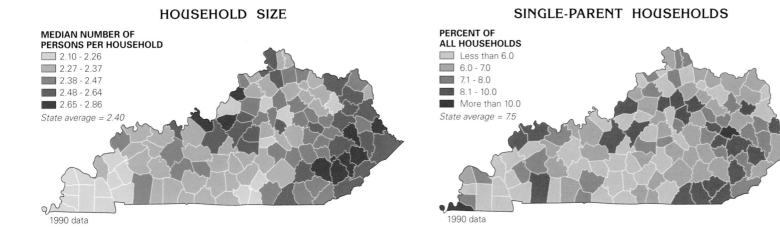

SINGLE-PARENT HOUSEHOLDS

PERCENT OF ALL HOUSEHOLDS
- Less than 6.0
- 6.0 - 7.0
- 7.1 - 8.0
- 8.1 - 10.0
- More than 10.0

State average = 7.5

1990 data

HOUSEHOLDS WITH ONE PERSON

PERCENT OF ALL HOUSEHOLDS
- 13.3 - 17.0
- 17.1 - 20.0
- 20.1 - 23.0
- 23.1 - 26.0
- 26.1 - 29.0

State average = 23.2

1990 data

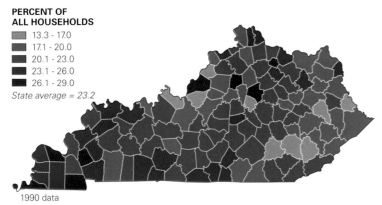

SINGLE-PARENT HOUSEHOLDS HEADED BY WOMEN

PERCENT OF ALL SINGLE-PARENT HOUSEHOLDS
- Less than 60
- 60 - 75
- 76 - 80
- 81 - 85
- More than 85

State average = 84

1990 data

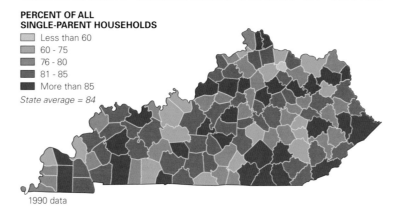

KENTUCKY'S CHANGING HOUSEHOLD SIZE

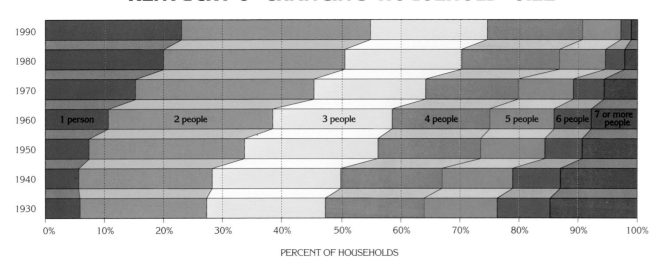

PERCENT OF HOUSEHOLDS

people by 1960, caused in part by a strong national economy, levels have gradually declined. Kentuckians now are somewhat less likely to be married than Americans in general. The sex structure of marriage has also evolved over this period; while females were slightly more likely than males to be married in 1930, just the opposite is true today, with a margin of about 3 percent now separating the sexes at both the state and the national level. Of the unmarried population, there is an increasing probability today that these individuals are divorced, and divorce in Kentucky and in the United States has increased dramatically, from levels of 1–2 percent in 1930 to 7–9 percent in 1990. The gap between male and female divorce has been widening over time, with a greater proportion of females than males now being divorced.

A view of the temporal changes in the number of persons per household is in many ways related to both the postponement of marriage to a later age and the increasing prevalence of divorce. Single-person households in Kentucky have increased in both number and share of all households, from about 6 percent in 1930 to over 23 percent in 1990. The proportion of the population living alone tends to be much higher in urban areas, and this can be attributed to both higher educational levels and the tendency for single professionals to move away from their parental families well before marriage. Also responsible for the urban trend of single-person households is the growing number of elderly persons, especially older women, who rely on the service availability offered in cities. Two-person households—most often involving married or cohabiting couples, but also increasingly comprising a single parent and child—have also increased, rising from about 21 percent of all households in 1930 to almost 32 percent in 1990. The shares of three- and four-person households have remained remarkably stable over time, while larger households—especially with six or more people—have become rare, with percentages falling from over 23 percent in 1930 to under 3 percent in 1990.

Changes in marriage, divorce, and family structure have many impacts that immediately involve population change at the county level. The traditions of early marriage and large extended families in Kentucky's rural areas, for example, clearly appear to have weakened over the past several decades. Marriage rates have fallen slightly but,

NATURAL POPULATION INCREASE

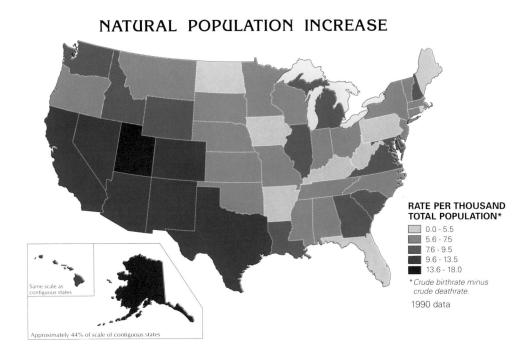

RATE PER THOUSAND
TOTAL POPULATION*

- 0.0 - 5.5
- 5.6 - 7.5
- 7.6 - 9.5
- 9.6 - 13.5
- 13.6 - 18.0

*Crude birthrate minus crude deathrate.

1990 data

Same scale as contiguous states

Approximately 44% of scale of contiguous states

NATURAL POPULATION INCREASE

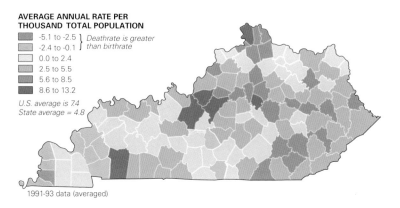

AVERAGE ANNUAL RATE PER
THOUSAND TOTAL POPULATION

- -5.1 to -2.5 ⎫ Deathrate is greater
- -2.4 to -0.1 ⎭ than birthrate
- 0.0 to 2.4
- 2.5 to 5.5
- 5.6 to 8.5
- 8.6 to 13.2

U.S. average is 7.4
State average = 4.8

1991-93 data (averaged)

GENERAL FERTILITY RATE

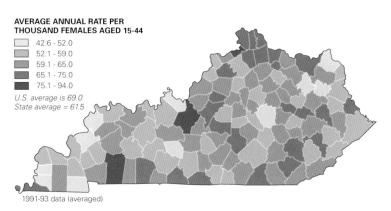

AVERAGE ANNUAL RATE PER
THOUSAND FEMALES AGED 15-44

- 42.6 - 52.0
- 52.1 - 59.0
- 59.1 - 65.0
- 65.1 - 75.0
- 75.1 - 94.0

U.S. average is 69.0
State average = 61.5

1991-93 data (averaged)

TEENAGE FERTILITY RATE

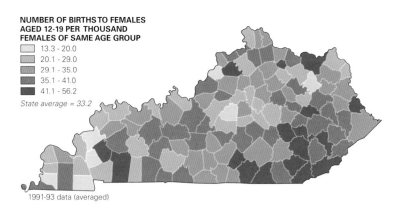

NUMBER OF BIRTHS TO FEMALES
AGED 12-19 PER THOUSAND
FEMALES OF SAME AGE GROUP

- 13.3 - 20.0
- 20.1 - 29.0
- 29.1 - 35.0
- 35.1 - 41.0
- 41.1 - 56.2

State average = 33.2

1991-93 data (averaged)

more important, more marriages end in divorce, which causes a separation of family members and the postponement or cessation of further reproduction. Indeed, the largest clustering of single-parent households can be found in the rural southeastern portion of the state. Such patterns take on further salience when one considers patterns of unemployment, welfare, and housing.

Components of Population Change

Topics such as marriage and the family serve as an illustrative transition to the processes that cause population change. These basic processes include fertility, mortality, and mobility, and marriage and family structure can influence each process. Such processes can help to explain, in demographic terms at least, the changing characteristics of Kentucky's population.

Fertility and Mortality. A summary measure of the impact of fertility and mortality is the rate of natural increase, a figure that remains popular for comparing areas at all scales—from local to international—by examining the balance between birth and death. Among the states, Kentucky ranks as one of the lowest in this measure, with an annual gain in population of fewer than five persons per one thousand residents of the state. Although falling fertility is partially responsible for this ranking, regionally high mortality is perhaps more influential, and a comparison of these two levels over about twenty years shows a gradual convergence in the crude rates of birth and death. A county-level map showing natural increase, for example, identifies eleven counties that had a negative rate of natural increase in the early 1990s, which means that there were more deaths than births in these areas. Recalling the age structure maps earlier in this chapter, we see that these counties are characterized by their relatively old populations and thus relatively high mortality and low fertility compared with other counties. This is confirmed in maps of general fertility rates (a fertility measure specific for females of childbearing age), crude birthrates, and crude deathrates by county.

The counties in the southeastern part of the state centered around Knox County have a fairly high rate of natural increase—about seven per thousand population and above. Whereas crude deathrates here also hover around seven deaths per thousand, crude birthrates are nearly twice as high in these counties. Furthermore, a large number of babies

CRUDE BIRTHRATE

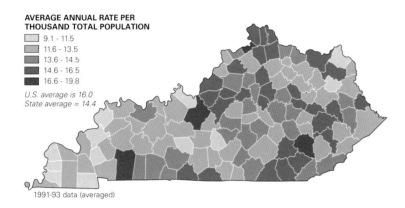

AVERAGE ANNUAL RATE PER
THOUSAND TOTAL POPULATION
- 9.1 - 11.5
- 11.6 - 13.5
- 13.6 - 14.5
- 14.6 - 16.5
- 16.6 - 19.8

U.S. average is 16.0
State average = 14.4

1991-93 data (averaged)

CRUDE DEATHRATE

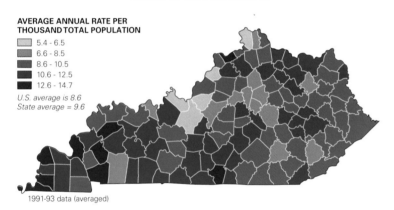

AVERAGE ANNUAL RATE PER
THOUSAND TOTAL POPULATION
- 5.4 - 6.5
- 6.6 - 8.5
- 8.6 - 10.5
- 10.6 - 12.5
- 12.6 - 14.7

U.S. average is 8.6
State average = 9.6

1991-93 data (averaged)

BIRTHS AND DEATHS IN KENTUCKY
1975-1993

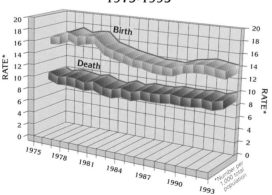

here are born to teenage mothers, and this portion of the state has been shown to have high levels of single-parent households as well.

Infant mortality rate, one noneconomic measure often used as an indicator of socioeconomic development, is comparatively low in Kentucky. The Kentucky average is about the same as that for the nation, and lower than that of any of the adjacent states. The state's infant mortality rates have witnessed a steady decline in recent decades, from nearly sixteen infant deaths per thousand live births in 1975 to just over eight deaths per thousand births in the early 1990s. Whereas Kentucky ranks favorably overall, such rates are of course not uniform across the state, and the county with the highest infant mortality rate in 1988, Carroll County, ranked seventh in this measure among all counties in the United States. Such high infant mortality rates are often related to both high fertility and poor access to health care clinics and hospitals, although low income also plays a role through its influence on diet and health hazards in the home.

Morbidity. The leading cause of death in Kentucky, as for the nation as a whole, is heart disease, although since the early 1970s the gap between it and the second leading cause of death, cancer, has narrowed. Together, these two leading causes accounted for nearly 60 percent of all deaths in the state in 1992, compared with 55 percent nationally. The next four leading causes of death were stroke, accidents, influenza and pneumonia, and pulmonary diseases—including emphysema, bronchitis, and asthma—which together accounted for a further 20 percent of all deaths. As is the case for the nation, motor vehicle accidents cause the majority of all accidental deaths in the state. Accidental deaths resulting from coal mining have declined significantly during the twentieth century, and these numbers pale in comparison with the number of deaths resulting from motor vehicle accidents. In 1990, 1,634 persons died in Kentucky from motor vehicle accidents, whereas just 20 persons died in coal mining accidents. Indeed, between 1980 and 1992, a total of 288 persons died in coal mining accidents; this is an unacceptable number but small compared with deaths by motor vehicles. Overall, changes in causes of

INFANT MORTALITY RATE

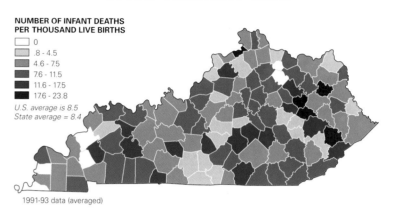

NUMBER OF INFANT DEATHS
PER THOUSAND LIVE BIRTHS
- 0
- .8 - 4.5
- 4.6 - 7.5
- 7.6 - 11.5
- 11.6 - 17.5
- 17.6 - 23.8

U.S. average is 8.5
State average = 8.4

1991-93 data (averaged)

INFANT MORTALITY RATE
(KENTUCKY AND ADJACENT STATES)

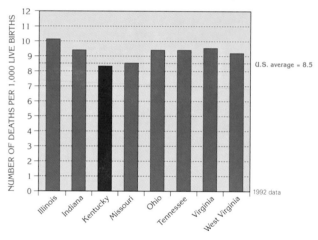

KENTUCKY INFANT MORTALITY RATE

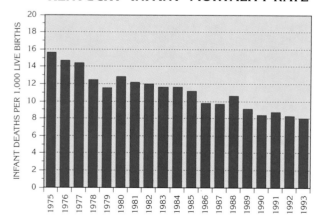

LEADING CAUSES OF DEATH IN KENTUCKY

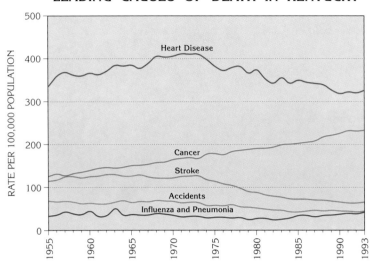

LEADING CAUSES OF DEATH

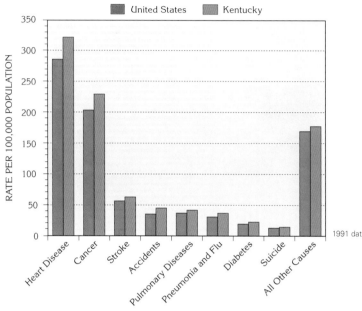

1991 data

LEADING CAUSES OF ACCIDENTAL DEATHS IN THE UNITED STATES

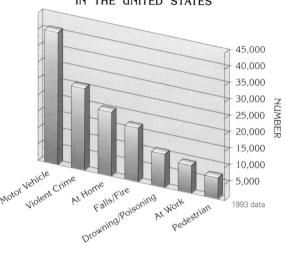

1993 data

death in Kentucky have followed the "typical" epidemiological transition: that is, a shift from occupational accidents and infectious diseases to chronic degenerative diseases such as heart disease, cancer, and stroke.

One of the nation's newest killers, death from infection by the human immunodeficiency virus (HIV), is significant in the state. Between 1982 and mid-1995, 1,639 cases of acquired immunodeficiency syndrome (AIDS) were reported, and 1,058 (65 percent) of these people died. Most cases were reported in urban areas: two-thirds of the cases reported in Kentucky were in the Louisville, Lexington, and Cincinnati–Northern Kentucky urban areas, and 52 percent of all cases were found in just two counties, Jefferson and Fayette. Nonetheless, AIDS cases were found throughout the state. The deathrate for AIDS in Kentucky, 2.6 per one hundred thousand residents in 1990, was among the lowest in the nation—only eleven states, none of which are adjacent to Kentucky, had lower AIDS deathrates in 1990—and the state rate is about one-quarter of that of the United States, which was 10.1 per hundred thousand people in 1990.

Holy Cross Cemetery and Church, Washington County.

LEADING CAUSES OF ACCIDENTAL DEATHS IN KENTUCKY

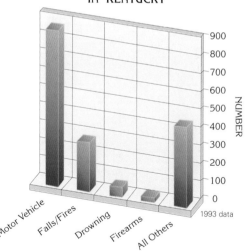

1993 data

LEADING CAUSES OF DEATH IN KENTUCKY

AGE GROUP

- 65 and over
- 55-64
- 45-54
- 35-44
- 25-34
- 15-24
- 6-14
- Under 6
- All ages

0% 10% 20% 30% 40% 50% 60% 70% 80% 90% 100%

CAUSE

- Heart Disease
- Cancer
- Stroke
- Accidents
- Influenza and Pneumonia
- Diabetes
- Suicide
- Nephritis and Nephrosis
- Septicemia
- Cirrhosis of Liver
- Emphysema
- Homicide
- Perinatal Conditions
- All other causes

1992 data

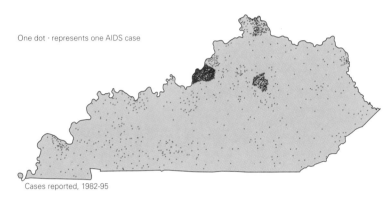

AIDS CASES BY COUNTY OF RESIDENCE

One dot · represents one AIDS case

Cases reported, 1982-95

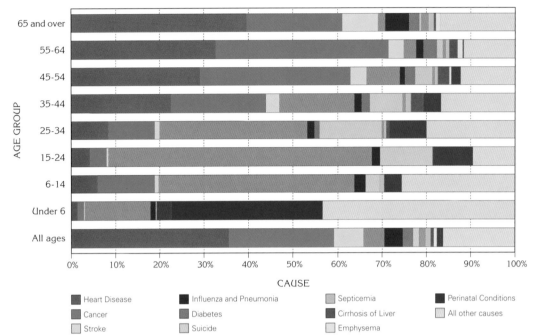

The Lucille Parker Markey
Cancer Center at the University
of Kentucky, Lexington.

STROKE DEATHRATE

AVERAGE ANNUAL RATE
PER 100,000 TOTAL POPULATION

- 28.4 - 40.0
- 40.1 - 60.0
- 60.1 - 75.0
- 75.1 - 90.0
- 90.1 - 124.0

U.S. average is 57.9
State average = 64.0

1991-93 data (averaged)

HEART DISEASE DEATHRATE

AVERAGE ANNUAL RATE
PER 100,000 TOTAL POPULATION

- 170.2 - 255.5
- 255.6 - 340.5
- 340.6 - 420.5
- 420.6 - 500.5
- 500.6 - 636.8

U.S. average is 289.5
State average = 322.6

1991-93 data (averaged)

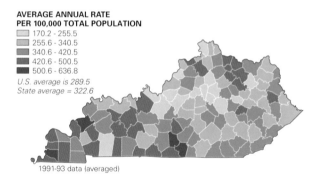

CANCER DEATHRATE

AVERAGE ANNUAL RATE
PER 100,000 TOTAL POPULATION

- 137.8 - 190.5
- 190.6 - 232.5
- 232.6 - 264.5
- 264.6 - 301.5
- 301.6 - 349.2

U.S. average is 203.2
State average = 231.6

1991-93 data (averaged)

ACCIDENT DEATHRATE

AVERAGE ANNUAL RATE
PER 100,000 TOTAL POPULATION

- 20.4 - 36.5
- 36.6 - 49.5
- 49.6 - 63.5
- 63.6 - 78.5
- 78.6 - 102.5

U.S. average is 37.0
State average = 40.4

1991-93 data (averaged)

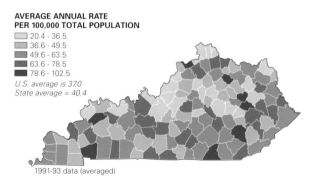

ORIGIN OF IMMIGRANTS LIVING IN KENTUCKY

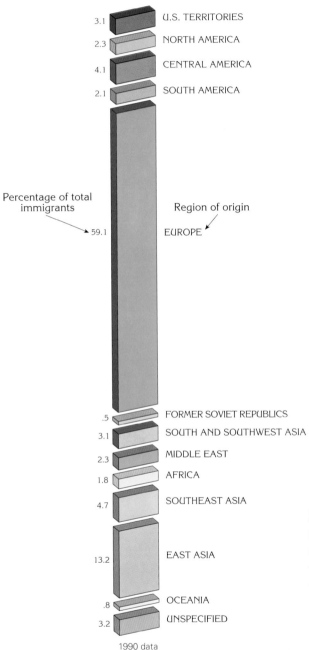

3.1 U.S. TERRITORIES
2.3 NORTH AMERICA
4.1 CENTRAL AMERICA
2.1 SOUTH AMERICA

Percentage of total immigrants

Region of origin

59.1 EUROPE

.5 FORMER SOVIET REPUBLICS
3.1 SOUTH AND SOUTHWEST ASIA
2.3 MIDDLE EAST
1.8 AFRICA
4.7 SOUTHEAST ASIA

13.2 EAST ASIA

.8 OCEANIA
3.2 UNSPECIFIED

1990 data

NET MIGRATION BY STATE, 1985-1990

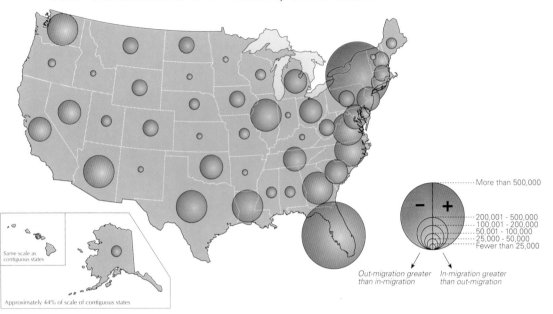

Same scale as contiguous states

Approximately 44% of scale of contiguous states

More than 500,000
200,001 - 500,000
100,001 - 200,000
50,001 - 100,000
25,000 - 50,000
25,000 - 50,000
Fewer than 25,000

Out-migration greater than in-migration

In-migration greater than out-migration

NET MIGRATION RATE, 1975-1980

RATE PER 1,000 POPULATION

Less than -50
-50 to -25
-24 to -1
0 to 50
51 to 200
More than 200

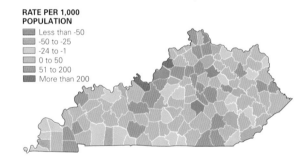

NET MIGRATION RATE, 1985-1990

RATE PER 1,000 POPULATION

Less than -100
-100 to -50
-49 to -1
0 to 50
51 to 150
More than 150

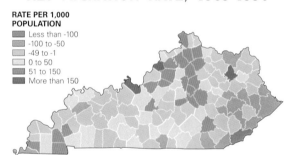

Mobility. Although some recent estimates suggest that Kentucky experienced a net in-migration (U.S. domestic migration) of about ten thousand persons and a net immigration (international migration) of four thousand persons between 1990 and 1993, the state with few exceptions has in the twentieth century been a net "exporter" of people. One exception was the decade of the 1970s, when Kentucky's population grew by 13.7 percent (compared with 11.4 percent for the United States). During this period Kentucky's rate of natural population increase approximated that of the United States. More typically, the 1950s, 1960s, and 1980s witnessed population increases of 3.2 percent, 6.0 percent, and 0.7 percent, respectively; all of these figures were well below the national averages, and all imply significant net out-migration. During the 1970s, however, a reversal occurred, and a number of eastern Kentucky counties actually experienced net in-migration between 1975 and 1980. The period 1985–90 again witnessed net out-migration from the mountains and coalfields.

The county with the highest out-migration rate during the 1985–90 period was Harlan County, where the population declined from 41,889 in 1980 to 36,574 in 1990. The net out-migration was more than 7,500. Rural counties that witnessed net in-migration during both the 1975–80 and the 1985–90 periods included those in the Land Between the Lakes and Lake Cumberland regions, areas typified by retirement living, leisure activities, and purchases of second homes. The Bluegrass Region also experienced significant net in-migration during the two periods. Jefferson County, like other counties with large central cities, witnessed out-migration—of more than 53,000 people between 1985 and 1990. The suburban,

KENTUCKY'S NET MIGRATION EXCHANGE, 1985-1990

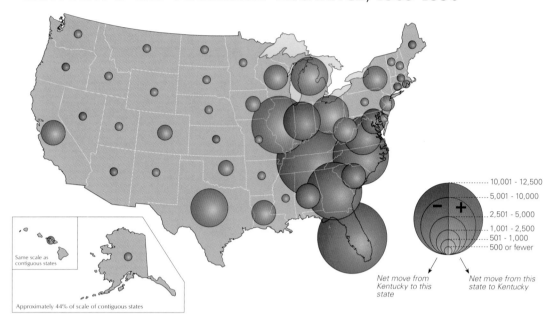

10,001 - 12,500
5,001 - 10,000
2,501 - 5,000
1,001 - 2,500
501 - 1,000
500 or fewer

Net move from
Kentucky to this
state

Net move from this
state to Kentucky

Same scale as
contiguous states

Approximately 44% of scale of contiguous states

IN-MIGRATION RATE

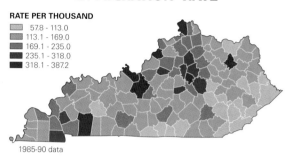

RATE PER THOUSAND
57.8 - 113.0
113.1 - 169.0
169.1 - 235.0
235.1 - 318.0
318.1 - 387.2

1985-90 data

OUT-MIGRATION RATE

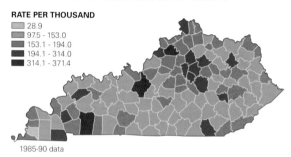

RATE PER THOUSAND
28.9
97.5 - 153.0
153.1 - 194.0
194.1 - 314.0
314.1 - 371.4

1985-90 data

MOVES WITHIN COUNTY

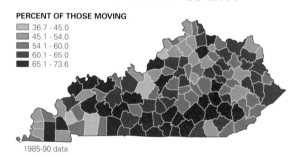

PERCENT OF THOSE MOVING
36.7 - 45.0
45.1 - 54.0
54.1 - 60.0
60.1 - 65.0
65.1 - 73.6

1985-90 data

CHANGE OF RESIDENCE

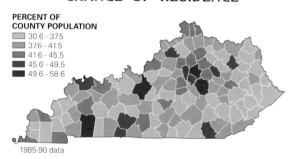

PERCENT OF
COUNTY POPULATION
30.6 - 37.5
37.6 - 41.5
41.6 - 45.5
45.6 - 49.5
49.6 - 58.6

1985-90 data

NO CHANGE OF RESIDENCE

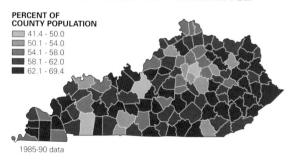

PERCENT OF
COUNTY POPULATION
41.4 - 50.0
50.1 - 54.0
54.1 - 58.0
58.1 - 62.0
62.1 - 69.4

1985-90 data

"bedroom" counties surrounding Jefferson County, however, experienced net in-migration during both periods. Oldham County had the second highest net migration rate in the 1985–90 period, following Grant County, which since mid-1993 has been considered part of the Cincinnati metropolitan area.

Recent Kentucky in-migrants come mostly from the North, especially Illinois, Ohio, Michigan, and, to a lesser extent, Wisconsin and New York. Certainly these in-migrants include many persons who are ready to retire. Many returning Kentuckians left the state in the 1950s and 1960s for job opportunities in cities such as Detroit, Chicago, Cleveland, Columbus, and Cincinnati. Such individuals maintain strong ties to their areas of origin and are returning to the eastern mountains, to the agricultural regions, or to amenity areas such as Lake Cumberland and the Land Between the Lakes area. Of course, some people are also moving to Kentucky in search of new economic opportunities, such as manufacturing jobs, available in the Bluegrass, in the northern Kentucky counties, and in other parts of the state. Kentucky lost migrants during the 1985–90 period, especially to states in the Southeast, and particularly, in order of the number of net out-migrants, to Tennessee, Florida, North Carolina, Georgia, and Virginia. This in part reflects the recent national trend of movement toward economic opportunities perceived to exist in the Sun Belt, as well as the retirement movement to Florida and other states with very mild climates.

A state's immigrant population, of course, reflects its degree of diversity. As we have already seen, Kentucky is among the more homogeneous and less diverse states. The majority of immigrants living in Kentucky in 1990 came from Europe (59 percent), and almost 90 percent of this group immigrated from Germany. The second largest region of origin, East Asia, accounted for 13 percent of the immigrant population. Within this group, the most significant countries of origin were Japan and Korea, reflecting in part the Japanese- and Korean-owned manufacturing industries that have recently located in the Commonwealth. Compared with other states, especially those in the

SIGNIFICANT COUNTY-TO-COUNTY MIGRANT FLOWS
(BETWEEN CONTIGUOUS COUNTIES)

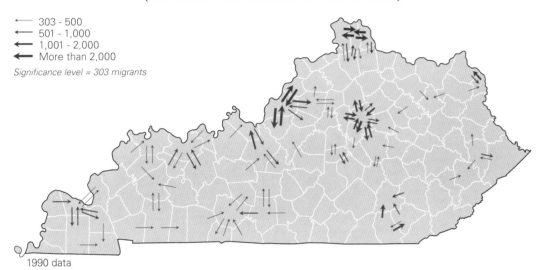

← 303 - 500
← 501 - 1,000
← 1,001 - 2,000
← More than 2,000

Significance level = 303 migrants

1990 data

SIGNIFICANT COUNTY-TO-COUNTY MIGRANT FLOWS
(BETWEEN NONCONTIGUOUS COUNTIES)

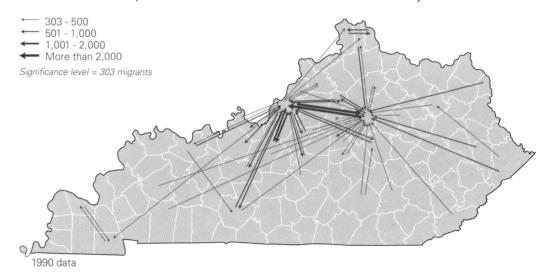

← 303 - 500
← 501 - 1,000
← 1,001 - 2,000
← More than 2,000

Significance level = 303 migrants

1990 data

MOVES FROM THIS KENTUCKY COUNTY TO ANOTHER KENTUCKY COUNTY

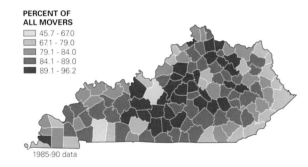

PERCENT OF
ALL MOVERS
□ 45.7 - 67.0
□ 67.1 - 79.0
□ 79.1 - 84.0
■ 84.1 - 89.0
■ 89.1 - 96.2

1985-90 data

MOVES FROM ANOTHER KENTUCKY COUNTY TO THIS KENTUCKY COUNTY

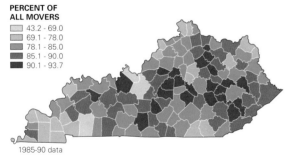

PERCENT OF
ALL MOVERS
□ 43.2 - 69.0
□ 69.1 - 78.0
□ 78.1 - 85.0
■ 85.1 - 90.0
■ 90.1 - 93.7

1985-90 data

southeastern and southwestern parts of the United States, Kentucky has few Hispanic immigrants. Hispanics accounted for just over 6 percent of total immigration. This may change in the future, however, as an increasing number of Hispanic agricultural workers enter the state to provide labor on farms.

Thus far we have examined migration to and from Kentucky, not moves within the state or county. Such local moves are significant because the largest share of Kentucky's population moves either locally or not at all. This level of local mobility or staying put is the reason Kentucky's nativity levels are among the highest in the nation. Longer-distance moves within the state are dominated by relocations to urban centers, especially Louisville and Lexington. Age selectivity of migrants is also important. Specifically, the labor force tends to the most mobile. Rural–urban moves are dominated by the young labor force (aged sixteen to thirty-five) and the very old (aged seventy-five or older), urban–urban moves are dominated by the middle and late labor force (thirty-five to sixty-five), and urban–rural moves are most typical of retirees (sixty-five and older). Social and economic selectivity of migrants is also important: for example, there is a positive relationship between distance of move and educational level or income; that is, those who move the longest distances are typically those with higher educational levels and higher incomes. Thus it follows that rural counties with high poverty levels typically have low mobility rates.

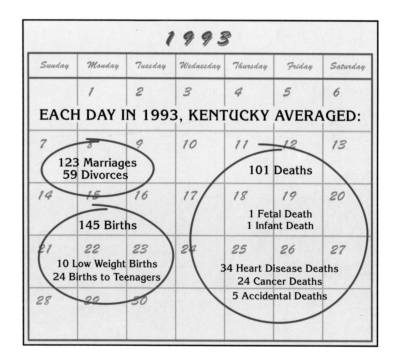

1993

EACH DAY IN 1993, KENTUCKY AVERAGED:

Sunday	Monday	Tuesday	Wednesday	Thursday	Friday	Saturday

123 Marriages
59 Divorces

101 Deaths

145 Births

1 Fetal Death
1 Infant Death

10 Low Weight Births
24 Births to Teenagers

34 Heart Disease Deaths
24 Cancer Deaths
5 Accidental Deaths

Social and Economic Characteristics

CHAPTER FIVE

Social and Economic Characteristics

K̲ENTUCKIANS HAVE LONG had to contend with stereotypes. Many people outside the state have made generalizations about Kentucky's residents: they are poorly educated and poverty-stricken, they mostly work as coal miners and tobacco farmers, they live in small houses and mobile homes situated in remote areas, and they have realized fewer benefits from social and technological advances than has the nation as a whole. Such stereotypes no doubt originated in part from historical observation. Kentucky has traditionally been a rural state. Agriculture and the extraction of natural resources have been economic mainstays, and participation in these activities, which requires little formal education, has continued from generation to generation across the Commonwealth. The absence of extensive industrialization, until recently, with its associated revenues, resulted in fewer state dollars that might have been spent to improve the general well-being of Kentuckians.

How accurate are the stereotypes of Kentucky's population? The information in this chapter sheds light on the question. Kentucky residents, in fact, exhibit much diversity in their socioeconomic characteristics, and many goods and services are readily available within the state. Some data support the stereotypes, while other material clearly refutes such generalizations. Unquestionably, certain areas within the Commonwealth have been—and remain—disadvantaged; income levels are low, and unemployment and poverty rates are high. Portions of Appalachian Kentucky, for example, consistently experience economic distress, while the Bluegrass Region tends to exhibit high incomes and educational levels and very low levels of distress.

A black neighborhood near Shawnee Park in west Louisville.

Primary and Secondary Education

Kentucky has 120 county school systems and 56 independent local school districts, each operating under the direction of a locally elected

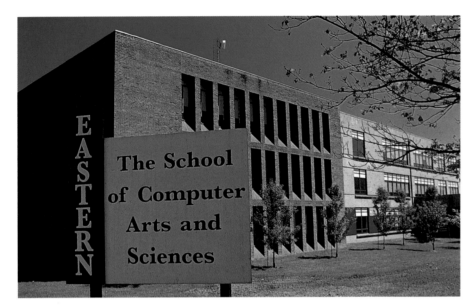

Eastern High School, Middletown.

LESS THAN NINTH GRADE EDUCATION

PERCENT OF PERSONS 25 YEARS OLD AND OLDER

- Less than 16
- 16 - 23
- 24 - 31
- 32 - 40
- More than 40

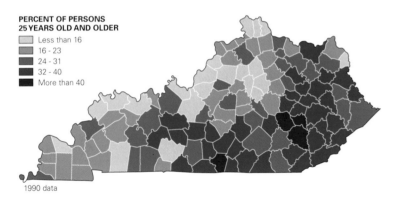

1990 data

NINTH GRADERS COMPLETING HIGH SCHOOL

PERCENT OF NINTH GRADERS

- Less than 50.0
- 50.0 - 60.0
- 60.1 - 70.0
- 70.1 - 85.0
- More than 85.0

County extremes:
Knox = 45.4
Hancock = 95.6

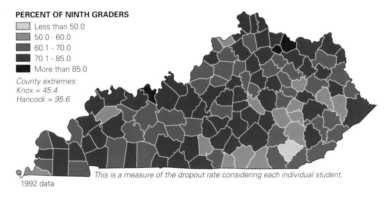

This is a measure of the dropout rate considering each individual student.

1992 data

HIGH SCHOOL EDUCATION OR HIGHER

PERCENT OF PERSONS 25 YEARS OLD AND OLDER

- Less than 45
- 45 - 54
- 55 - 64
- 65 - 75
- More than 75

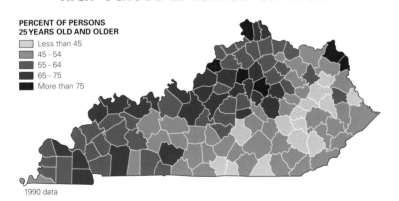

1990 data

HIGH SCHOOL GRADUATES ENTERING COLLEGE

PERCENT OF HIGH SCHOOL GRADUATES

- Less than 40
- 40 - 60
- 61 - 70
- 71 - 80
- More than 80

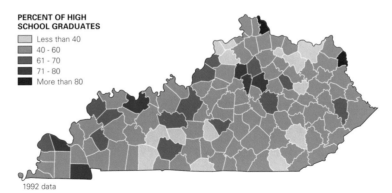

1992 data

five-member school board. The Kentucky General Assembly created the state's first system of universal free public education in 1838, providing for school taxes and local boards of trustees and establishing qualifications for teachers. In the same year the governor appointed Kentucky's first superintendent of public instruction. Seven years later, in 1845, the legislature created the State Board of Education. Kentucky's fourth constitution, adopted in 1891, called for "an efficient system of schools throughout the state" and placed the responsibility for this "efficient system" upon the legislature.

For nearly a century after the fourth state constitution, however, Kentuckians seemed to make only halting progress toward promoting quality education across the state. Indeed, in 1985 a suit filed by the Council for Better Education, a group of local boards of education, claimed that present educational funds were neither adequate nor equitable. As a result, in 1989 the Kentucky Supreme Court ruled that the entire existing Kentucky school system was unconstitutional. After lengthy examination, debate, and consultation with educational experts, the Kentucky legislature passed and the governor signed House Bill 940, the Kentucky Education Reform Act of 1990 (KERA). Underscoring the importance of educational reform, the 1990 census reported that Kentucky ranked last among all states in the percentage of its population aged twenty-five and over who had completed less than ninth grade (19 percent). Only 64.6 percent of adult Kentuckians had graduated from high school, well below the U.S. figure of 75.2 percent; the only state ranking lower than Kentucky was Mississippi. Moreover, only 13.6 percent of Kentucky's population had completed college; the national average was 20.3 percent, and all but two states ranked higher than Kentucky.

Some of the effects of KERA, which addresses elementary and secondary education, can be seen in recent changes in the public schools of the Commonwealth. Average daily attendance increased from 93 percent in 1980-81 to 95 percent in 1991-92. During the same period the pupil-teacher ratio dropped from 21.1 students per teacher to 17.4 students, equivalent to the U.S. average. While the pupil-teacher ratio decreased, the percentage of teachers holding higher certification increased; both of these measures imply a higher-quality classroom environment. In the 1980-81 school year, only 67 percent of Kentucky's teachers held

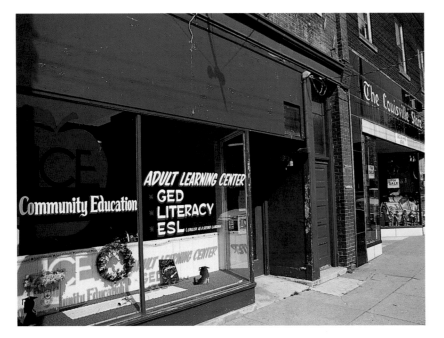

The Adult Learning Center in Lawrenceburg offers classes in literacy and English as a second language and prepares students for their GED exam.

HIGH SCHOOL GRADUATES, 1990
(KENTUCKY AND ADJACENT STATES)

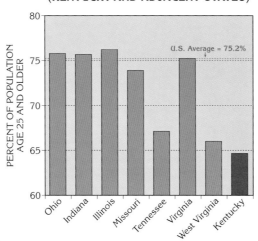

CLASSROOM TEACHER'S SALARY

AVERAGE ANNUAL SALARY ($)

- Less than 28,000
- 28,000 - 29,000
- 29,001 - 30,000
- 30,001 - 33,000
- More than 33,000

County extremes:
Gallatin = $26,657
Fayette = $34,379

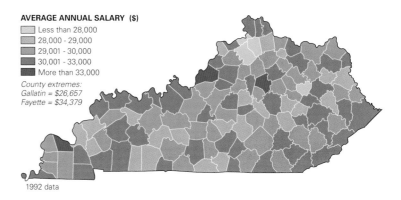

1992 data

COST OF INSTRUCTION

DOLLARS PER PUPIL

- Less than 2,300
- 2,300 - 2,500
- 2,501 - 2,700
- 2,701 - 3,000
- More than 3,000

County extremes:
Mercer = 2,188
Fayette = 3,383

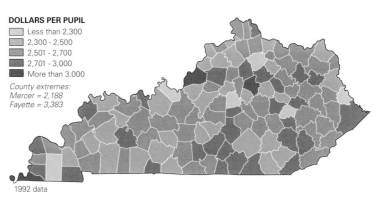

1992 data

a Rank II or higher certificate, indicating that they had earned at least a master's degree; by the 1991-92 school year, 81 percent of the state's teachers held such certification.

Total expenditures per pupil have increased from the 1980-81 average of $1,784 to the 1991-92 average of $4,719. While the state average fell below the national average in 1980-81 by $2,208, it had approached within $702 of the national average by 1991-92. In that school year Kentucky's average per-pupil expenditure was above the state averages in all but one of the seven states contiguous to the Commonwealth. What appears to be a positive trend in expenditures per student, however, is moderated by an examination of the way in which costs are allocated. Kentucky spends more of its funding on administration than the nation as a whole (9 percent versus 6 percent). The average annual salary for a teacher in Kentucky schools improved by $5,290 between the 1989-90 and 1993-94 school years (to $31,582), although it remained more than $4,000 shy of the national average. The cost per public school student solely for instructional purposes—that is, for books, computer equipment, and so on—during 1991-92 ranged from a low of $2,188 to a high of $3,383 across Kentucky counties.

Funds allocated to support local school districts are based on revenue from federal, state, and local sources, with small amounts coming from private sources. The percentage of revenue from these sources has not varied significantly during 1985-95. During the 1991-92 school year, the state provided the largest portion of funding, at 67 percent; local sources added another 23 percent, and the federal government contributed 10 percent. The proportions are quite different from national averages: local support amounts to 47 percent, states generally contribute about 46 percent, and federal additions total about 7 percent.

Higher Education

The Council on Higher Education coordinates the fourteen community colleges and eight universities that constitute the system of public higher education in Kentucky. This council also oversees the licensing of twenty-one independent institutions that are part of the Council on Independent Kentucky Colleges and Universities. The roots of higher education in Kentucky can be traced back to 1780, when Transylvania Seminary (later Transylvania Univer-

sity) was established. Transylvania is the oldest college west of the Allegheny Mountains. The University of Louisville dates back to 1837, when the city of Louisville established a medical institute and a collegiate institute; in 1846 the two were combined, and with the addition of a law department, the new institution became the University of Louisville. The passage of the federal Morrill Act in 1862 initiated the development of Kentucky's public land-grant colleges. The first of these, founded in 1865, was the Agricultural and Mechanical College, which later became the University of Kentucky.

Kentucky's Community College System began in 1964, in an effort to promote the accessibility of postsecondary education. The fourteen community colleges now offer courses not only at their campuses but also at more than one hundred other locations, including public schools, libraries, work sites, and even shopping malls. The demand for community college education

Built in 1820, Old Centre is the original main building of Centre College, Danville. Established in 1819 by the Kentucky Legislature, the college became affiliated with the Presbyterian Church in 1824.

PUBLIC TECHNICAL AND VOCATIONAL SCHOOLS

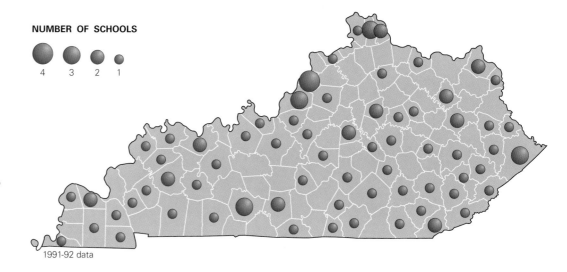

NUMBER OF SCHOOLS

4 3 2 1

1991-92 data

ASSOCIATE DEGREE OR HIGHER

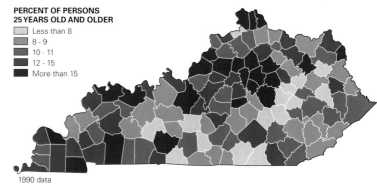

PERCENT OF PERSONS
25 YEARS OLD AND OLDER

Less than 8
8 - 9
10 - 11
12 - 15
More than 15

1990 data

PUBLIC UNIVERSITY SYSTEM

TYPE OF INSTITUTION
● University
● Community college

ENROLLMENT
18,000
6,000
1,000 or fewer

Extremes:
University of Kentucky = 18,336
Henderson Community College = 113

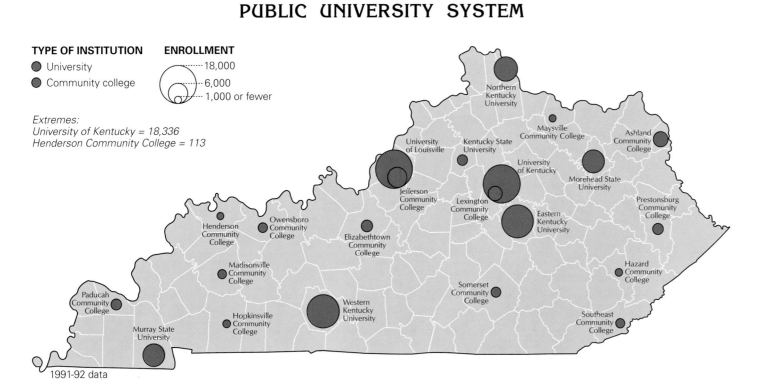

Northern Kentucky University
Maysville Community College
University of Louisville
Kentucky State University
Ashland Community College
University of Kentucky
Morehead State University
Jefferson Community College
Lexington Community College
Eastern Kentucky University
Prestonsburg Community College
Henderson Community College
Owensboro Community College
Elizabethtown Community College
Hazard Community College
Madisonville Community College
Paducah Community College
Somerset Community College
Hopkinsville Community College
Western Kentucky University
Southeast Community College
Murray State University

1991-92 data

BACHELOR'S DEGREE OR HIGHER

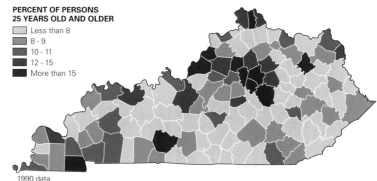

PERCENT OF PERSONS
25 YEARS OLD AND OLDER

Less than 8
8 - 9
10 - 11
12 - 15
More than 15

1990 data

ENROLLMENT IN STATE-SUPPORTED INSTITUTIONS OF HIGHER EDUCATION

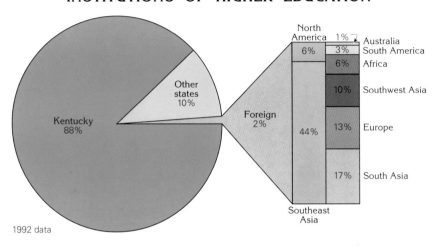

Kentucky 88%
Other states 10%
Foreign 2%

North America 6%
Australia 1%
South America 3%
Africa 6%
Southwest Asia 10%
Europe 13%
South Asia 17%
Southeast Asia 44%

1992 data

ORIGIN OF COLLEGE STUDENTS

State	Fall 1992 Enrollment	Percent of Total
Kentucky	138,576	88.1
Ohio	5,308	3.4
Tennessee	3,246	2.1
Indiana	2,304	1.5
Illinois	1,244	0.8
West Virginia	719	0.5
Florida	501	0.3
Michigan	394	0.3
Virginia	392	0.3
Missouri	350	0.2

TUITION AT STATE-SUPPORTED INSTITUTIONS

School Year	SEMESTER TUITION FOR RESIDENT STUDENTS				
	Doctoral Institutions		Masters Institutions		Community Colleges
	Undergraduate	Graduate	Undergraduate	Graduate	
1993-94	$980	$1,081	$750	$830	$420
1992-93	840	920	670	740	350
1991-92	810	890	650	720	340
1990-91	750	830	590	650	320
1989-90	690	760	530	580	300
1988-89	680	750	520	570	290
1987-88	660	730	500	550	280
1986-87	620	680	470	520	270
1985-86	572	630	442	486	260
1984-85	520	572	415	457	234
1983-84	467	514	388	427	207
1982-83	406	447	337	371	195

Authorized as a teacher training institution in 1922, Murray State Normal School and Teachers College became Murray State University in 1966. The campus has grown to include nearly eighty major buildings on 234 acres and enrolls more than 6,000 full-time undergraduates and more than 1,200 graduate students.

FUNDING SOURCES FOR PUBLIC HIGHER EDUCATION IN KENTUCKY

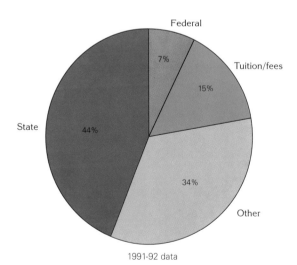

Federal 7%
Tuition/fees 15%
State 44%
Other 34%

1991-92 data

has grown dramatically: enrollment more than doubled between 1988 and 1992, and more than forty-eight thousand students were enrolled in the state's community colleges in 1993. Vocational and technical schools have also become an important element of higher education in the state, emphasizing opportunities for those who do not wish to pursue a traditional college degree. In 1986 the Kentucky Commission on Vocational and Technical Education issued a report to the General Assembly, recommending the use of interactive television to instruct students in rural areas, the establishment of pilot programs for dislocated workers, and the funding of Advanced Technology Centers to offer associate degrees in automated industrial systems technology with options in robotics and computer applications.

The enrollment in all of Kentucky's institutions of higher learning, both public and private, totaled 188,320 students in the fall of 1992. Female students accounted for 58 percent of this number; Kentucky thus had a greater proportion of female students than any of the seven contiguous states. (The national average of female students in 1992 was 55 percent.) Of the Commonwealth's neighboring states, Illinois had the highest percentage of black students in 1992, at 12.5 percent, and West Virginia had the lowest level, at 3.7 percent. Kentucky reported a black student population of 6.4 percent. The national figure was 9.6 percent. (Blacks accounted for 7.1 percent of Kentucky's total population in 1990.)

PRIVATE POSTSECONDARY INSTITUTIONS

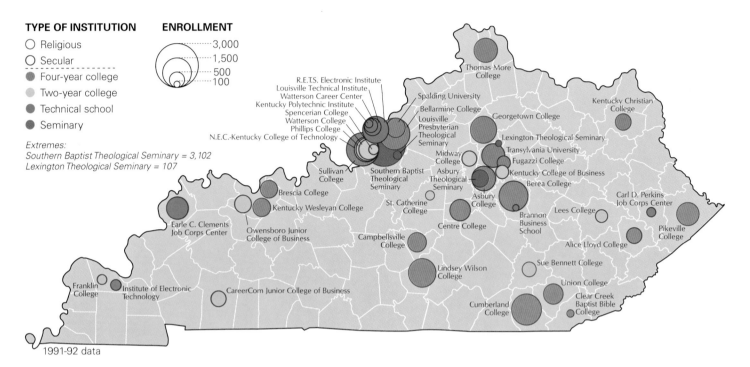

TYPE OF INSTITUTION

○ Religious
○ Secular
● Four-year college
○ Two-year college
● Technical school
● Seminary

ENROLLMENT

- 3,000
- 1,500
- 500
- 100

Extremes:
Southern Baptist Theological Seminary = 3,102
Lexington Theological Seminary = 107

R.E.T.S. Electronic Institute
Louisville Technical Institute
Watterson Career Center
Kentucky Polytechnic Institute
Spencerian College
Watterson College
Phillips College
N.E.C.-Kentucky College of Technology

Spalding University
Bellarmine College
Louisville Presbyterian Theological Seminary

Thomas More College

Kentucky Christian College

Georgetown College
Lexington Theological Seminary
Transylvania University
Fugazzi College
Kentucky College of Business
Berea College

Midway College
Asbury Theological Seminary
Asbury College

Sullivan College
Southern Baptist Theological Seminary

Brescia College
Kentucky Wesleyan College

St. Catherine College

Carl D. Perkins Job Corps Center

Lees College

Brannon Business School

Earle C. Clements Job Corps Center

Owensboro Junior College of Business

Centre College

Campbellsville College

Pikeville College

Alice Lloyd College

Lindsey Wilson College

Sue Bennett College

Union College

Franklin College
Institute of Electronic Technology

CareerCom Junior College of Business

Cumberland College

Clear Creek Baptist Bible College

1991-92 data

In the fall of 1992 foreign students at Kentucky's state-supported institutions totaled 3,054, or 2 percent of total enrollment. China, India, Malaysia, and Japan led the list of origin areas of foreign students. Persons from other states accounted for 10 percent of total enrollment, and Kentucky residents constituted 88 percent of the student population at Kentucky's public colleges and universities. About 59 percent of all students were between the ages of eighteen and twenty-four, and the remaining 41 percent were "nontraditional" students, aged twenty-five years or over.

The average annual salary for a full professor in the state's public universities was $60,096 during the 1991-92 academic year. Although this figure was above the average salary of a professor in Indiana, Missouri, Tennessee, and West Virginia, it fell almost $2,000 below the national average. In comparison, professors at Kentucky's private institutions earned an average of $40,118. Following national trends, the cost of tuition in state-supported schools has increased dramatically in recent years, more than doubling between 1983 and 1995. Average undergraduate tuition per semester, for example, increased from $428 in 1983-84 to $940 in 1994-95, and graduate tuition for Kentucky residents increased, on average, from $471 to $1,035 over the same period.

The Dominican Sisters established St. Catherine College near Springfield in 1920 as a two-year liberal arts college.

Income, Poverty, and Welfare

The economic well-being of a population can be assessed in a variety of ways. When applied to Kentucky, many of these measures reveal the disadvantaged state of the Appalachian region, especially the Eastern Kentucky Coal Field Region. Personal income is one measure that comprises all sources of income, such as wages and salaries, employer contributions to retirement funds, and noncash benefits such as food stamps. Average personal income, per person, is lowest in Owsley County ($9,466 in 1992) and a large group of contiguous counties in eastern Kentucky. High-paying jobs are scarce in this portion of the state, and the mechanization of coal production has elevated unemployment levels. Indeed, most rural nonagricultural counties of the state are conspicuous for low per capita incomes. The large urban areas generally enjoy the highest income levels, above the state average of $16,528 per capita in 1992. Family and household income are additional measures commonly used to assess well-being, and patterns revealed by mapping these measures are quite similar to those shown for personal per capita income.

There seems to be a consensus that across the United States, the rich are getting richer, and the poor are getting poorer. Kentucky might appear to exhibit evidence of this trend as well, since the difference between the highest and lowest county median family incomes expanded by an average of 14 percent per year between 1959 and 1989. The poorest counties, however, have seen more rapid income advances than the richest counties. In 1959, for example, the highest median family income by county was more than 4.8 times greater than that in the poorest county; by 1989 the richest county had a median family income about 3.8 times greater than that in the poorest county, indicating a slow but noticeable convergence. This trend is partly reflected in poverty rates: the proportion of the state population living in poverty fell from 22.9 percent in 1969 to 18.8 percent in 1989. This decline represents a significantly more positive trend in Kentucky than in the nation as a whole. In fact, the actual number of persons living below the poverty level in Kentucky dropped by nearly 13 percent between 1969 and 1979, while the number nationwide increased. Although the number in Kentucky increased during 1979-89, it did so only about half as fast as the national figure.

A working-class neighborhood on Milton Avenue in Louisville. Nineteenth-century developers often used the "shotgun" house plan—one room wide, three or more rooms deep—to keep lot sizes and housing prices low.

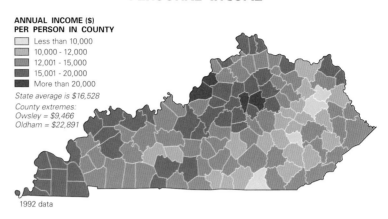

PERSONAL INCOME

ANNUAL INCOME ($)
PER PERSON IN COUNTY

- Less than 10,000
- 10,000 - 12,000
- 12,001 - 15,000
- 15,001 - 20,000
- More than 20,000

State average is $16,528
County extremes:
Owsley = $9,466
Oldham = $22,891

1992 data

MEDIAN HOUSEHOLD INCOME IN KENTUCKY

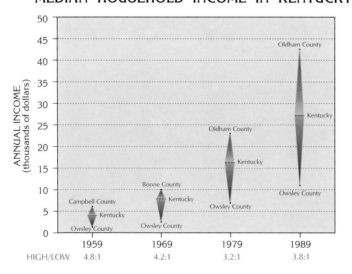

ANNUAL INCOME (thousands of dollars)

	1959	1969	1979	1989
HIGH/LOW	4.8:1	4.2:1	3.2:1	3.8:1

HOUSEHOLDS WITH ANNUAL INCOME BELOW $10,000

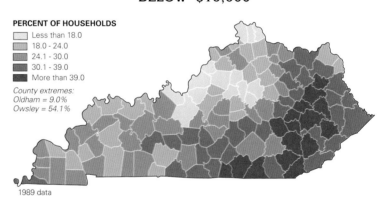

PERCENT OF HOUSEHOLDS

- Less than 18.0
- 18.0 - 24.0
- 24.1 - 30.0
- 30.1 - 39.0
- More than 39.0

County extremes:
Oldham = 9.0%
Owsley = 54.1%

1989 data

HOUSEHOLDS WITH ANNUAL INCOME OVER $100,000

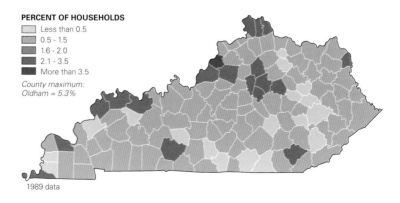

PERCENT OF HOUSEHOLDS

- Less than 0.5
- 0.5 - 1.5
- 1.6 - 2.0
- 2.1 - 3.5
- More than 3.5

County maximum:
Oldham = 5.3%

1989 data

PERSONS BELOW POVERTY LEVEL

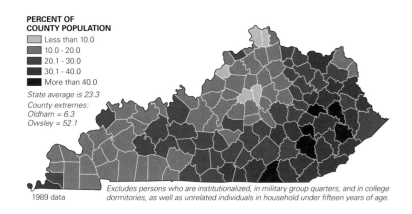

PERCENT OF
COUNTY POPULATION
- Less than 10.0
- 10.0 - 20.0
- 20.1 - 30.0
- 30.1 - 40.0
- More than 40.0

State average is 23.3
County extremes:
Oldham = 6.3
Owsley = 52.1

1989 data

Excludes persons who are institutionalized, in military group quarters, and in college dormitories, as well as unrelated individuals in household under fifteen years of age.

POPULATION FIVE YEARS OLD OR YOUNGER BELOW POVERTY LEVEL

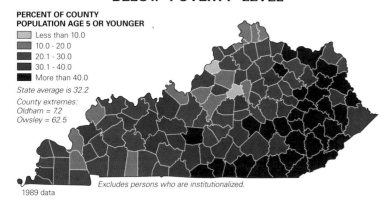

PERCENT OF COUNTY
POPULATION AGE 5 OR YOUNGER
- Less than 10.0
- 10.0 - 20.0
- 20.1 - 30.0
- 30.1 - 40.0
- More than 40.0

State average is 32.2
County extremes:
Oldham = 7.2
Owsley = 62.5

1989 data

Excludes persons who are institutionalized.

POPULATION SIXTY-FIVE OR OLDER BELOW POVERTY LEVEL

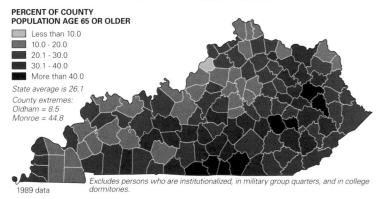

PERCENT OF COUNTY
POPULATION AGE 65 OR OLDER
- Less than 10.0
- 10.0 - 20.0
- 20.1 - 30.0
- 30.1 - 40.0
- More than 40.0

State average is 26.1
County extremes:
Oldham = 8.5
Monroe = 44.8

1989 data

Excludes persons who are institutionalized, in military group quarters, and in college dormitories.

UNITED STATES POVERTY THRESHOLDS
(WEIGHTED AVERAGES)

Family Size	Poverty Threshold
1 person	
Under 65 years	$7,518
65 and older	6,930
2 persons	
Head under 65	9,676
Head 65 and older	8,734
3 persons	11,303
4 persons	14,904
5 persons	17,974
6 persons	20,673
7 persons	23,787
8 persons	26,604
9 persons	32,003

1993 data

POPULATION BELOW POVERTY LEVEL

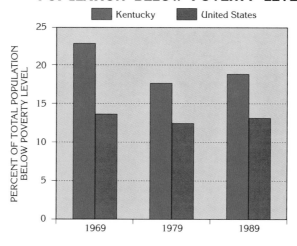

■ Kentucky ■ United States

PERCENT OF TOTAL POPULATION BELOW POVERTY LEVEL

1969   1979   1989

Regardless of improvements, though, more than seven hundred thousand Kentuckians live below the poverty level. In six eastern counties in 1989, over 40 percent of the total population lived in poverty, and in twelve counties more than half of all children under the age of five lived in poverty. Poverty level is based on annual federal calculations of the amount of money necessary for adequate personal and family survival. Although this measure provides a rough indicator of economic well-being, it cannot account for local realities associated with the cost of living.

Low income immediately affects society, most noticeably through state and federal government contributions to needy individuals and families. Approximately 12 percent of Kentuckians receive some form of direct public assistance in the form of Supplementary Security Income (SSI) payments, Aid to Families with Dependent Children (AFDC), or food stamps. Low family income is a central criterion for receiving aid, and single-parent families, blind and disabled individuals, and the aged are especially prominent among program recipients. During the 1994-95 fiscal year, for example, almost 194,000 persons in Kentucky received AFDC payments amounting to $188,164,304, about 520,000 Kentuckians received food stamps worth $407,058,186, and 509,319 individuals were eligible for Medicaid. In each of these cases, participation rates were highest in Appalachian Kentucky. Considering SSI payments in general and retirement income in particular, however, results in a very different spatial pattern. Supplemental Security Income, which targets both the elderly and the blind and disabled, is more concentrated in southern and western portions of Kentucky. This region is characterized by retirement settlement, agriculture, and mining in the Western Kentucky Coal Field.

MEDICAID ELIGIBILITY

PERCENT OF COUNTY POPULATION
- Less than 10.0
- 10.0 - 15.0
- 15.1 - 20.0
- 20.1 - 30.0
- More than 30.0

County extremes:
Trigg = 0.7
Owsley = 44.0

1995 data

AFDC RECIPIENTS

PERCENT OF COUNTY POPULATION
- Less than 3.50
- 3.50 - 5.50
- 5.51 - 8.50
- 8.51 - 12.50
- More than 12.50

County extremes:
Oldham = 0.95
Owsley = 18.43

1995 data   *AFDC - Aid to Families with Dependent Children*

AFDC PAYMENTS

DOLLARS PER RECIPIENT
- Less than 925
- 925 - 950
- 951 - 975
- 976 - 1,000
- More than 1,000

County extremes:
Edmonson = $1,152
Trimble = $881

1995 data   *AFDC - Aid to Families with Dependent Children*

FOOD STAMP RECIPIENTS

PERCENT OF COUNTY POPULATION
- Less than 12.5
- 12.5 - 20.0
- 20.1 - 27.5
- 27.6 - 35.0
- More than 35.0

County extremes:
Oldham = 3.0
Owsley = 48.4

1995 data

FOOD STAMP PAYMENTS

DOLLARS (IN MILLIONS)
- Less than 1.0
- 1.0 - 2.5
- 2.6 - 4.0
- 4.1 - 8.0
- More than 8.0

County extremes:
Oldham = 0.1
Jefferson = 31

1995 data

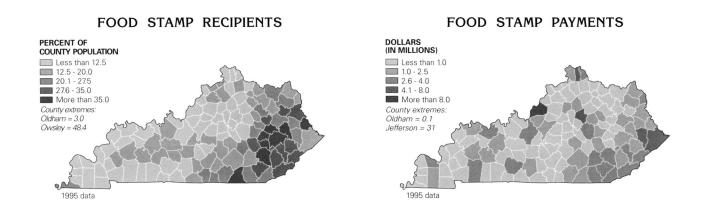

HOUSEHOLDS WITH RETIREMENT INCOME

PERCENT OF ALL HOUSEHOLDS
- Less than 11.5
- 11.5 - 13.5
- 13.6 - 15.5
- 15.6 - 18.5
- More than 18.5

County extremes:
Magoffin = 8.4
Lyon = 22.6

1989 data

Includes retirement pensions and survivors' benefits from a former employer, labor union, or federal, state, county, or other governmental agency; disability income from sources such as workers' compensation, companies or unions, federal, state or local government, and the U.S. military; periodic receipts from annuities and insurance; and regular income from IRA and KEOGH plans.

HOUSEHOLDS RECEIVING SUPPLEMENTAL SECURITY INCOME

PERCENT OF ALL HOUSEHOLDS
- Less than 23.0
- 23.0 - 29.0
- 29.1 - 32.0
- 32.1 - 35.0
- More than 35.0

County extremes:
Meade = 17.3
Fulton = 39.4

1989 data

Includes Social Security pensions and survivors' benefits and permanent disability insurance payments made by the Social Security Administration. Excludes Medicare reimbursements.

HOUSEHOLDS WITH PUBLIC ASSISTANCE INCOME

PERCENT OF ALL HOUSEHOLDS
- Less than 7.5
- 7.5 - 10.5
- 10.6 - 15.5
- 15.6 - 20.5
- More than 20.5

State average = 12.0
County extremes:
Oldham = 3.1
Wolfe = 26.8

1989 data

Includes Supplementary Security Income payments made by federal or state welfare agencies to low-income persons age sixty-five and over, blind, or disabled; Aid to Families with Dependent Children; and general assistance. Excludes payments received for hospital or medical care.

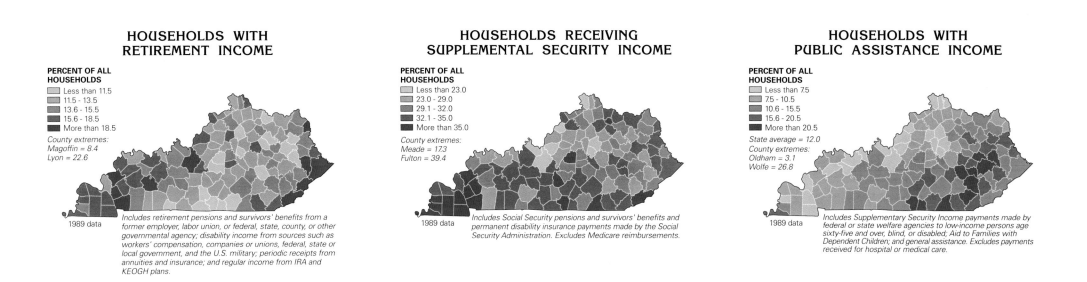

UNEMPLOYMENT

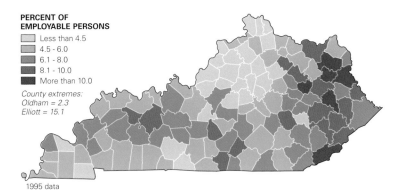

PERCENT OF
EMPLOYABLE PERSONS

- Less than 4.5
- 4.5 - 6.0
- 6.1 - 8.0
- 8.1 - 10.0
- More than 10.0

County extremes:
Oldham = 2.3
Elliott = 15.1

1995 data

LABOR SURPLUS

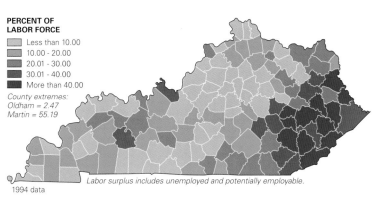

PERCENT OF
LABOR FORCE

- Less than 10.00
- 10.00 - 20.00
- 20.01 - 30.00
- 30.01 - 40.00
- More than 40.00

County extremes:
Oldham = 2.47
Martin = 55.19

Labor surplus includes unemployed and potentially employable.

1994 data

FEMALE UNEMPLOYMENT

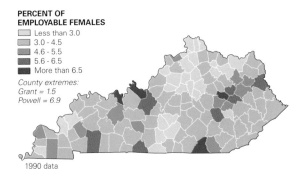

PERCENT OF
EMPLOYABLE FEMALES

- Less than 3.0
- 3.0 - 4.5
- 4.6 - 5.5
- 5.6 - 6.5
- More than 6.5

County extremes:
Grant = 1.5
Powell = 6.9

1990 data

MALE UNEMPLOYMENT

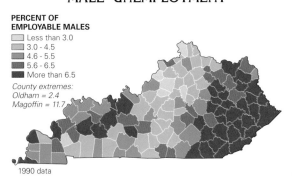

PERCENT OF
EMPLOYABLE MALES

- Less than 3.0
- 3.0 - 4.5
- 4.6 - 5.5
- 5.6 - 6.5
- More than 6.5

County extremes:
Oldham = 2.4
Magoffin = 11.7

1990 data

Labor Force Activity and Occupations

Employment, along with income and poverty, is a measure often used to assess economic well-being. The civilian labor force in Kentucky grew significantly during the 1970s and 1980s, primarily as a result of the aging of individuals born during the post-World War II Baby Boom. This growth stimulated competition for jobs, and unemployment rates waxed and waned as employment opportunities changed. The Kentucky Cabinet for Human Resources estimated that unemployment rates in Kentucky during the 1970s were below national figures. The state seems to have been hit especially hard by the recession of the 1980s, however, and unemployment rose to levels that exceeded the national average. Yet another reversal has occurred during the 1990s, and state unemployment rates are again lower than U.S. levels of about 7 percent. Unemployment is most severe in eastern Kentucky and is especially prominent among males as compared with females. Measures of unemployment are somewhat misleading, however, as they do not consider the population formally defined as not being in the labor force. Such individuals include students, homemakers, retirees, institutionalized persons, and persons doing only incidental unpaid family work. It can be argued that many of these people are potentially employable; adding them to the numbers of unemployed workers would substantially increase the measured labor surplus.

Agriculture and mining, historically considered the foundations of the state's economy, have offered fewer and fewer jobs during the past twenty years. In 1975 about 85,400 persons were employed in full-time agriculture, and mining provided another 46,600 jobs. By 1993 employment in these two occupations had fallen to 47,329 and 27,700, respectively. Manufacturing employment has increased

Drive-in restaurants such as this one in Catlettsburg employ predominantly young people in entry-level service jobs.

EMPLOYMENT IN AGRICULTURE

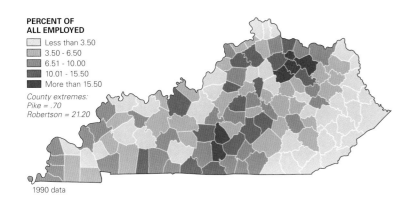

PERCENT OF ALL EMPLOYED
- Less than 3.50
- 3.50 - 6.50
- 6.51 - 10.00
- 10.01 - 15.50
- More than 15.50

County extremes:
Pike = .70
Robertson = 21.20

1990 data

EMPLOYMENT IN MINING

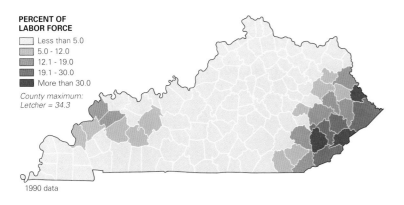

PERCENT OF LABOR FORCE
- Less than 5.0
- 5.0 - 12.0
- 12.1 - 19.0
- 19.1 - 30.0
- More than 30.0

County maximum:
Letcher = 34.3

1990 data

EMPLOYMENT IN GOVERNMENT

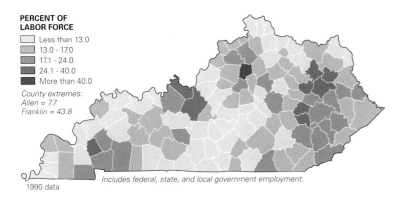

PERCENT OF LABOR FORCE
- Less than 13.0
- 13.0 - 17.0
- 17.1 - 24.0
- 24.1 - 40.0
- More than 40.0

County extremes:
Allen = 7.7
Franklin = 43.8

Includes federal, state, and local government employment.

1990 data

EMPLOYMENT IN SERVICES

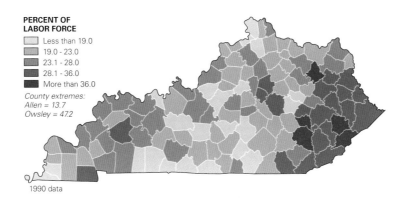

PERCENT OF LABOR FORCE
- Less than 19.0
- 19.0 - 23.0
- 23.1 - 28.0
- 28.1 - 36.0
- More than 36.0

County extremes:
Allen = 13.7
Owsley = 47.2

1990 data

PERSONS WHO WORK AT HOME

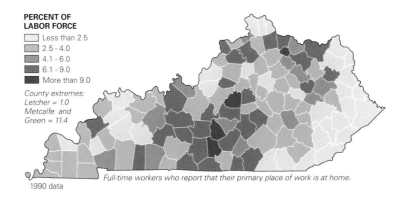

PERCENT OF LABOR FORCE
- Less than 2.5
- 2.5 - 4.0
- 4.1 - 6.0
- 6.1 - 9.0
- More than 9.0

County extremes:
Letcher = 1.0
Metcalfe and
Green = 11.4

Full-time workers who report that their primary place of work is at home.

1990 data

NONAGRICULTURAL EMPLOYMENT
(By Industry Division)

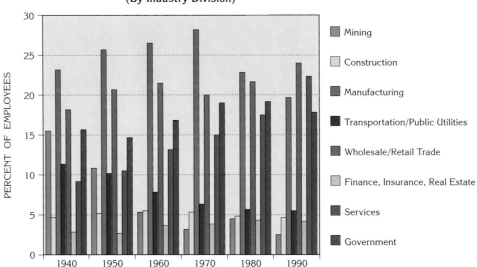

- Mining
- Construction
- Manufacturing
- Transportation/Public Utilities
- Wholesale/Retail Trade
- Finance, Insurance, Real Estate
- Services
- Government

slightly. Having dominated nonagricultural employment in the state from 1940 through 1970, manufacturing barely edged out the wholesale/retail sector in the number of jobs in 1980. By 1990 manufacturing had dropped to third place, behind wholesale/retail trade and services, which together provided nearly 690,000 jobs in Kentucky, or about 47 percent of all nonfarm employment. Another changing element of employment is the number of persons who work principally at home. Advances in computer and communications technology now allow individuals to be productive and to remain in contact with employers without having to be physically present at their place of employment.

AGE OF HOUSING

**MEDIAN YEAR RESIDENCES
IN THE COUNTY WERE BUILT**
- ☐ 1953-59
- ☐ 1960-64
- ☐ 1965-69
- ■ 1970-74
- ■ 1975-79

*County extremes:
Campbell = 1953
Oldham = 1976*

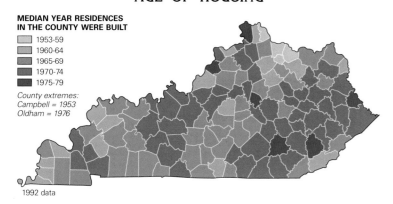

1992 data

NUMBER OF ROOMS IN RESIDENTIAL STRUCTURES

**AVERAGE NUMBER
OF ROOMS**
- ☐ Fewer than 5.0
- ☐ 5.0 - 5.2
- ■ 5.3 - 5.5
- ■ 5.6 - 5.7
- ■ More than 5.7

*County extremes:
Menifee = 4.9
Oldham = 6.4*

Excludes bathrooms.

1990 data

Much of the housing stock near Covington's downtown area dates from the nineteenth century. Typically built of brick on narrow lots, the houses have small yards, and space for parking is limited.

HOUSING UNITS IN KENTUCKY

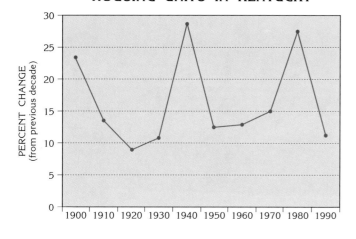

CHANGE IN NUMBER OF HOUSEHOLDS
(KENTUCKY AND ADJACENT STATES)

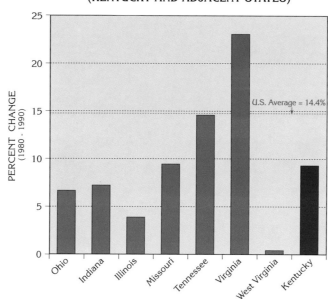

U.S. Average = 14.4%

DWELLINGS WITH PUBLIC SEWER

**PERCENT OF ALL
DWELLINGS IN COUNTY**
- ☐ Less than 20.0
- ☐ 20.0 - 35.0
- ■ 35.1 - 50.0
- ■ 50.1 - 65.0
- ■ More than 65.0

*County extremes:
Leslie = 3.9
Fayette = 95.2*

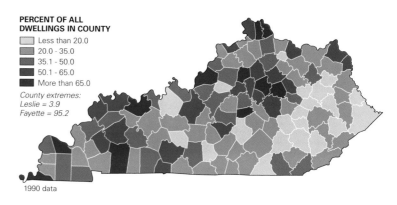

1990 data

DWELLINGS WITH PUBLIC WATER

**PERCENT OF ALL
DWELLINGS IN COUNTY**
- ☐ Less than 35.0
- ☐ 35.0 - 50.0
- ■ 50.1 - 70.0
- ■ 70.1 - 85.0
- ■ More than 85.0

*County extremes:
Knott = 10.4
Fayette = 99.7*

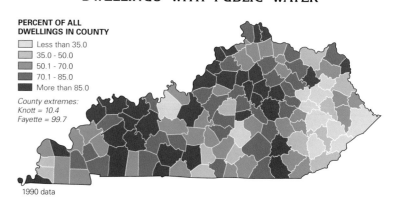

1990 data

Housing

Housing tends to be a visible manifestation of a population's well-being; it is essential for survival and often consumes the largest share of household income. Although older structures may eventually be destroyed, such subtractions from the housing stock are nearly always overshadowed by a much higher number of new housing starts, intended to accommodate a growing population. New housing is strongly tied not only to population growth but also to economic conditions. In the late 1930s, for example, Kentucky's housing pool grew substantially, corresponding to a revival after the Depression and a rapid growth in the state's population. Lower levels of housing starts reported from 1950 through 1970 correspond to an extended period of out-migration from Kentucky, lessening the pace of population growth. Current trends suggest a slow growth in housing, perhaps related to a slow increase in population and stability in statewide economic conditions.

SEWAGE DISPOSAL
(BY AREA DEVELOPMENT DISTRICT)

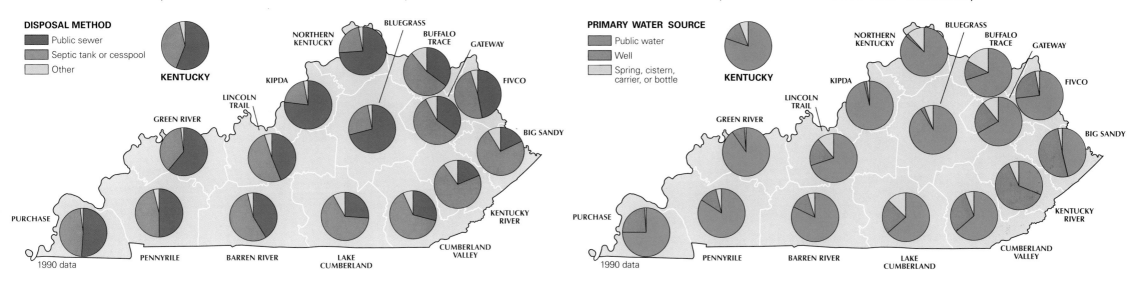

DISPOSAL METHOD
- Public sewer
- Septic tank or cesspool
- Other

KENTUCKY

NORTHERN KENTUCKY
BLUEGRASS
BUFFALO TRACE
GATEWAY
KIPDA
LINCOLN TRAIL
FIVCO
GREEN RIVER
BIG SANDY
PURCHASE
KENTUCKY RIVER
PENNYRILE
BARREN RIVER
LAKE CUMBERLAND
CUMBERLAND VALLEY

1990 data

WATER SOURCE
(BY AREA DEVELOPMENT DISTRICT)

PRIMARY WATER SOURCE
- Public water
- Well
- Spring, cistern, carrier, or bottle

KENTUCKY

NORTHERN KENTUCKY
BLUEGRASS
BUFFALO TRACE
GATEWAY
KIPDA
LINCOLN TRAIL
FIVCO
GREEN RIVER
BIG SANDY
PURCHASE
KENTUCKY RIVER
PENNYRILE
BARREN RIVER
LAKE CUMBERLAND
CUMBERLAND VALLEY

1990 data

The quality of housing is often difficult to assess. Size provides one very rough estimate of quality. Larger houses naturally will demand a higher relative price, and such structures are most often found within the commuter zones of major urban centers like Louisville and Lexington. Better indicators of quality include the presence or absence of complete kitchens, plumbing facilities, and telephones, the method of domestic waste disposal, the source of water, and the means of home heating. A state-level view of these indicators confirms first a rural influence on Kentucky housing. In 1990 fully 63 percent of residential structures in Kentucky were not connected to public sewers, compared with a U.S. average of 25 percent, and about 32 percent of Kentucky's residences lacked access to public water, compared with a national figure of 16 percent. Indicators also suggest the influence of low income, especially in Appalachian Kentucky, where in some counties more than 10 percent of all structures lacked complete kitchens and plumbing in 1990. Finally, nearly 23 percent of Kentucky homes did not have a telephone; this figure was over four times the national figure.

Regardless of quality, homeownership commonly plays a role in promoting personal well-being. Ownership has the potential of increasing personal net worth as equity builds over time. There is also an emotional benefit associated with

DWELLINGS WITH PUBLIC ELECTRIC OR GAS HEAT

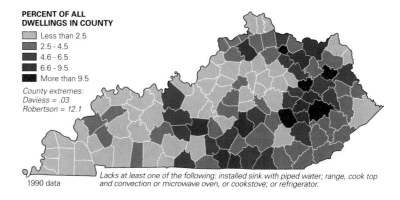

PERCENT OF ALL DWELLINGS IN COUNTY
- Less than 35.0
- 35.0 - 50.0
- 50.1 - 60.0
- 60.1 - 75.0
- More than 75.0

County extremes:
Elliott = 22.7
Jefferson = 89.2

1990 data

DWELLINGS WITHOUT A COMPLETE KITCHEN

PERCENT OF ALL DWELLINGS IN COUNTY
- Less than 2.5
- 2.5 - 4.5
- 4.6 - 6.5
- 6.6 - 9.5
- More than 9.5

County extremes:
Daviess = .03
Robertson = 12.1

1990 data

Lacks at least one of the following: installed sink with piped water; range, cook top and convection or microwave oven, or cookstove; or refrigerator.

DWELLINGS WITHOUT COMPLETE PLUMBING

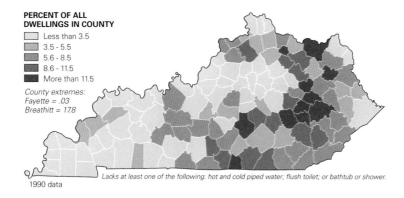

PERCENT OF ALL DWELLINGS IN COUNTY
- Less than 3.5
- 3.5 - 5.5
- 5.6 - 8.5
- 8.6 - 11.5
- More than 11.5

County extremes:
Fayette = .03
Breathitt = 17.8

1990 data

Lacks at least one of the following: hot and cold piped water; flush toilet; or bathtub or shower.

KENTUCKY HOUSING QUALITY INDICATORS	
Indicator	Percent of Structures
Not on Public Sewer	63.0
Not on Public Gas or Electric	45.7
Not on Public Water	31.8
Without Phone	22.7
Without Full Kitchen	3.6
Without Full Plumbing	3.4

1990 data

FUEL SOURCE
(BY AREA DEVELOPMENT DISTRICT)

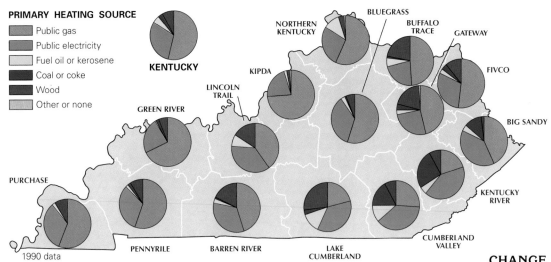

PRIMARY HEATING SOURCE
- Public gas
- Public electricity
- Fuel oil or kerosene
- Coal or coke
- Wood
- Other or none

KENTUCKY

NORTHERN KENTUCKY

BLUEGRASS

BUFFALO TRACE

GATEWAY

KIPDA

FIVCO

LINCOLN TRAIL

GREEN RIVER

BIG SANDY

PURCHASE

KENTUCKY RIVER

PENNYRILE

BARREN RIVER

LAKE CUMBERLAND

CUMBERLAND VALLEY

1990 data

Suburban expansion into agricultural land allows developers to erect large homes on very large lots, as seen here in Jefferson County.

Many rural residences, like this home in Madison County, are heated with LP gas.

CHANGE IN NUMBER OF OCCUPIED DWELLINGS, 1980-1990

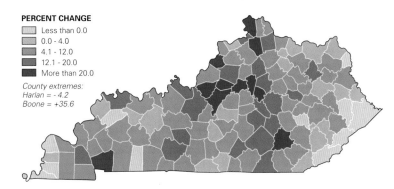

PERCENT CHANGE
- Less than 0.0
- 0.0 - 4.0
- 4.1 - 12.0
- 12.1 - 20.0
- More than 20.0

County extremes:
Harlan = - 4.2
Boone = +35.6

RESIDENTIAL VACANCY RATE

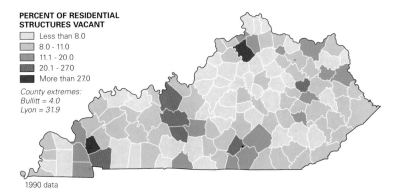

PERCENT OF RESIDENTIAL STRUCTURES VACANT
- Less than 8.0
- 8.0 - 11.0
- 11.1 - 20.0
- 20.1 - 27.0
- More than 27.0

County extremes:
Bullitt = 4.0
Lyon = 31.9

1990 data

claiming a house as one's family home, and well-being is increased through a tangible connection to a specific place of residence and its family traditions, social networks, and familiar routines. Rural areas in Kentucky generally have higher percentages of homeownership than do urban areas. The greatest increases in ownership between 1980 and 1990 were found in eastern Kentucky. Housing values in this portion of the state, however, are relatively low, with few structures valued at over $40,000 (the U.S. median value in 1990 was just over $79,000). Relative increases in ownership in eastern Kentucky and other rural counties are primarily a reflection of a small housing stock base. Homeownership in such areas is, in fact, a more elusive goal than in urban areas. Besides the prevalence of substandard quality housing, rural residents often encounter difficulty in acquiring home loans. This phenomenon is most likely tied to the low incomes and poverty conditions that prevail in rural—and especially eastern—Kentucky, and to the high percentage of personal incomes necessarily used for housing expenses.

Housing and rental costs in Appalachian Kentucky consume a larger share of family income than in other portions of the state. Despite the financial hardships imposed by homeownership, emotional benefits can be great, and the traditional Appalachian ties to land and home should not be discounted when considering the overall well-being of Kentucky's population.

Timber-framed, clapboard-sided Victorian-style homes were popular as single-family residences around the turn of the century. This view is on Griffith Street, Owensboro.

OWNER-OCCUPIED DWELLINGS

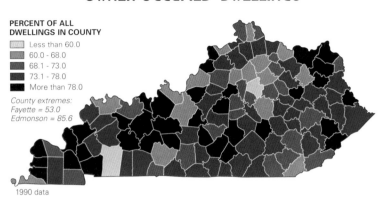

PERCENT OF ALL
DWELLINGS IN COUNTY

- Less than 60.0
- 60.0 - 68.0
- 68.1 - 73.0
- 73.1 - 78.0
- More than 78.0

County extremes:
Fayette = 53.0
Edmonson = 85.6

1990 data

CHANGE IN NUMBER OF
OWNER-OCCUPIED DWELLINGS, 1980-1990

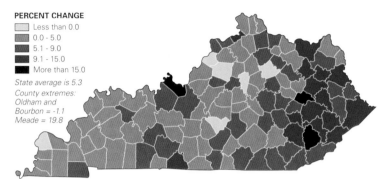

PERCENT CHANGE

- Less than 0.0
- 0.0 - 5.0
- 5.1 - 9.0
- 9.1 - 15.0
- More than 15.0

State average is 5.3
County extremes:
Oldham and
Bourbon = -1.1
Meade = 19.8

OWNER-OCCUPIED HOUSING IN KENTUCKY

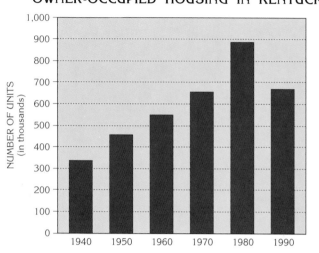

NUMBER OF UNITS (in thousands)

1,000 — 900 — 800 — 700 — 600 — 500 — 400 — 300 — 200 — 100 — 0

1940 1950 1960 1970 1980 1990

MEDIAN OWNER-OCCUPIED
HOME VALUE

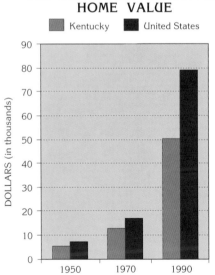

■ Kentucky ■ United States

DOLLARS (in thousands)

90 — 80 — 70 — 60 — 50 — 40 — 30 — 20 — 10 — 0

1950 1970 1990

VALUE OF OWNED RESIDENTIAL STRUCTURES

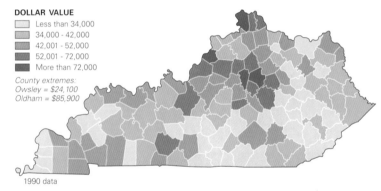

DOLLAR VALUE

- Less than 34,000
- 34,000 - 42,000
- 42,001 - 52,000
- 52,001 - 72,000
- More than 72,000

County extremes:
Owsley = $24,100
Oldham = $85,900

1990 data

MONTHLY GROSS RENT

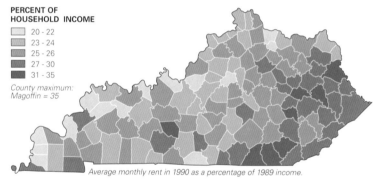

PERCENT OF
HOUSEHOLD INCOME

- 20 - 22
- 23 - 24
- 25 - 26
- 27 - 30
- 31 - 35

County maximum:
Magoffin = 35

Average monthly rent in 1990 as a percentage of 1989 income.

MONTHLY OWNER COSTS

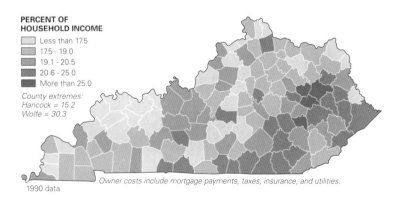

PERCENT OF
HOUSEHOLD INCOME

- Less than 17.5
- 17.5 - 19.0
- 19.1 - 20.5
- 20.6 - 25.0
- More than 25.0

County extremes:
Hancock = 15.2
Wolfe = 30.3

Owner costs include mortgage payments, taxes, insurance, and utilities.

1990 data

MONTHLY GROSS RENTAL COSTS

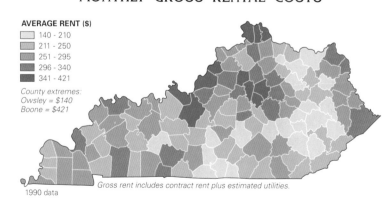

AVERAGE RENT ($)

- 140 - 210
- 211 - 250
- 251 - 295
- 296 - 340
- 341 - 421

County extremes:
Owsley = $140
Boone = $421

Gross rent includes contract rent plus estimated utilities.

1990 data

APPROVAL RATE FOR HOME PURCHASE LOAN APPLICATIONS

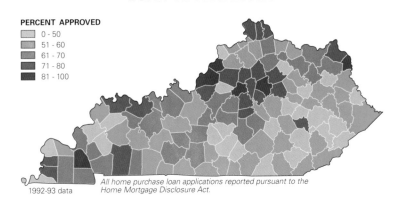

PERCENT APPROVED
- 0 - 50
- 51 - 60
- 61 - 70
- 71 - 80
- 81 - 100

All home purchase loan applications reported pursuant to the Home Mortgage Disclosure Act.

1992-93 data

MOBILE HOMES IN KENTUCKY

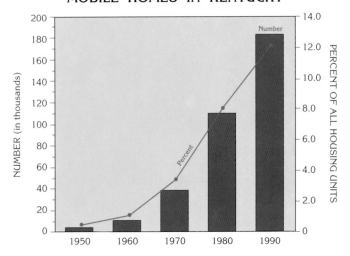

MOBILE HOMES

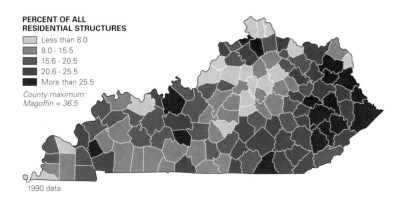

PERCENT OF ALL RESIDENTIAL STRUCTURES
- Less than 8.0
- 8.0 - 15.5
- 15.6 - 20.5
- 20.6 - 25.5
- More than 25.5

County maximum: Magoffin = 36.5

1990 data

CHANGE IN NUMBER OF MOBILE HOMES, 1980-1990

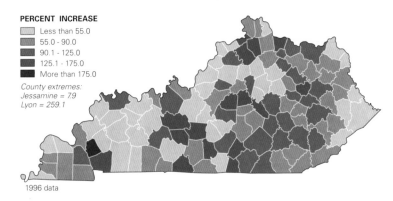

PERCENT INCREASE
- Less than 55.0
- 55.0 - 90.0
- 90.1 - 125.0
- 125.1 - 175.0
- More than 175.0

County extremes: Jessamine = 7.9 Lyon = 259.1

1996 data

MOBILE HOMES IN THE UNITED STATES

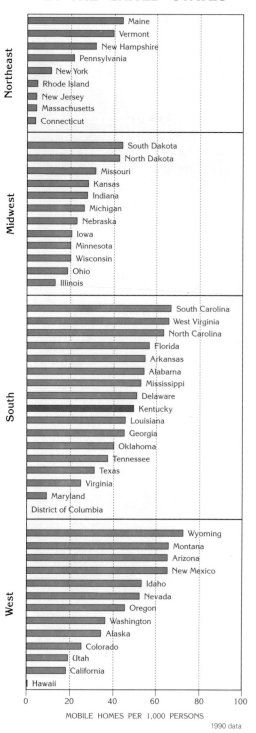

MOBILE HOMES PER 1,000 PERSONS

1990 data

Mobile homes are an interesting American phenomenon. Even with a reputation of being poor-quality boxes that seem to attract tornadoes like a magnet, these structures have gained in popularity over time. Mobile homes likely owe their allure to their relatively low cost as a housing alternative, to recent dramatic improvements in quality, and to the advantages of movability. Mobile homes can be placed without the owner's having to contend with—and pay for—clearing and leveling land or excavating a foundation. Mobile homes are part of Kentucky's rural landscape, and in some counties of the state, particularly in eastern Kentucky, mobile homes account for one-fifth or more of all housing structures. Kentucky is not unique in its substantial number of mobile homes, however; many southern and western states have significantly higher concentrations of such residential structures.

Mobile homes like these in Grant County are popular "starter" homes for young couples and serve as replacement for aging housing stock.

Some people in Kentucky neither rent nor own their homes. In 1990 more than one hundred thousand persons, or 2.7 percent of all Kentucky residents, lived in group quarters; this state percentage is identical to the proportion of individuals living in group quarters across the nation. College dormitories, nursing homes, and military quarters predominated among the types of group housing in the state.

Military Areas

Fort Knox, about thirty-five miles south of Louisville, is the home of the U.S. Army's armored warfare training center and the nation's gold bullion depository. The post is 109,000 acres in size. In addition to the 10,600 military personnel assigned to the post, civil service employees and military dependents bring the total population to about 33,000, making Fort Knox the state's sixth-largest urban community and its largest employer. Fort Campbell, another army post, is the home of the 101st Airborne Division. It sits astride the Kentucky-Tennessee border south of Hopkinsville. This post includes 105,668 acres, of which 36,000 are in Kentucky, in Trigg and Christian Counties. The post population includes about 23,000 military personnel and 10,000 dependents. The army depot near Avon was recently demilitarized, but the 14,000-acre Lexington-Bluegrass Army Depot near Richmond continues to serve as a munitions storage site. A Naval Ordnance Depot is maintained in Louisville, the 100th Army Division offers training in Lexington, and more than fifty National Guard units are scattered across the state.

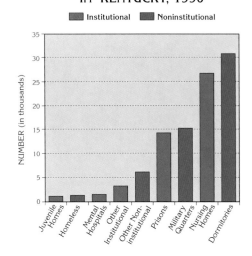

PERSONS LIVING IN GROUP QUARTERS IN KENTUCKY, 1990

Institutional ■ Noninstitutional

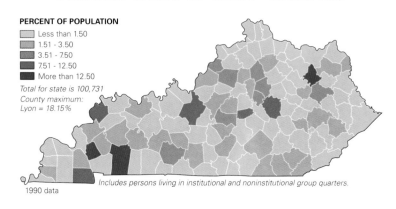

PERSONS LIVING IN GROUP QUARTERS

PERCENT OF POPULATION
- Less than 1.50
- 1.51 - 3.50
- 3.51 - 7.50
- 7.51 - 12.50
- More than 12.50

Total for state is 100,731
County maximum:
Lyon = 18.15%

Includes persons living in institutional and noninstitutional group quarters.

1990 data

MILITARY AREAS

• National Guard Unit(s)

▨ Military post

1996 data

BANKS

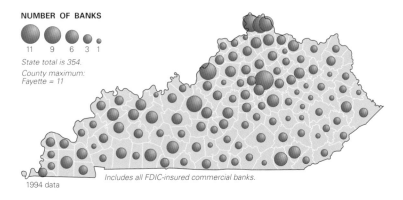

NUMBER OF BANKS

11　9　6　3　1

State total is 354.
County maximum:
Fayette = 11

Includes all FDIC-insured commercial banks.

1994 data

BANK DEPOSITS

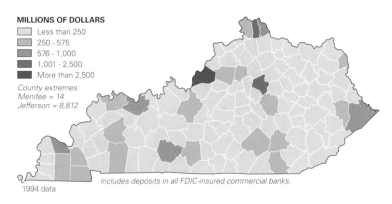

MILLIONS OF DOLLARS

- Less than 250
- 250 - 575
- 576 - 1,000
- 1,001 - 2,500
- More than 2,500

County extremes:
Menifee = 14
Jefferson = 8,812

Includes deposits in all FDIC-insured commercial banks.

1994 data

BANK DEPOSITS PER CAPITA

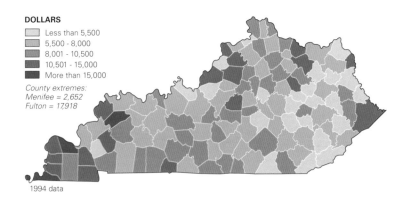

DOLLARS

- Less than 5,500
- 5,500 - 8,000
- 8,001 - 10,500
- 10,501 - 15,000
- More than 15,000

County extremes:
Menifee = 2,652
Fulton = 17,918

1994 data

This bank remains a viable business on Main Street in Cloverport, an early port town on the Ohio River in Breckinridge County.

Commerce and Services

Examining the local institutional systems that support personal and family livelihoods can offer much information about a population. These systems provide the goods and services necessary to maintain life, and the diversity of such systems reflects both the demands of the local population and the financial ability of local areas to support the systems. It should not be assumed, however, that a certain mix and magnitude of commercial and service institutions are always the results of the characteristics of a resident population. Institutions may instead cause particular population characteristics to emerge, as migrants are drawn to specific elements of places. A variety of shopping opportunities, eating establishments, or transportation services, for example, can strongly influence migrants in choosing a new place of residence.

Banking tends to be a central institution in regional and local economies. Banks serve not only as repositories for personal, commercial, and industrial funds but also as sources of loans that help support continued development in an area. Total deposits in Kentucky banks in 1994 amounted to nearly 34 billion dollars, almost 9 billion of which were deposited in Jefferson County banks alone. The actual number of large commercial banks, in Kentucky and elsewhere, has declined in recent years because of mergers and buyouts. According to Kentucky government sources, the Commonwealth had 354

banks in 1994. These institutions, as one would expect, were concentrated in large population centers. The number of branch banks, however, has recently increased, making banking more convenient and accessible to the population.

Wholesale and retail trade establishments are also found in greater numbers within and around major cities, which have a large enough market to support such enterprises. These enterprises provide finished goods for purchase by consumers. Patterns of retail trade are particularly revealing because they offer a partial indicator of general consumer behavior. Per person retail sales, for example, show that individuals in or near cities and towns such as Paducah, Bowling Green, Covington, Maysville, and Lexington may be likely to buy more goods or to spend more on goods than persons in rural areas. Such an observation must be qualified, however, by the fact that shopping opportunities in cities also attract people from outlying and even remote areas. Rural dwellers may make weekly or monthly trips to cities for a portion of their purchases. Appalachian Kentucky has a large pocket of counties where per person retail sales are low. Although economic conditions in these counties suggest that the resident populations spend less on their purchases than individuals elsewhere in the state, there is strong evidence that families often make day trips to cities such as Lexington to shop for goods, including food.

Suburban shopping centers such as this one on U.S. 60 in Ashland account for a large proportion of retail sales in Kentucky's medium-sized towns and cities.

WHOLESALE SALES

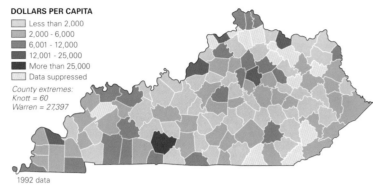

DOLLARS PER CAPITA
- Less than 2,000
- 2,000 - 6,000
- 6,001 - 12,000
- 12,001 - 25,000
- More than 25,000
- Data suppressed

County extremes:
Knott = 60
Warren = 27,397

1992 data

FOOD SALES

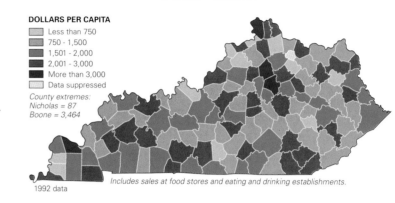

DOLLARS PER CAPITA
- Less than 750
- 750 - 1,500
- 1,501 - 2,000
- 2,001 - 3,000
- More than 3,000
- Data suppressed

County extremes:
Nicholas = 87
Boone = 3,464

1992 data   *Includes sales at food stores and eating and drinking establishments.*

RETAIL SALES

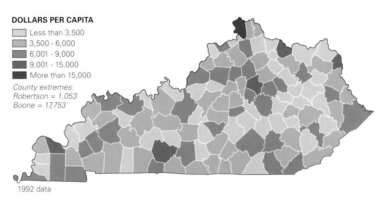

DOLLARS PER CAPITA
- Less than 3,500
- 3,500 - 6,000
- 6,001 - 9,000
- 9,001 - 15,000
- More than 15,000

County extremes:
Robertson = 1,053
Boone = 17,753

1992 data

SERVICE INDUSTRY RECEIPTS

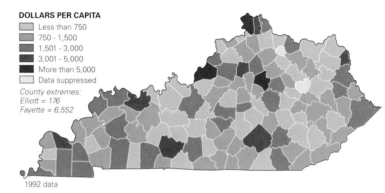

DOLLARS PER CAPITA
- Less than 750
- 750 - 1,500
- 1,501 - 3,000
- 3,001 - 5,000
- More than 5,000
- Data suppressed

County extremes:
Elliott = 176
Fayette = 6,552

1992 data

RETAIL SALES BY TYPE

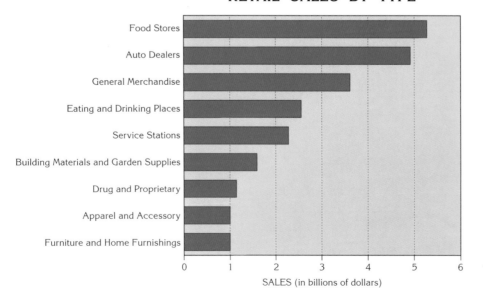

Food Stores
Auto Dealers
General Merchandise
Eating and Drinking Places
Service Stations
Building Materials and Garden Supplies
Drug and Proprietary
Apparel and Accessory
Furniture and Home Furnishings

0   1   2   3   4   5   6
SALES (in billions of dollars)

SERVICE INDUSTRY RECEIPTS BY TYPE

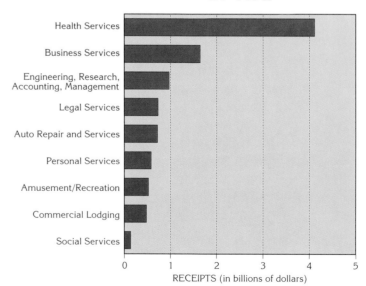

Health Services
Business Services
Engineering, Research, Accounting, Management
Legal Services
Auto Repair and Services
Personal Services
Amusement/Recreation
Commercial Lodging
Social Services

0   1   2   3   4   5
RECEIPTS (in billions of dollars)

Indeed, food accounts for the largest single element of retail sales statewide, valued at more than five billion dollars per year.

Kentucky's service industry recorded more than ten billion dollars' worth of receipts in 1992. The service sector can be considered as including those commercial activities that do not provide a tangible thing at purchase, like food, clothes, automobiles, or appliances. Instead, services generally comprise actions by individuals, like health care, auto repair, personal grooming, and legal advising, or the temporary use of facilities, as in hotels and motels, cinemas, golf courses, and amusement parks. Urban areas are service-rich, and rural areas tend to have less active service sectors. Certain service establishments, however, concentrate in recreational areas, such as the Lake Cumberland area and the Kentucky Lake and Lake Barkley area. Health care is by far the largest single component of the service industry in Kentucky, a fact that relates not only to the need for health care for basic survival but also to escalating medical costs in the United States.

Unfortunately, local, regional, and state development is not supported solely by sales revenues from businesses and industries. Construction and maintenance of public infrastructure and the support of public services are gained through contributions from residents in the form of taxes on purchases and on land. Total tax payments are not, of course, equitable across space; they are influenced by income and personal expenditures, landownership, and local tax laws. Residents of rural eastern and southern Kentucky, for instance, pay less in taxes than residents of northern and more urbanized counties. Persons in rural counties, however, pay a larger share of their taxes on property than do persons in urban areas.

With generally lower local tax payments in rural areas, such areas generate less funding to support infrastructure and public services than do urban areas. In some cases more monies are spent locally than are generated through taxes. State tax allocations, as well as federal allocations and grants, provide supplemental sources of revenue for local development.

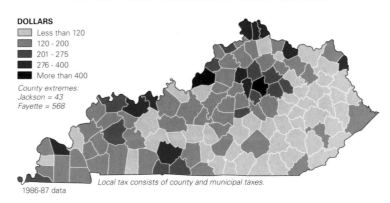

LOCAL TAX PAYMENTS PER CAPITA

DOLLARS
- Less than 120
- 120 - 200
- 201 - 275
- 276 - 400
- More than 400

County extremes:
Jackson = 43
Fayette = 568

Local tax consists of county and municipal taxes.

1986-87 data

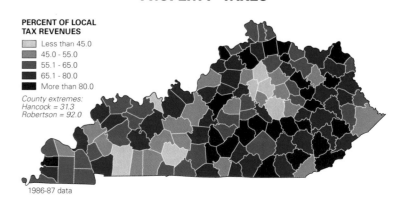

PROPERTY TAXES

PERCENT OF LOCAL TAX REVENUES
- Less than 45.0
- 45.0 - 55.0
- 55.1 - 65.0
- 65.1 - 80.0
- More than 80.0

County extremes:
Hancock = 31.3
Robertson = 92.0

1986-87 data

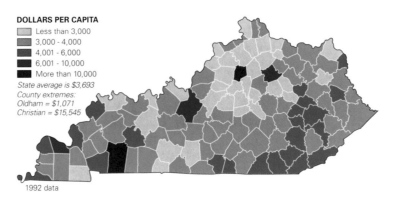

FEDERAL FUNDING / GRANTS

DOLLARS PER CAPITA
- Less than 3,000
- 3,000 - 4,000
- 4,001 - 6,000
- 6,001 - 10,000
- More than 10,000

State average is $3,693
County extremes:
Oldham = $1,071
Christian = $15,545

1992 data

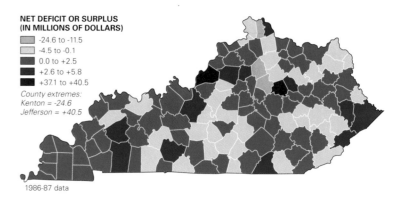

LOCAL GOVERNMENT GENERAL FINANCES

NET DEFICIT OR SURPLUS (IN MILLIONS OF DOLLARS)
- -24.6 to -11.5
- -4.5 to -0.1
- 0.0 to +2.5
- +2.6 to +5.8
- +37.1 to +40.5

County extremes:
Kenton = -24.6
Jefferson = +40.5

1986-87 data

HOSPITAL ADMISSIONS

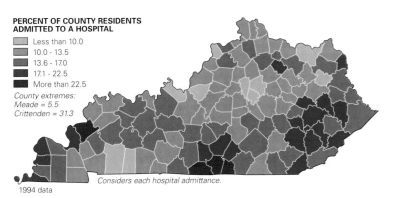

PERCENT OF COUNTY RESIDENTS
ADMITTED TO A HOSPITAL
- Less than 10.0
- 10.0 - 13.5
- 13.6 - 17.0
- 17.1 - 22.5
- More than 22.5

County extremes:
Meade = 5.5
Crittenden = 31.3

Considers each hospital admittance.

1994 data

HEALTH SERVICE RECEIPTS

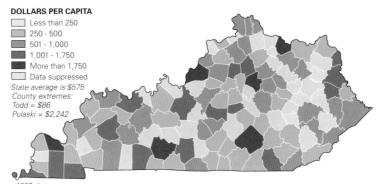

DOLLARS PER CAPITA
- Less than 250
- 250 - 500
- 501 - 1,000
- 1,001 - 1,750
- More than 1,750
- Data suppressed

State average is $575
County extremes:
Todd = $86
Pulaski = $2,242

1992 data

King's Daughters Medical Center on Lexington Street in Ashland serves patients from across much of northeastern Kentucky.

Health Care

Health care is both necessary and controversial. Modern society almost demands that individuals ensure proper health maintenance by seeking regular medical attention, from periodic physical examinations for school and work to biannual dental checkups. Recent medical advances have resulted in the introduction of new drugs that fight disease and reduce the effects of disease, as well as new procedures that allow outpatient and short-term hospital visits for operations that once required lengthy hospitalizations. New regional hospitals and local clinics, and even the introduction of telemedicine, have brought medical care to rural areas that have long been poorly served.

The downside of health care is cost. In 1992 each Kentuckian paid, on average, about $1,100 for health care services. Urban areas, where health care is both more accessible and more expensive, generally had the highest per person payments. Counties with older populations also reported higher health care expenditures. A report by the Families USA Foundation demonstrated the dramatic increase in the cost of family health care over recent years. In 1980, for example, about $2,250 was spent per family for health care in Kentucky. Just over 67 percent of this amount was paid directly by fami-

lies in the form of premiums, taxes, and out-of-pocket expenses. The remaining 33 percent was paid through business supplements in health care plans. By 1993 the total payment per family had increased to about $7,750; business supplements had increased slightly as a share of total payments, but families paid much more in the form of general taxes that were routed toward health care.

Examining the distribution of physicians across the state reveals quite dramatic inequities. Urban areas, especially Louisville and Lexington, are generally well supplied with doctors, primarily because of their medical training facilities. Yet the number of doctors, and particularly the number of doctors compared with the population size, drops dramatically as one moves away from major cities. Rural Kentucky counties have long experienced an undersupply of practicing doctors and dentists. The draw of large incomes and prestige associated with practices in large cities accounts for some of this undersupply. More important factors, however, include the lack of suitable support facilities in rural areas—such as modern hospitals and equipment—and the difficulties facing doctors trying to assimilate into and be accepted by local rural cultures and social networks.

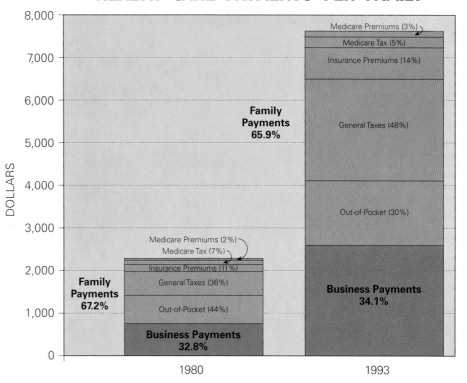

HEALTH CARE PAYMENTS PER FAMILY

1980
- Family Payments 67.2%
  - Medicare Premiums (2%)
  - Medicare Tax (7%)
  - Insurance Premiums (11%)
  - General Taxes (36%)
  - Out-of-Pocket (44%)
- Business Payments 32.8%

1993
- Family Payments 65.9%
  - Medicare Premiums (3%)
  - Medicare Tax (5%)
  - Insurance Premiums (14%)
  - General Taxes (48%)
  - Out-of-Pocket (30%)
- Business Payments 34.1%

(DOLLARS axis: 0 to 8,000)

NUMBER OF PHYSICIANS

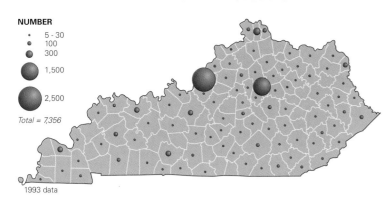

NUMBER

· 5 - 30
· 100
● 300
● 1,500
● 2,500

Total = 7,356

1993 data

AVAILABILITY OF PHYSICIANS

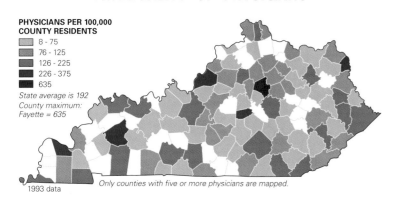

PHYSICIANS PER 100,000
COUNTY RESIDENTS

8 - 75
76 - 125
126 - 225
226 - 375
635

State average is 192
County maximum:
Fayette = 635

1993 data   Only counties with five or more physicians are mapped.

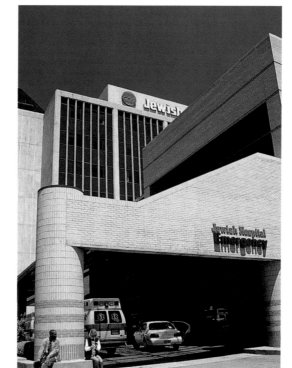

Jewish Hospital on Chestnut Street is part of a large medical complex in downtown Louisville that serves patients from across the state.

GENERAL ACUTE CARE HOSPITALS

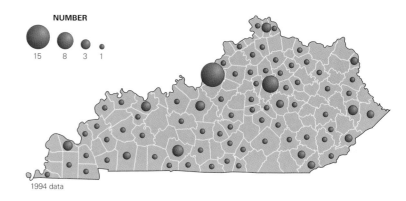

NUMBER

● 15   ● 8   ● 3   · 1

1994 data

GENERAL ACUTE CARE HOSPITAL BEDS

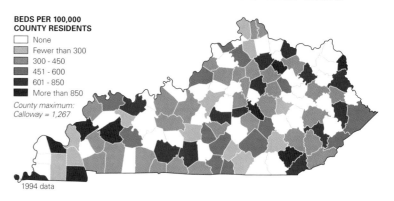

BEDS PER 100,000
COUNTY RESIDENTS

None
Fewer than 300
300 - 450
451 - 600
601 - 850
More than 850

County maximum:
Calloway = 1,267

1994 data

GENDER OF KENTUCKY PHYSICIANS

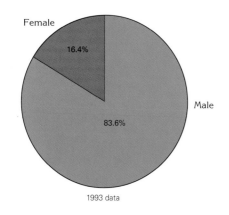

Female
16.4%

Male
83.6%

1993 data

STATUS OF KENTUCKY PHYSICIANS

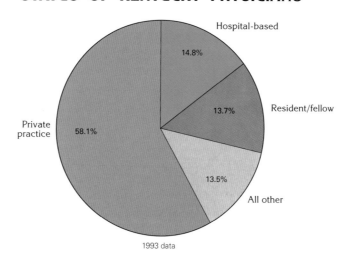

Hospital-based
14.8%

Resident/fellow
13.7%

All other
13.5%

Private
practice
58.1%

1993 data

ORIGIN OF KENTUCKY PHYSICIANS' MEDICAL DEGREES

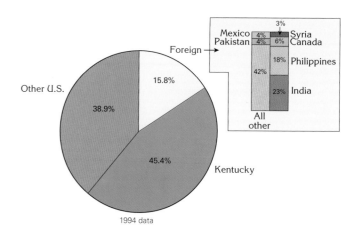

Foreign

Mexico 4%
Pakistan 4%
Syria 6%
Canada 3%
Philippines 18%
India 23%
All other 42%

Other U.S.
38.9%

Foreign
15.8%

Kentucky
45.4%

1994 data

PHYSICIANS WITH MEDICAL DEGREES FROM KENTUCKY INSTITUTIONS

PERCENT OF PHYSICIANS IN COUNTY
- Less than 25.0
- 25.0 - 40.0
- 40.1 - 55.0
- 55.1 - 75.0
- More than 75.0

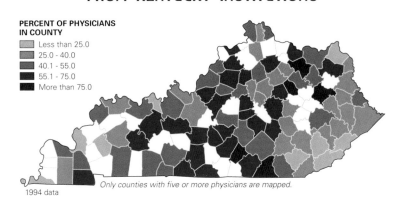

Only counties with five or more physicians are mapped.

1994 data

PHYSICIANS WITH MEDICAL DEGREES FROM FOREIGN INSTITUTIONS

PERCENT OF PHYSICIANS IN COUNTY
- Less than 10.0
- 10.0 - 20.0
- 20.1 - 40.0
- 40.1 - 60.0
- More than 60.0

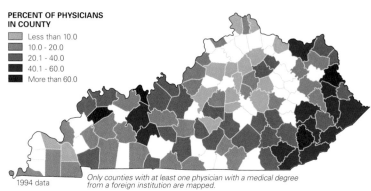

Only counties with at least one physician with a medical degree from a foreign institution are mapped.

1994 data

SPECIALTIES OF KENTUCKY PHYSICIANS

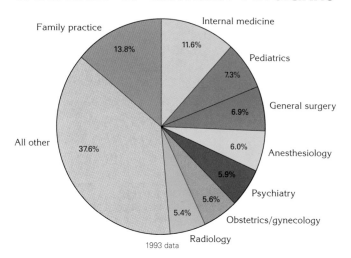

- Family practice 13.8%
- Internal medicine 11.6%
- Pediatrics 7.3%
- General surgery 6.9%
- Anesthesiology 6.0%
- Psychiatry 5.9%
- Obstetrics/gynecology 5.6%
- Radiology 5.4%
- All other 37.6%

1993 data

ORIGIN OF PATIENTS HOSPITALIZED IN LEXINGTON AND LOUISVILLE

One • represents 300 patients.
- • Lexington hospitals
- • Louisville hospitals

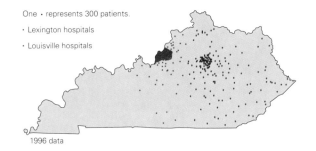

1996 data

ORIGIN OF PATIENTS HOSPITALIZED IN REGIONAL MEDICAL CENTERS

One • represents 50 patients.
- · Highland Regional Medical Center in Prestonsburg
- · Regional Medical Center in Madisonville

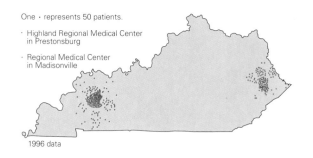

1996 data

DENTISTS

DENTISTS PER 100,000 COUNTY RESIDENTS
- Fewer than 25
- 25 - 45
- 46 - 65
- 66 - 95
- More than 95

County extremes:
Edmonson = 10
Fayette = 114

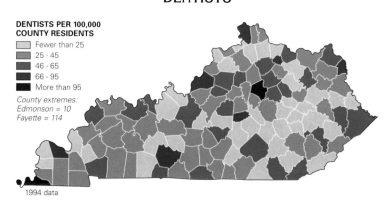

1994 data

INPATIENTS BY TYPE OF CARE
(BY AREA DEVELOPMENT DISTRICT)

NUMBER OF INPATIENTS
- 100,000
- 50,000
- 25,000
- 5,000

PERCENT OF ALL PATIENTS

KENTUCKY
Not to scale

TYPE OF CARE
- Medical/surgical
- Pediatric
- Obstetric
- Psychiatric
- Other

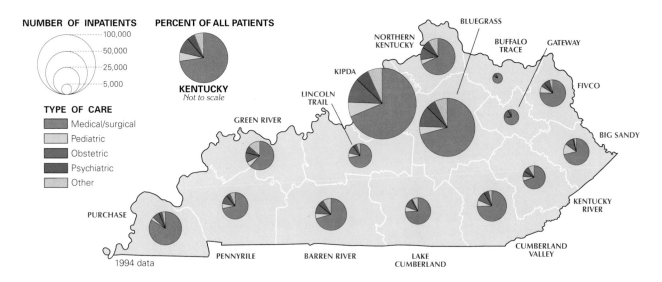

1994 data

The problem of retaining physicians is an issue throughout rural America, and the predicament is especially acute in Kentucky. Because few U.S.-trained doctors prefer to continue practicing in rural counties, the demand has largely been filled by physicians holding degrees from foreign medical schools. Specialized medicine continues to be generally an urban phenomenon, although specialization is increasing in some rural counties that have large regional hospitals. Most commonly, though, Kentuckians requiring specialized medical care, including cardiovascular surgery, neurosurgery, and psychiatric or rehabilitative services, must seek treatment at large hospitals located in Lexington and Louisville, or at specialized hospitals in other cities.

Kentucky in 1990 had nearly half a million persons aged sixty-five and over. Although many of these individuals are quite healthy, a growing number of elderly people require some sort of assistance. Home care of elders is increasingly difficult in today's society; families have become more nuclear in nature, with generations living in physically distinct households, and time pressures allow little opportunity for adult children to care for their aging parents. One alternative is to seek care in a nursing home, and the number of nursing homes in Kentucky has increased significantly since 1980, primarily in response to the rapid aging of the state's population. Despite the increasing number of nursing homes and of nursing home beds, the supply has barely met demand. Some counties have no nursing homes whatsoever, and few vacant beds are available statewide. Consequently, families often must place their elderly relatives in homes located quite far away, thus increasing the social distance between family members and potentially reducing the quality of life of both the elderly persons and their younger relatives.

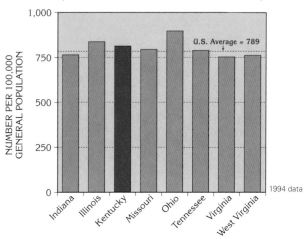

REGISTERED NURSES
(KENTUCKY AND ADJACENT STATES)

NUMBER PER 100,000 GENERAL POPULATION

U.S. Average = 789

Indiana · Illinois · Kentucky · Missouri · Ohio · Tennessee · Virginia · West Virginia

1994 data

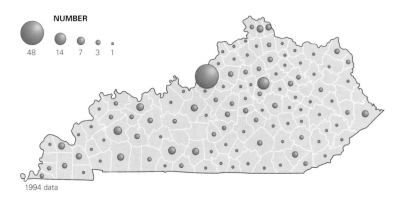

NURSING HOMES

NUMBER

48 14 7 3 1

1994 data

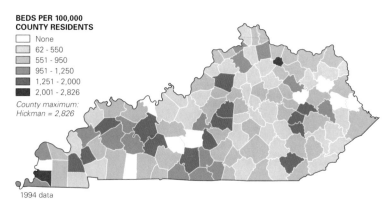

NURSING HOME BEDS

BEDS PER 100,000
COUNTY RESIDENTS

None
62 - 550
551 - 950
951 - 1,250
1,251 - 2,000
2,001 - 2,826

County maximum: Hickman = 2,826

1994 data

Shriner's Hospital in Lexington offers specialized treatment for crippled children.

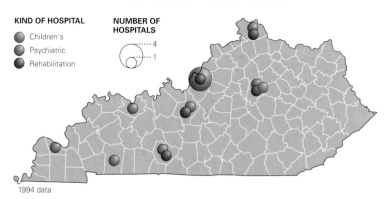

SPECIALIZED HOSPITALS

KIND OF HOSPITAL
Children's
Psychiatric
Rehabilitation

NUMBER OF HOSPITALS
4
1

1994 data

Law Enforcement and Crime

A key element of public service provision in Kentucky is law enforcement and, along with that, the prosecution and imprisonment of criminals. State and local taxes primarily support these services, and the extent of such services within counties is governed by local revenues and residents' perceived need. Nearly seven thousand full-time law enforcement officers were employed in the state in 1993, and nearly half of those worked for municipal police departments. Approximately 13 percent were officers of the Kentucky State Police; these officers, as well as others, such as those employed by the law enforcement divisions of the Kentucky Department of Fish and Wildlife Resources and the Kentucky Water Patrol, could not be allocated to individual counties and therefore are not reflected on the county-level map.

Urban counties tend to employ a greater number of law enforcement officers per county resident than do rural counties. On average, Kentucky in 1993 hired about 140 local law enforcement officers for every 100,000 residents of the state, although coverage ranged from 14 officers per 100,000 persons in Wolfe County to 257 officers per 100,000 persons in Fulton County.

Prisoners in state and federal correctional institutions are supported by tax dollars. Kentucky's state and federal prison population numbered 10,440 in 1993, or just over 1 percent of the nation's prison population of nearly 1 million. Kentucky's prisoners are housed in fifteen state and three federal correctional institutions statewide. About one-third of the state's prison population is black, compared with the national figure of nearly one-half. The state has one maximum-security facility, the Kentucky State Penitentiary, located on Lake Barkley in Eddyville. Recurring public issues in Kentucky, and elsewhere in the nation, concern not only the adequacy of law enforcement but also the increasing financial burden of supporting a growing prison population.

POLICE OFFICERS

NUMBER PER 100,000 RESIDENT POPULATION
- Fewer than 175
- 175 - 200
- 201 - 225
- 226 - 300
- More than 300

U.S. average is 237
State average = 162

1992 data

Data comprise full-time sworn police officers in state and local law enforcement agencies.

Same scale as contiguous states

Approximately 44% of scale of contiguous states

JURISDICTION OF LAW ENFORCEMENT OFFICERS IN KENTUCKY

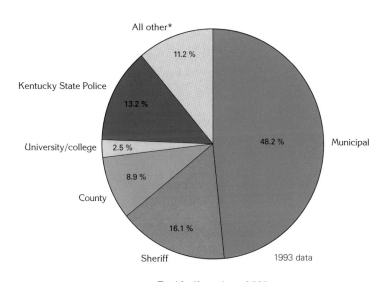

All other* 11.2 %
Kentucky State Police 13.2 %
University/college 2.5 %
County 8.9 %
Sheriff 16.1 %
Municipal 48.2 %

1993 data

Total for Kentucky is 6,935

Other includes Division of Forestry, Department of Fish and Wildlife Resources, Department of Parks, Airport Police, and Water Patrol.

LAW ENFORCEMENT OFFICERS

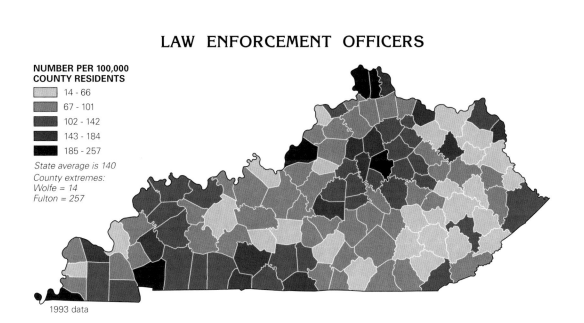

NUMBER PER 100,000 COUNTY RESIDENTS
- 14 - 66
- 67 - 101
- 102 - 142
- 143 - 184
- 185 - 257

State average is 140
County extremes:
Wolfe = 14
Fulton = 257

1993 data

PRISONS

OWNERSHIP

Adjacent Camp — Federal prison

State prison

Privately contracted by state

SECURITY LEVEL

Minimum

Low

Medium

Maximum

Administrative (all levels)

CAPACITY

1,500
1,000
500
200

Kentucky Correctional Institution for Women

Roederar Correctional Complex

Luther Luckett Correctional Complex

Kentucky State Reformatory

Frankfort Career Development Center

Federal Medical Center Lexington

Blackburn Correctional Complex

Northpoint Training Center

Federal Correctional Institution Ashland

Eastern Kentucky Correctional Complex

Lee Adjustment Center

Otter Creek Correctional Complex

Marion Adjustment Center

Western Kentucky Correctional Complex

Green River Correctional Complex

Kentucky State Penitentiary

Federal Correctional Institution Manchester

Bell County Forestry Camp

1995 data

STATE AND FEDERAL PRISON POPULATION

Black 48.2% | 33.6%

White 45.6% | 66.4%

Kentucky

United States

Native American (0.9%)
Asian/Pacific Islander (0.6%)

Unknown race (4.7%)

1993 data

Kentucky State Penitentiary at Eddyville, in Lyon County, has a capacity of 800 maximum-security prisoners.

PRISON POPULATION

NUMBER PER 100,000 RESIDENT POPULATION

Fewer than 150
150 - 225
226 - 350
351 - 500
More than 500

U.S. average is 367
State average = 276

1993 data — *Data comprise prisoners under jurisdiction of state and federal correctional authorities.*

Same scale as contiguous states

Approximately 44% of scale of contiguous states

CHANGE IN KENTUCKY'S PRISON POPULATION

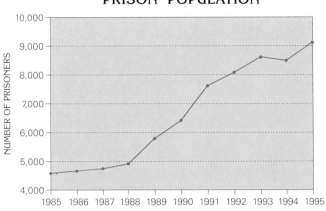

Data comprises prisoners under jurisdiction of state correctional authorities.

VIOLENT CRIMES

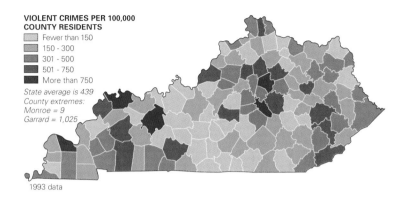

VIOLENT CRIMES PER 100,000 COUNTY RESIDENTS
- Fewer than 150
- 150 - 300
- 301 - 500
- 501 - 750
- More than 750

State average is 439
County extremes:
Monroe = 9
Garrard = 1,025

1993 data

SERIOUS CRIMES

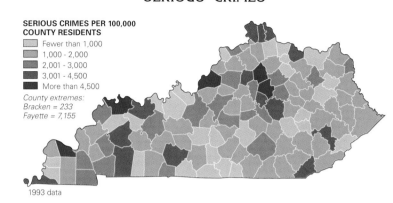

SERIOUS CRIMES PER 100,000 COUNTY RESIDENTS
- Fewer than 1,000
- 1,000 - 2,000
- 2,001 - 3,000
- 3,001 - 4,500
- More than 4,500

County extremes:
Bracken = 233
Fayette = 7,155

1993 data

DRUG ARRESTS

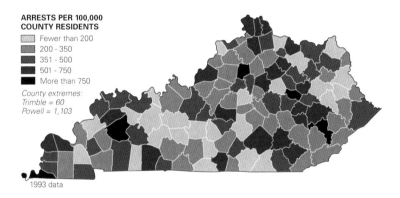

ARRESTS PER 100,000 COUNTY RESIDENTS
- Fewer than 200
- 200 - 350
- 351 - 500
- 501 - 750
- More than 750

County extremes:
Trimble = 60
Powell = 1,103

1993 data

DUI ARRESTS

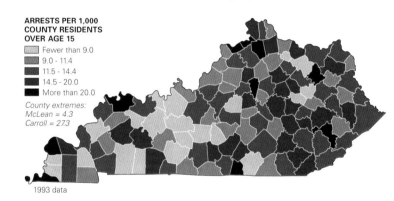

ARRESTS PER 1,000 COUNTY RESIDENTS OVER AGE 15
- Fewer than 9.0
- 9.0 - 11.4
- 11.5 - 14.4
- 14.5 - 20.0
- More than 20.0

County extremes:
McLean = 4.3
Carroll = 27.3

1993 data

DUI FATALITIES

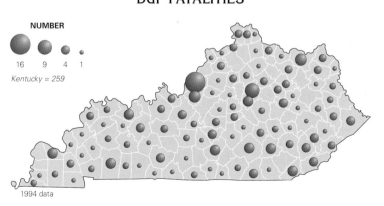

NUMBER

16   9   4   1

Kentucky = 259

1994 data

DUI ACCIDENTS

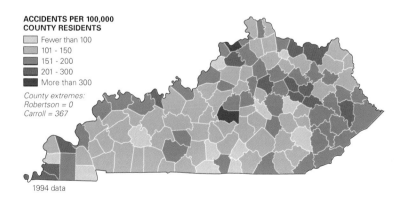

ACCIDENTS PER 100,000 COUNTY RESIDENTS
- Fewer than 100
- 101 - 150
- 151 - 200
- 201 - 300
- More than 300

County extremes:
Robertson = 0
Carroll = 367

1994 data

Kentucky's law enforcement officers made more than seventeen thousand arrests for violations of drug laws in 1993. Fifty-eight percent of these involved the use of marijuana. Portions of the state that had high rates of drug-related arrests included metropolitan counties and the counties of eastern Kentucky, where considerable amounts of marijuana are grown.

Despite financial pressures affecting the support of law enforcement, crime rates in Kentucky for serious crimes (violent crimes plus burglary, larceny, theft, and arson) are among the lowest in the nation. In 1991 Kentucky's serious crime rate of 3,375 per 100,000 residents compared favorably with the national average of 5,928 per 100,000. Although the underreporting of crime accounts for some of the differences in crime rates among counties, considerable variation does exist across the state. Serious crime rates tend to be highest in metropolitan counties and in counties served by interstate highways.

In 1994 approximately one of every twelve Kentucky drivers was involved in a traffic accident in the state, and 791 fatalities occurred because of traffic accidents. Kentucky's death rate from traffic accidents compares with that of the nation (2.0 deaths in Kentucky versus 1.8 deaths nationwide per 100 million miles traveled). Also like the nation as a whole, Kentucky has seen a decline in traffic fatalities in recent years. In 1979 the rate was 3.3 deaths per 100 million miles traveled, nearly two-thirds

WET AND DRY COUNTIES

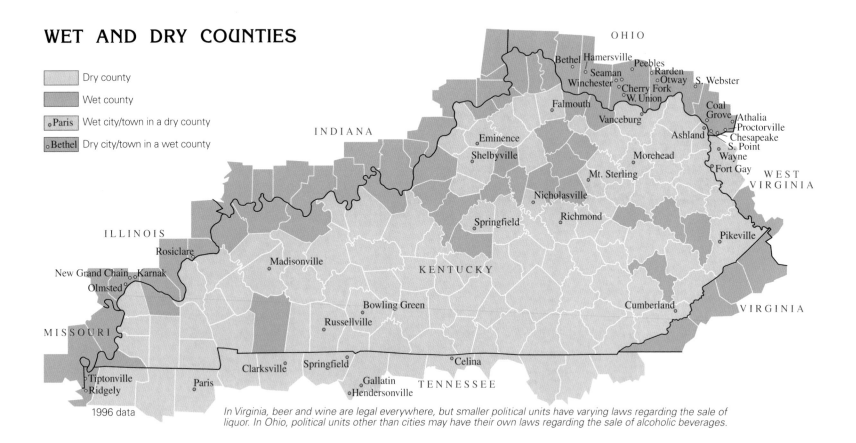

Dry county

Wet county

○ Paris Wet city/town in a dry county

Bethel Dry city/town in a wet county

1996 data

In Virginia, beer and wine are legal everywhere, but smaller political units have varying laws regarding the sale of liquor. In Ohio, political units other than cities may have their own laws regarding the sale of alcoholic beverages.

higher than the rate for 1994. Traffic accidents involving drivers under the influence of alcohol have also decreased, from a total of 8,014 in 1988 to 5,995 in 1994. The highest rates prevail in metropolitan counties, in the Bluegrass Region, in the eastern portion of the state, and in counties along the Ohio River.

Although voting in dry counties on the issue of the sale of alcohol has been common in recent decades, the majority of Kentucky's counties remain completely or partially dry. Seventy-five of the 120 counties are totally dry, and 15 counties are split. In cases of a split, the largest city or town is usually wet, and the remainder of the county is dry. Only one-quarter of the state's counties are completely wet. Debates over the wet-dry issue commonly pit arguments for increased revenues—both direct receipts from sales and indirect profits, such as those from increased tourism—against moral concerns. A comparison of the map of wet and dry counties with that of DUI accidents suggests a possible correspondence between dry counties and low DUI rates.

County liquor laws that prohibit the sale of alcoholic beverages often stimulate liquor sales in adjoining counties or states. Last Chance liquor store is located on U.S. 641 at the border between Calloway County, Kentucky, and Henry County, Tennessee.

Mineral, Energy, and Timber Resources

Previous page: *A coal mining operation removes a mountaintop on the rim of the Middlesboro basin. Cumberland Mountain trends from left to the distant horizon, while Middlesboro lies beneath a layer of morning fog.*

CHAPTER SIX

Mineral, Energy, and Timber Resources

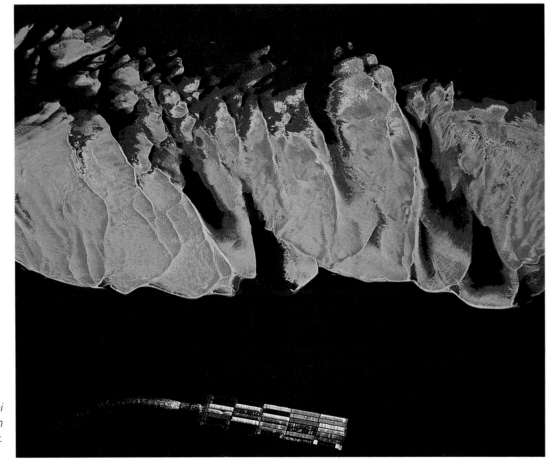

A sandbar in the Mississippi River west of Bardwell, in Carlisle County.

KENTUCKY'S PRINCIPAL NATURAL RESOURCES, in order of value, are bituminous coal, wood and wood products, crushed stone, natural gas, and petroleum. Together these resources account for annual revenues in excess of $5.5 billion.

Kentucky's coal is mined from deposits found in two distinct coalfields, one in the eastern part of the state and the other in the region commonly called the Western Kentucky Coal Field. The eastern part, a portion of the Appalachian Basin Coal Field, accounts for more than three-quarters of the state's annual production. The coalfield of western Kentucky is part of the extensive Interior Basin Coal Field. Coal produced there has a higher sulfur content than that mined in the Appalachian field. These two regions also generate most of Kentucky's natural gas and petroleum. Crushed stone is produced at quarries in areas rich in limestone or dolomite, especially along the Dripping Springs and Pottsville Escarpments and in the Inner Bluegrass Region. The state's Appalachian region is by far the most heavily forested part of the state, and thus it offers the greatest potential for the production of timber. Timberland accounts for at least 80 percent of the areas of most counties in southeastern Kentucky. If the state has a resource cornucopia, it is found in the eastern Appalachian counties, an area that has experienced chronic poverty and yet contains and produces a disproportionate share of the Commonwealth's natural resource wealth.

Coal

By far the state's most important natural resource is bituminous coal, which in 1993 accounted for 85 percent of the Commonwealth's mineral production value. Although commercial coal production began as early as the mid-1700s in the United States, it was not until 1820 that the first commercial mine opened in Kentucky, near Paradise in Muhlenberg

133

MINING AND MINERALS

MINERAL SYMBOLS

- B   Brick plant
- DIM   Dimension stone
- Fe   Iron
- Fu   Iron ore furnace
- K   Kiln
- Pc   Penny Wain conglomerate
- ⤬⤬   Phosphatic area mines
- SG   Sand/gravel
- SST   Sandstone deposit
- QT   Terrace gravel/sand
- ▽   Zinc deposits

- Clay deposits within clay region
- Commercial limestone
- Dolomite
- Glacial sand/gravel
- Industrial limestone
- Iron deposit
- New Providence shale
- Olive Hill clay
- Riverine sand/gravel
- Rockcastle sandstone

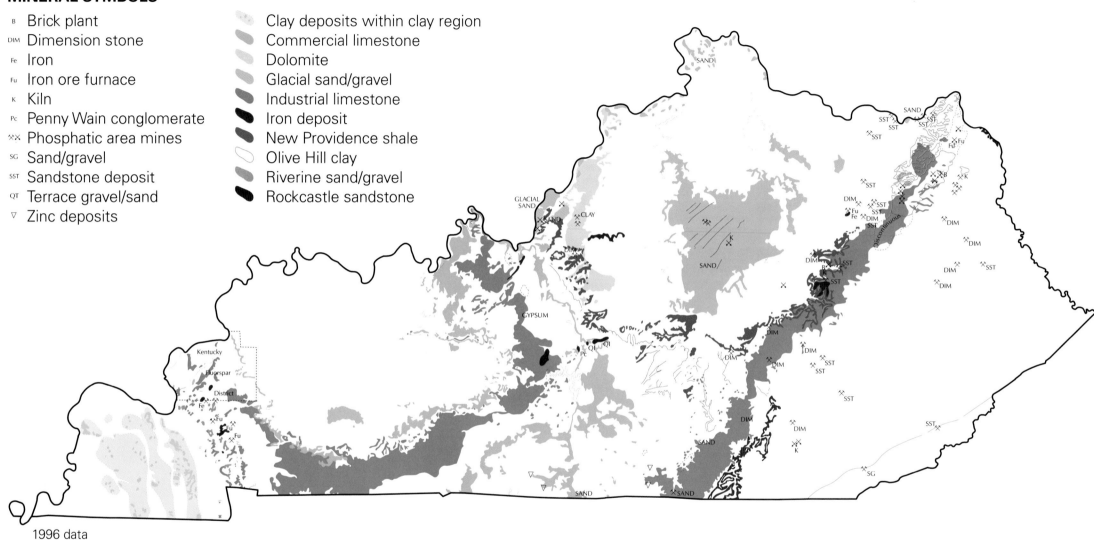

1996 data

COALFIELDS OF THE UNITED STATES

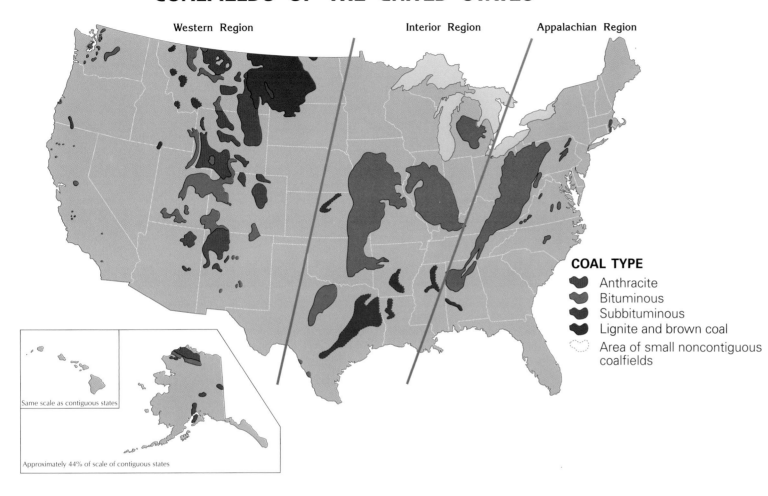

Western Region | Interior Region | Appalachian Region

COAL TYPE
- Anthracite
- Bituminous
- Subbituminous
- Lignite and brown coal
- Area of small noncontiguous coalfields

Same scale as contiguous states

Approximately 44% of scale of contiguous states

KENTUCKY'S MINERAL PRODUCTION VALUE

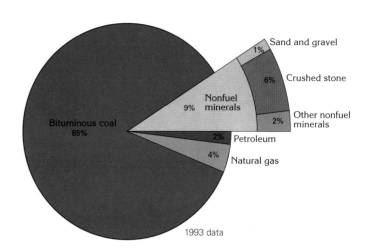

Bituminous coal 85%

Nonfuel minerals 9%

Sand and gravel 1%

Crushed stone 6%

Other nonfuel minerals 2%

Petroleum 2%

Natural gas 4%

1993 data

Total value = $4.56 billion; other nonfuel minerals include clays, lime, cement, gemstones, silver, zinc, and lead.

County. Since then coal production has increased steadily: by the early 1840s the state was producing 100,000 tons of coal annually, in 1879 the state's mines produced 1 million tons for the first time, and state coal production first exceeded 100 million tons in the mid-1960s. From 1973 to 1987 Kentucky was the nation's leading coal-producing state; since 1987 it has ranked second or third, after Wyoming (ranked first since 1987) and, in some years, West Virginia. Annual production in 1994 exceeded 160 million tons and accounted for 15.6 percent of the total national production.

During the early 1970s most coal mined in Kentucky came from surface mines. In 1994, however, about three-fifths of Kentucky's coal production was derived from 425 underground mines and two-fifths from 250 surface mines; in comparison, less than two-fifths of the nation's coal came from underground

mines. Today, the proportions coming from underground and surface mines in the coalfields of eastern and western Kentucky are almost identical. Coal that comes from the gently sloped lands of western Kentucky is produced in large surface mines, whereas coal from surface mines in the much steeper terrain of eastern Kentucky is produced through mountaintop removal or through the use of large augers in a method called auger mining. Most coal produced in Kentucky comes from large, modern, and efficient underground mines of various entry types (shaft, slope, or drift). These mines have become increasingly mechanized—through the use, for example, of continuous mining machines and longwall mining panels—thereby reducing the number of miners employed. Most underground coal is mined by the room-and-pillar method. In this process, coal is extracted by cutting a

series of "rooms" into the coal bed and leaving "pillars," or columns, of coal to help support the mine roof. A single series of rooms is typically about 400 feet wide and one-half mile long. When the end of the area being mined is reached, the direction of mining can be reversed, and additional coal can be recovered from the pillars that were supporting the roof. Generally, 50 to 60 percent of the minable coal in the area is recovered using this method.

To avoid the expense of reclamation, companies that extract coal have abandoned more than 100,000 acres of surface mines in Kentucky. A further 270,000 acres (about 2.5 percent of the total land in the state's coal-producing counties) have been disturbed by mining since 1977, the year that the federal Surface Mining Reclamation and Enforcement Act was signed into law.

LEADING COAL-PRODUCING STATES

Tons of coal produced (in millions)

Only states that produced more than 10 million tons of coal in 1994 are shown.

Underground mines in Harlan County move bituminous coal to tipples by way of covered conveyor belts. In the tipples the coal is cleaned of rock and other foreign materials and loaded onto rail cars or trucks. The water used in the cleaning process contains fine particles of rock and coal and must be stored to avoid stream pollution. Here a dam across a side valley holds a pond of coal-cleaning water for evaporation.

Nearly three-quarters of the disturbed land is in eastern Kentucky. Contour mining and mountaintop removal have been extensively employed there, and land has also settled as a result of underground mining. Pike County has the greatest number of acres disturbed (25,677, or more than 5 percent of the county's area), and Martin County has the largest percentage of its area disturbed (16,576 disturbed acres account for 11.2 percent of the county's area). Abandoned mine lands and disturbed lands are especially prone to landslides, and between 1982 and 1992 almost 2,500 such landslides created a threat to public safety and property.

More than three-quarters of Kentucky's coal is produced in the eastern part of the state. Pike County is the state's largest coal producer. In 1994 the county produced more than 34 million tons of coal, or more than one-fifth of the state's total. Of the thirty-two counties that produced some coal in 1994, the four leading producers—Pike, Harlan, Martin, and Perry Counties, all in eastern Kentucky—accounted for 46 percent of the state's total production. An estimated ninety billion tons of minable coal remain in the state; this figure represents 87 percent of the original coal deposits. Sixty percent of the estimated coal remaining is located in eastern Kentucky, 10 percent in Pike County alone.

Similar to the situation at the national level, more than four-fifths of the coal mined in Kentucky is used for electric utilities, and industry and exports are the second and third leading markets. Approximately 17.5 percent of Kentucky's total 1994 coal production went to the electric utilities located in the state; 63.5 percent was sold to electric utility companies in other states. The largest purchasers of Kentucky coal for use by utility companies in 1994 were Kentucky (27.33 million tons), Tennessee (15.58), Georgia (14.40), North Carolina (10.27), South Carolina (10.05), Ohio (9.82), and Michigan (7.03). The largest electric utility consumers of Kentucky coal were the Tennessee Valley Authority

COAL MINING METHODS

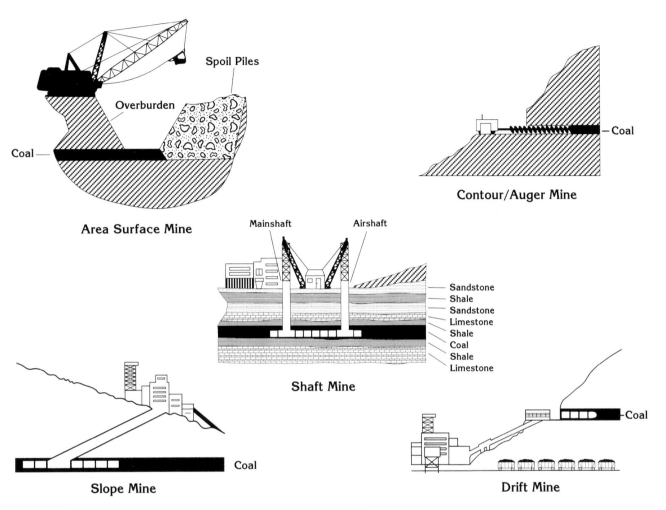

Area Surface Mine

Spoil Piles

Overburden

Coal

Contour/Auger Mine

Coal

Shaft Mine

Mainshaft   Airshaft

Sandstone
Shale
Sandstone
Limestone
Shale
Coal
Shale
Limestone

Slope Mine

Coal

Drift Mine

Coal

A surface coal mine in the Western Kentucky Coal Field's Muhlenberg County. After the coal is removed, the disturbed surface is reclaimed by grading and dirt fill. The dragline's scale is suggested by the maintenance trucks parked next to the bucket.

Mountaintop removal utilizes large-scale equipment to surface mine extensive areas by removing overlying rock to expose several coal seams. By reducing ridges and filling valleys the process creates a rolling surface in otherwise mountainous terrain. This Perry County mine near Hazard illustrates the process. A drill bores blasting holes (to the left of the mine cut); a large dragline removes broken rock from the cut to expose the coal seam and stacks the rock, or spoil, in long heaps or piles. The new surface is then regraded and planted (to the right of the spoil heaps). The wooded mountains in the distance represent an unmined landscape.

Large tipples process coal from a Johnson County surface mine and load waiting coal cars in the valley below. Much of the coal mined in eastern Kentucky is purchased by electrical utilities such as the Tennessee Valley Authority and Georgia Power Company.

KENTUCKY COAL PRODUCTION 1890-1989

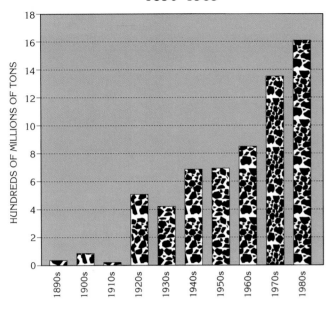

KENTUCKY COAL PRODUCTION 1950-1994

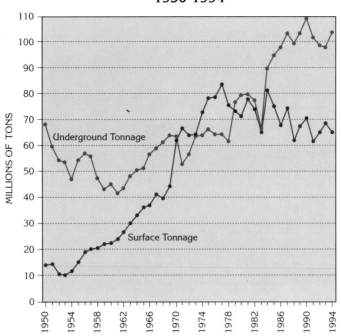

COAL PRODUCTION

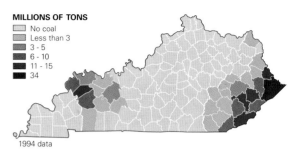

MILLIONS OF TONS
- No coal
- Less than 3
- 3 - 5
- 6 - 10
- 11 - 15
- 34

1994 data

SURFACE-MINED COAL

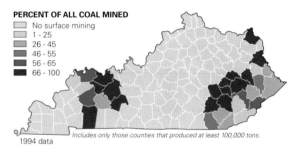

PERCENT OF ALL COAL MINED
- No surface mining
- 1 - 25
- 26 - 45
- 46 - 55
- 56 - 65
- 66 - 100

1994 data   Includes only those counties that produced at least 100,000 tons.

LAND DISTURBED BY COAL MINING

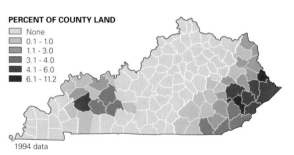

PERCENT OF COUNTY LAND
- None
- 0.1 - 1.0
- 1.1 - 3.0
- 3.1 - 4.0
- 4.1 - 6.0
- 6.1 - 11.2

1994 data

ESTIMATED COAL RESOURCES

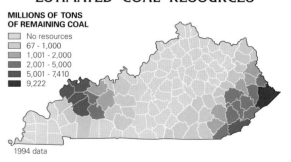

MILLIONS OF TONS OF REMAINING COAL
- No resources
- 67 - 1,000
- 1,001 - 2,000
- 2,001 - 5,000
- 5,001 - 7,410
- 9,222

1994 data

The 1977 federal Surface Mining Reclamation and Enforcement Act stipulated that mined land had to be restored to its approximate original contour. This coal mine in Johnson County has been reclaimed and planted. Stone-covered channels direct precipitation runoff.

COAL USAGE

United States

Kentucky

Electric utilities 82.2%
80.7%

Exports 7.2%
4.5%

Coke plants 2.9%
3.2%

Other industrial 6.9%
9.8%

All other 1%
1.6%

1994 data

COAL INTO KILOWATTS

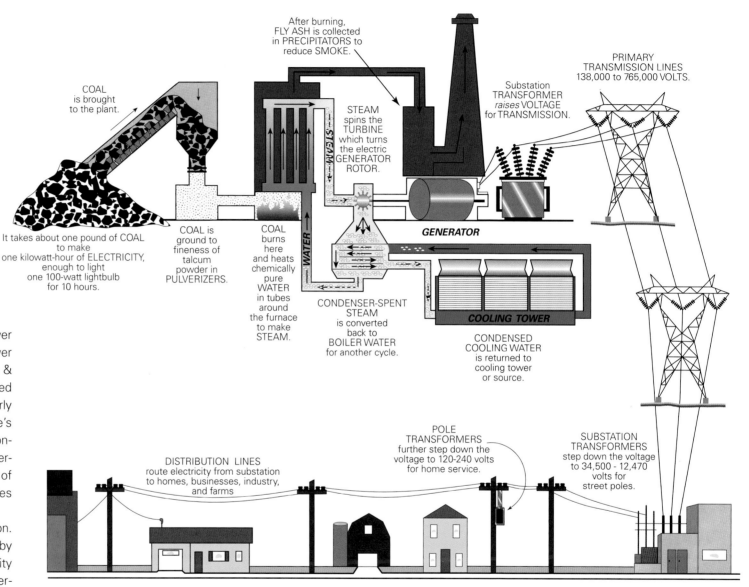

After burning, FLY ASH is collected in PRECIPITATORS to reduce SMOKE.

COAL is brought to the plant.

PRIMARY TRANSMISSION LINES 138,000 to 765,000 VOLTS.

Substation TRANSFORMER *raises* VOLTAGE for TRANSMISSION.

STEAM spins the TURBINE which turns the electric GENERATOR ROTOR.

GENERATOR

It takes about one pound of COAL to make one kilowatt-hour of ELECTRICITY, enough to light one 100-watt lightbulb for 10 hours.

COAL is ground to fineness of talcum powder in PULVERIZERS.

COAL burns here and heats chemically pure WATER in tubes around the furnace to make STEAM.

CONDENSER-SPENT STEAM is converted back to BOILER WATER for another cycle.

COOLING TOWER

CONDENSED COOLING WATER is returned to cooling tower or source.

DISTRIBUTION LINES route electricity from substation to homes, businesses, industry, and farms

POLE TRANSFORMERS further step down the voltage to 120-240 volts for home service.

SUBSTATION TRANSFORMERS step down the voltage to 34,500 - 12,470 volts for street poles.

(TVA) (28.03 million tons), Georgia Power Company (14.13), Duke Power Company (7.12), South Carolina Public Service (5.40), and Dayton Power & Light (4.49), South Carolina Electric & Gas (4.17), Louisville Gas & Electric (4.14), and Kentucky Utilities Company (4.11). The TVA used more than half of all the coal produced in western Kentucky. Nearly two-thirds of the coal consumed by the TVA came from the state's Western Kentucky Coal Field; its coal is noted for its higher sulfur content than that mined in eastern Kentucky (on average, 3.1 percent versus 1.1 percent). Western Kentucky coal also has a higher content of ash than coal produced in eastern Kentucky. Both of these impurities can be reduced through a process called coal washing.

Electricity costs in Kentucky are among the lowest in the nation. Only Washington and Idaho, where power production is augmented by low-cost hydroelectric dams, have lower costs. Kentucky's electricity costs for all sectors, 4.3 cents per kilowatt-hour in 1994, are considerably less than the national average of 6.9 cents. Cost per kilowatt-hour for the industrial sector only is also among the lowest nationally, 3.2 cents per kilowatt-hour compared with the national average of 4.6 cents. Again, only Washington and Idaho boast lower costs than Kentucky. The low cost of electricity in the state is not surprising, since coal, Kentucky's most abundant mineral resource, offers a much less costly form of electrical energy than either petroleum, natural gas, or nuclear power. Kentucky Utilities Company, Kentucky Power Company,

ELECTRIC GENERATING STATIONS

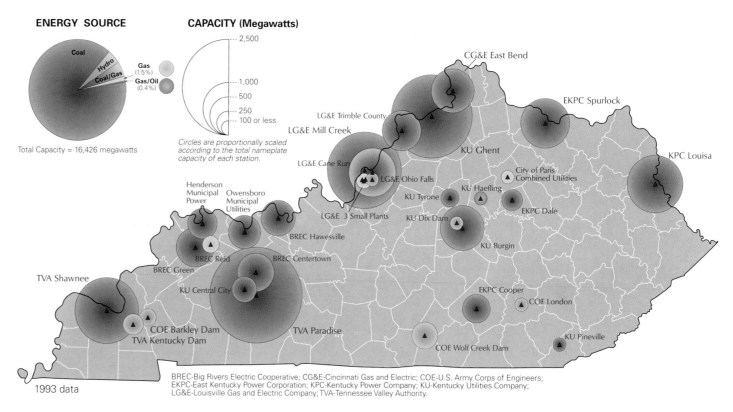

ENERGY SOURCE

Coal
Hydro
Coal/Gas
Gas (1.5%)
Gas/Oil (0.4%)

Total Capacity = 16,426 megawatts

CAPACITY (Megawatts)

2,500
1,000
500
250
100 or less

Circles are proportionally scaled according to the total nameplate capacity of each station.

CG&E East Bend
EKPC Spurlock
LG&E Trimble County
LG&E Mill Creek
KU Ghent
KPC Louisa
LG&E Cane Run
LG&E Ohio Falls
City of Paris Combined Utilities
KU Haefling
KU Tyrone
EKPC Dale
Henderson Municipal Power
Owensboro Municipal Utilities
LG&E 3 Small Plants
KU Dix Dam
BREC Hawesville
KU Burgin
BREC Reid
BREC Centertown
BREC Green
EKPC Cooper
COE London
TVA Shawnee
KU Central City
EKPC Cooper
COE Barkley Dam
TVA Kentucky Dam
TVA Paradise
COE Wolf Creek Dam
KU Pineville

BREC-Big Rivers Electric Cooperative; CG&E-Cincinnati Gas and Electric; COE-U.S. Army Corps of Engineers; EKPC-East Kentucky Power Corporation; KPC-Kentucky Power Company; KU-Kentucky Utilities Company; LG&E-Louisville Gas and Electric Company; TVA-Tennessee Valley Authority.

1993 data

Electrical power generating plants stand along the Ohio River, capitalizing on close proximity to Appalachian and interior coalfields in western Kentucky, Indiana, and Illinois. River tows move coal via barge to power plants such as this one in Carroll County. Pollution control devices on the smokestacks trap fly ash that must then be stored so as to prevent pollution of air or water.

UTILITY COST

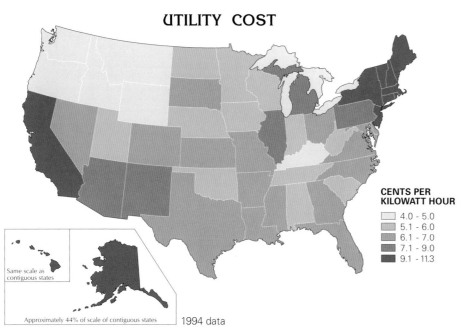

CENTS PER KILOWATT HOUR
- 4.0 - 5.0
- 5.1 - 6.0
- 6.1 - 7.0
- 7.1 - 9.0
- 9.1 - 11.3

Same scale as contiguous states

Approximately 44% of scale of contiguous states

1994 data

INDUSTRIAL UTILITY COST

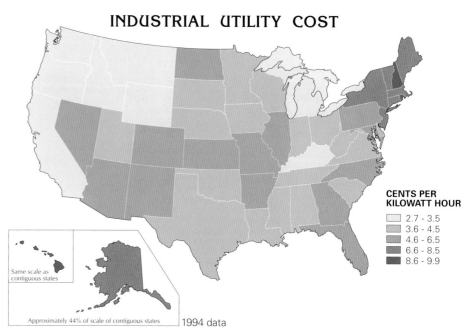

CENTS PER KILOWATT HOUR
- 2.7 - 3.5
- 3.6 - 4.5
- 4.6 - 6.5
- 6.6 - 8.5
- 8.6 - 9.9

Same scale as contiguous states

Approximately 44% of scale of contiguous states

1994 data

ELECTRICITY IN KENTUCKY

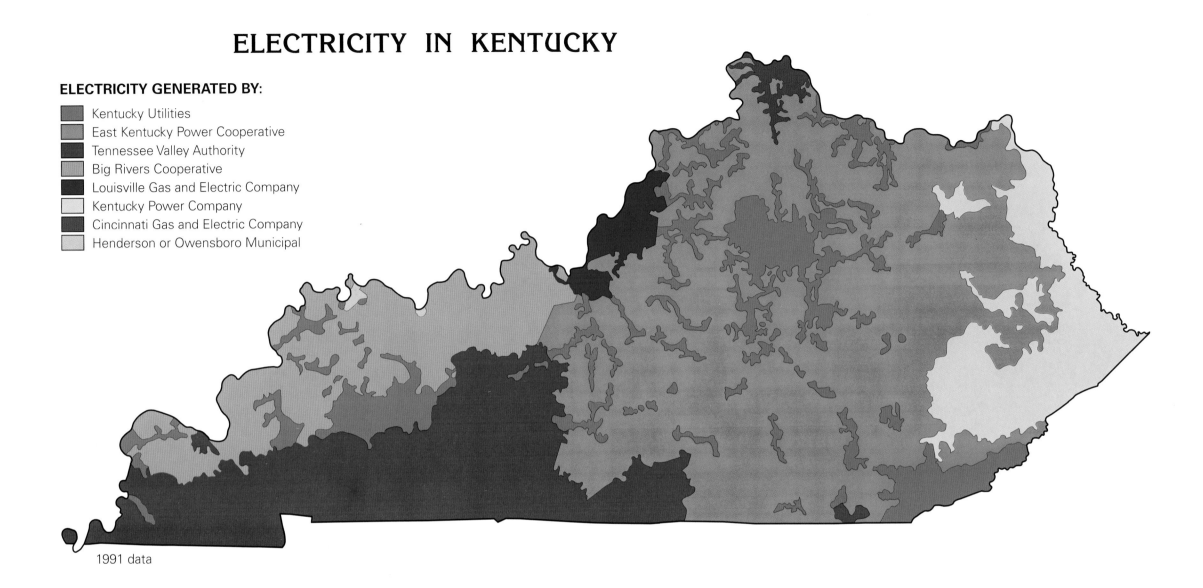

ELECTRICITY GENERATED BY:

- Kentucky Utilities
- East Kentucky Power Cooperative
- Tennessee Valley Authority
- Big Rivers Cooperative
- Louisville Gas and Electric Company
- Kentucky Power Company
- Cincinnati Gas and Electric Company
- Henderson or Owensboro Municipal

1991 data

and Louisville Gas & Electric Company were the sixth, fourteenth, and twenty-second least expensive utility companies in the U.S. in 1994, respectively.

Four major investor-owned power companies, twenty-nine municipal electric systems, the Tennessee Valley Authority, Berea College, and twenty-seven rural electric cooperative corporations (RECCs) distribute electric power to Kentucky's households, industries, and businesses. Of the four major companies, the most widespread geographically is the Kentucky Utilities Company, which serves Lexington and seventy-seven counties across the state. Kentucky Power Company serves all or parts of twenty

counties in eastern Kentucky; Louisville Gas and Electric Company serves metropolitan Louisville; and Union Light, Heat, and Power Company, a subsidiary of the Cincinnati Gas and Electric Company, serves portions of five northern Kentucky counties. Almost 50 percent of the state's electric power capacity is produced by power plants owned by these four investor-owned power companies. The TVA power facilities account for another 25 percent of capacity, and two electric cooperatives (Big Rivers Electric Corporation and East Kentucky Power Cooperative) for 18 percent. The remaining 7 percent of the state's capacity is produced by three U.S. Army Corps of Engineers hydroelectric

plants and three municipally owned plants. The largest single generating station in the state, accounting for 15 percent of the state's total electric power capacity, is TVA's Paradise plant near Drakesboro in Muhlenberg County. In all, more than 95 percent of the state's electric power is produced by coal-fired plants, and about 5 percent by hydroelectric dams. Kentucky has no nuclear-powered electric plants, and none are currently in the approval process. In 1997 a merger was proposed between two of Kentucky's largest power companies, Kentucky Utilities and Louisville Gas and Electric. By late 1997, the proposal had passed most of the regulatory hurdles and appeared to be close to final approval.

COAL EXPORTS

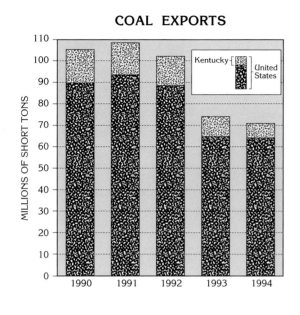

MILLIONS OF SHORT TONS

Kentucky / United States

1990 1991 1992 1993 1994

COAL MINING EMPLOYMENT

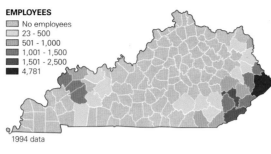

EMPLOYEES
- No employees
- 23 - 500
- 501 - 1,000
- 1,001 - 1,500
- 1,501 - 2,500
- 4,781

1994 data

COAL MINING EMPLOYMENT CHANGE 1990-1994

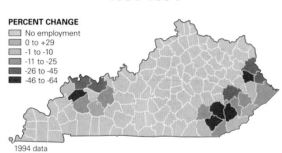

PERCENT CHANGE
- No employment
- 0 to +29
- -1 to -10
- -11 to -25
- -26 to -45
- -46 to -64

1994 data

COAL EMPLOYMENT AND PRODUCTION

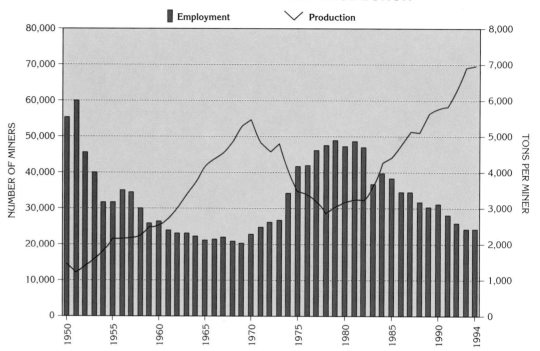

Employment · Production

NUMBER OF MINERS

TONS PER MINER

1950 1955 1960 1965 1970 1975 1980 1985 1990 1994

COAL EXPORT DESTINATIONS

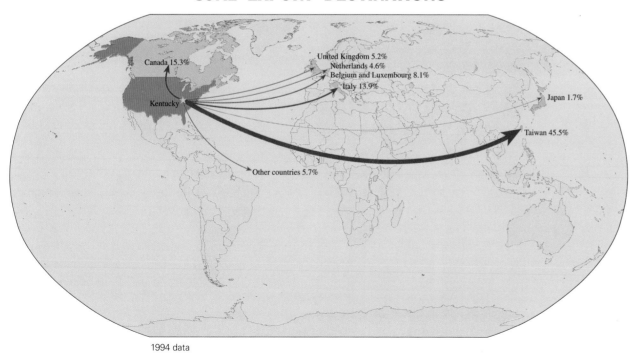

Canada 15.3%
United Kingdom 5.2%
Netherlands 4.6%
Belgium and Luxembourg 8.1%
Italy 13.9%
Japan 1.7%
Kentucky
Taiwan 45.5%
Other countries 5.7%

1994 data

In 1994 Kentucky exported more than seven million tons of coal, approximately 10 percent of the nation's total coal exports. Reflecting the national trend, Kentucky's coal exports have declined in recent years; in 1994 exports stood at less than half the 1990 figure. The single most important destination was Taiwan, which accounted for 45 percent of the tonnage of Kentucky coal exported in 1994 and 45 percent of the nearly $275 million total value exported.

Since 1950, trends in coal mining employment have reflected the demand for coal and, more recently, modernization and new techniques in coal mining. Since the early 1980s the number of miners employed has steadily declined. In 1994, 24,133 persons were directly employed in mining. (Only one county, Webster County in western Kentucky, experienced an increase in coal mining employment during the first half of the 1990s.) Yet total coal production as well as coal production per miner has increased significantly. In the early 1980s, for example, each miner produced about 3,000 tons of coal annually. In 1994 per-miner production had increased to 7,000 tons. New coal mining technologies and machinery developed for coal production are the primary reasons for increased production. The use of new technology and huge machines including continuous miners, longwall mining panels, and immense power shovels has meant that fewer workers can now produce more coal.

WEEKLY MINING EARNINGS

EARNINGS IN DOLLARS
- Non-disclosure
- No mining earnings
- 356 - 500
- 501 - 650
- 651 - 800
- 801 - 950
- 951 - 1,041

1994 data

MINING WAGES

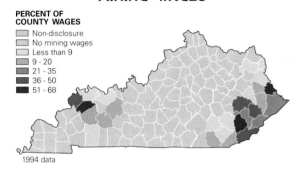

PERCENT OF COUNTY WAGES
- Non-disclosure
- No mining wages
- Less than 9
- 9 - 20
- 21 - 35
- 36 - 50
- 51 - 68

1994 data

MINING FATALITIES, 1890s-1994

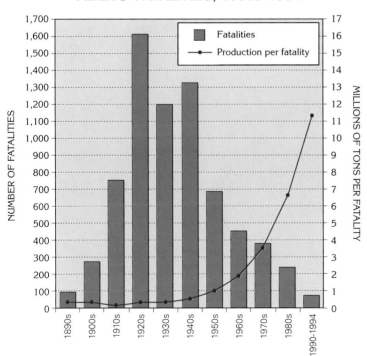

Legend: Fatalities; Production per fatality

COAL MINING FATALITIES, 1982-1994

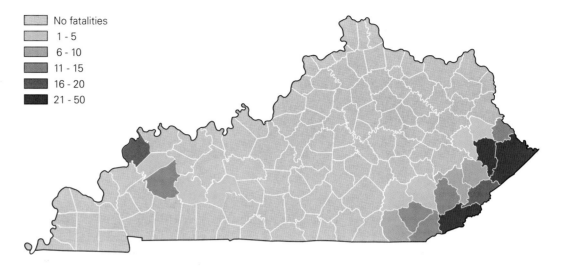

- No fatalities
- 1 - 5
- 6 - 10
- 11 - 15
- 16 - 20
- 21 - 50

Overall, coal mining has had a significant economic impact for the state. The University of Kentucky Center for Business and Economic Research estimates that in 1994 the state's mines produced and processed nearly four billion dollars' worth of coal, and this production in turn generated additional economic activity of nearly five billion dollars and 71,400 jobs. Thus, the total direct and indirect economic activity yielded by coal production and processing amounted to nearly nine billion dollars and about 95,500 jobs.

Mining in the coalfields of eastern Kentucky accounts for nearly four-fifths of the state's total coal mining employment. Pike County has far more employment in coal mining than any other county: in 1994, 4,781 miners accounted for nearly one-fifth of the county's total labor force. Indeed, nearly as many miners worked in Pike County mines in 1994 as worked in the entire Western Kentucky Coal Field (5,262). Wages from coal mining employment in Pike County totaled more than $175 million, representing one-third of total county wages. Pike County miners earned more than twice as much as the total mining wages of the second-ranked county, Harlan. Although Pike County led all counties in total employment and total wages, a number of counties depended more on the coal industry for employment and wages, as measured by the proportion contributed by coal. Martin County led all counties in

these categories. Coal miners represented two-fifths of this eastern Kentucky county's total labor force and earned more than two-thirds of the total wages. In western Kentucky, Webster County led, with nearly one-quarter of the labor force and more than half of the total wages coming from coal mining. Overall, the state's coal miners earned about $750 weekly in 1994, but there was significant disparity between the wages in western Kentucky and those in eastern Kentucky. In the west, the average wage was about $900 weekly; in the east, about $700. The state's highest mining wages were found in Webster County, where a miner's weekly pay averaged $1,040.

Since the 1890s, more than 7,100 work-related fatalities have occurred in Kentucky's coal mines. The number of fatalities has steadily declined, however: during the 1950s the average number of fatalities was sixty-nine per year; during 1990-94 fatalities stood at an average of fifteen per year. Certainly the enactment in 1969 of a comprehensive federal mine safety and health measure helped to further the decline in mining-related fatalities, which began in the 1950s. Within the state, a better-educated labor force, increased miner safety training, increased use of mining machinery, enhanced safety inspections, and union contracts that prohibit work under unsafe conditions have all contributed to improvements in the fatality rates.

POTENTIAL SULFUR DIOXIDE EMISSIONS

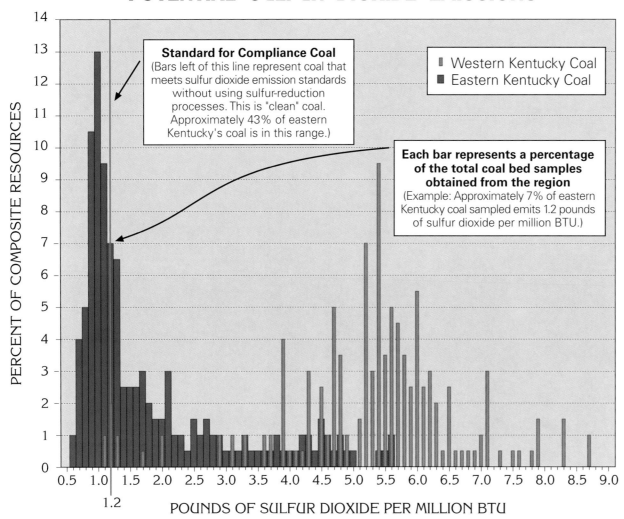

Standard for Compliance Coal
(Bars left of this line represent coal that meets sulfur dioxide emission standards without using sulfur-reduction processes. This is "clean" coal. Approximately 43% of eastern Kentucky's coal is in this range.)

▌ Western Kentucky Coal
▐ Eastern Kentucky Coal

Each bar represents a percentage of the total coal bed samples obtained from the region
(Example: Approximately 7% of eastern Kentucky coal sampled emits 1.2 pounds of sulfur dioxide per million BTU.)

PERCENT OF COMPOSITE RESOURCES

POUNDS OF SULFUR DIOXIDE PER MILLION BTU

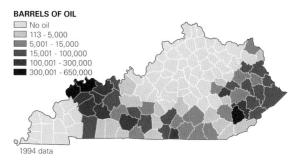

OIL PRODUCTION

BARRELS OF OIL
- No oil
- 113 - 5,000
- 5,001 - 15,000
- 15,001 - 100,000
- 100,001 - 300,000
- 300,001 - 650,000

1994 data

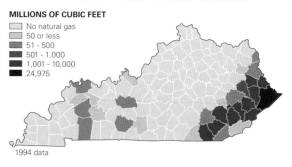

NATURAL GAS PRODUCTION

MILLIONS OF CUBIC FEET
- No natural gas
- 50 or less
- 51 - 500
- 501 - 1,000
- 1,001 - 10,000
- 24,975

1994 data

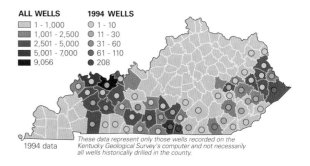

GAS AND OIL WELLS DRILLED

ALL WELLS
- 1 - 1,000
- 1,001 - 2,500
- 2,501 - 5,000
- 5,001 - 7,000
- 9,056

1994 WELLS
- ○ 1 - 10
- ○ 11 - 30
- ○ 31 - 60
- ○ 61 - 110
- ● 208

1994 data   *These data represent only those wells recorded on the Kentucky Geological Survey's computer and not necessarily all wells historically drilled in the county.*

Gas and Oil

Natural gas and oil production accounted for 4 percent and 2 percent, respectively, of the total value of mineral production in Kentucky in 1993. The state traces its petroleum industry to 1818, when the Martin Beatty well was drilled in what is today McCreary County. In the mid-1990s about 40,000 wells were producing gas and oil throughout the state; 883 wells were drilled in 1994 alone. In 1993 Kentucky ranked eighteenth in the nation in the volume of gas produced and twentieth in oil production. The average depth of oil and gas wells across the state is a relatively shallow 1,500 feet, which makes the state attractive to small, independent producers. The success ratio for gas wells

drilled in Kentucky is about 25 percent. Although the chief oil-producing county in the state, Leslie County, is located in Appalachian Kentucky, more than half the state's oil (54 percent in 1994) is produced in the counties of western Kentucky. On the other hand, most natural gas produced—98 percent of the state total—comes from wells in the Appalachian counties. The single largest producing county is Pike, which accounted for more than one-third of the state gas production in 1994. Only about 10 percent of the natural gas consumed in the state is produced in Kentucky. The distribution of natural gas production may change in the future, however, since the Kentucky Geological Survey

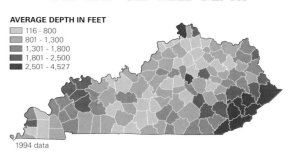

GAS AND OIL WELL DEPTH

AVERAGE DEPTH IN FEET
- 116 - 800
- 801 - 1,300
- 1,301 - 1,800
- 1,801 - 2,500
- 2,501 - 4,527

1994 data

OIL AND GAS FIELDS

KNOWN FIELD

- Oil
- Gas

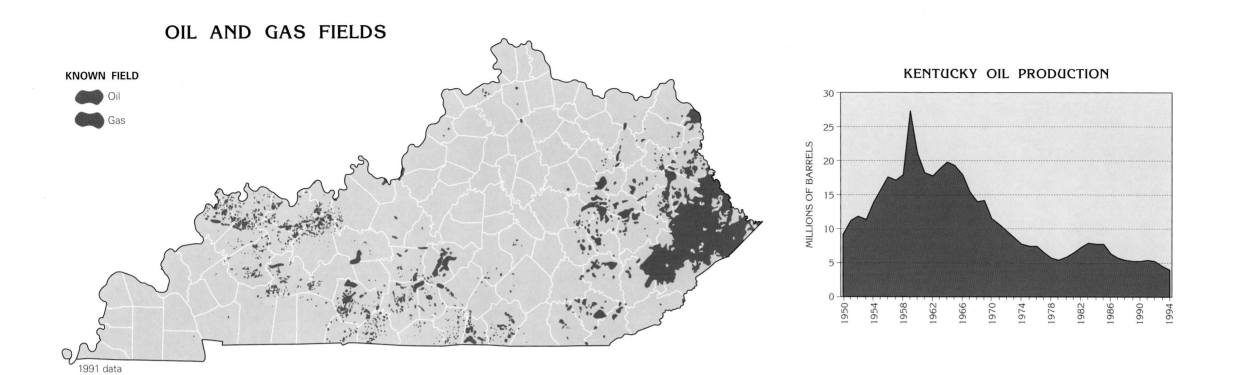

KENTUCKY OIL PRODUCTION

1991 data

recently reported that a major natural gas field may lie in western Kentucky, stretching from Union County east to Grayson and Edmonson Counties.

Kentucky oil production peaked in 1959 and has generally declined since the mid-1960s, except for a slight rise during the early 1980s. On the other hand, in recent years natural gas production has increased. A 1994 status report on Kentucky's environment attributes this increase to natural gas's "clean burning characteristics . . . , a slight increase in . . . prices, recently enacted federal tax credits, and [price] deregulation . . . by the federal Energy Regulatory Commission [that allowed] smaller independent operators access to natural gas transmission pipelines." The state's major cities and industries receive natural gas from the interstate pipeline system; less than one-quarter of the gas produced in the state is used in Kentucky, the major portion being transported out of state via pipelines. Kentucky lies astride the principal northeast-southwest corridor of gas transmission lines running between the major northeastern and midwestern markets and the gas- and oil-producing Gulf states. According to the Kentucky Cabinet for Economic Development, about three trillion cubic feet of gas is transported out of or through the state each year; Kentucky production accounts for only 3 percent of this volume. In 1990 eleven interstate transmission companies had more than 7,000 miles of pipelines crossing the state, and thirty-eight pipeline companies transported gas from the state's producing fields through 3,745 miles of local pipelines.

GAS TRANSMISSION AND DISTRIBUTION

GAS TRANSMISSION LINES

— Major pipelines (20" and larger)
— Other pipelines (10" - 19")
Local and/or less than 10" pipelines are not shown.

GAS DISTRIBUTION SYSTEMS

- Columbia Gas of Kentucky
- Delta Natural Gas Co.
- Louisville Gas & Electric Co.
- Union Light Heat and Power Co.
- Western Kentucky Gas Co.
- Unserved areas or areas served by municipal gas distribution utilities

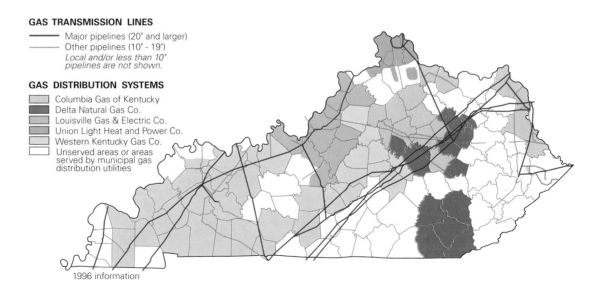

1996 information

NONFUEL MINERALS

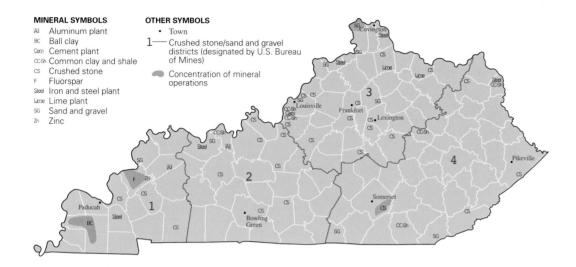

MINERAL SYMBOLS
Al Aluminum plant
BC Ball clay
Cem Cement plant
CC-Sh Common clay and shale
CS Crushed stone
F Fluorspar
Steel Iron and steel plant
Lime Lime plant
SG Sand and gravel
Zn Zinc

OTHER SYMBOLS
• Town
1— Crushed stone/sand and gravel districts (designated by U.S. Bureau of Mines)
⬤ Concentration of mineral operations

Nonfuel Minerals

In 1992 Kentucky ranked twenty-seventh among the states in the value of nonfuel mineral production. The most significant nonfuel mineral produced in the state is crushed stone (including limestone). Crushed stone accounts for more than half the value of all nonfuel minerals and 6 percent of the value of total mineral production (fuel and nonfuel). Kentucky ranked seventh nationally in the value of crushed stone production in 1992. The construction industry, notably the road construction business, is the state's principal consumer of crushed stone. Another use of crushed stone is in environmental applications; limestone, for example, is used as the scrubbing agent in coal-fired power generating stations. This application has recently increased in importance. Most of the state's crushed stone is produced at quarries in areas underlain with limestone or dolomite, including the Inner Bluegrass, the area stretching from Oldham County south to Nelson County, Pulaski County, and several counties in western Kentucky. After crushed stone, Kentucky's most valuable nonfuel minerals are lime (Kentucky ranks fourth nationally), cement, sand and gravel, and clays.

Timber Resources

Kentucky's large forested areas contain a diversity of commercially important tree species. Much of this diversity derives from the state's latitudinal positioning between the forests of the South, which feature yellow poplar, shortleaf pine, black gum, and sweet gum, and northern forests, which are characterized by such species as northern red oak, white pine, hemlock, hard maple, basswood, black walnut, and beech. Another reason for the diversity is the topography of eastern Kentucky: its varied aspects, slopes, and elevations create the conditions in which both northern and southern species can thrive. Just under one-half of Kentucky's land area is forested, and 97 percent of this forested area, more than 12 million acres, is classified as timberland (formerly called commercial forest land). Forty-one of the state's 120 counties each con-

Broken, crushed, or pulverized limestone has many commercial uses, especially in agriculture and construction. This large quarry in Livingston County lies along the Cumberland River near Smithland. Because stone is a bulky commodity with a low value per unit of weight, it must be transported by way of the cheapest mode available—water. Although limestone is ubiquitous across the Pennyroyal and could be quarried in many places, this location minimizes land transport costs by allowing large trucks to move stone directly to the river's edge and dump it into waiting barges (lower left).

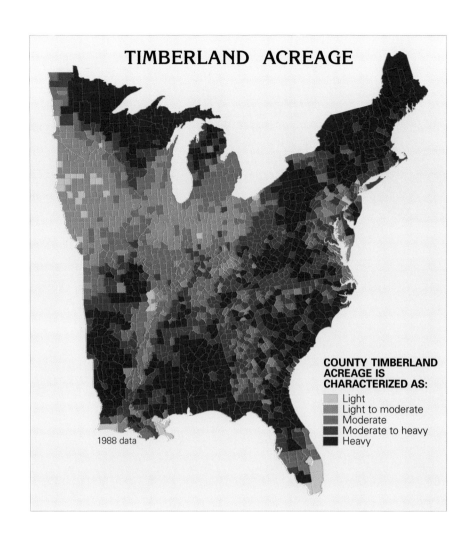

TIMBERLAND ACREAGE

COUNTY TIMBERLAND ACREAGE IS CHARACTERIZED AS:

- Light
- Light to moderate
- Moderate
- Moderate to heavy
- Heavy

1988 data

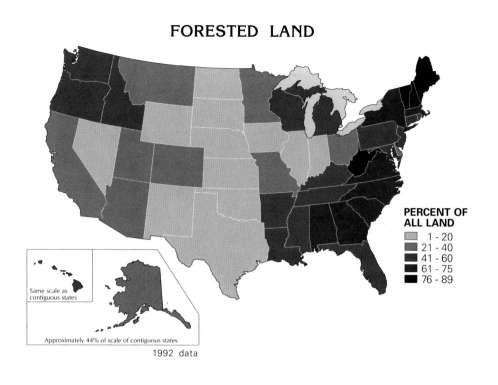

FORESTED LAND

PERCENT OF ALL LAND
- 1 - 20
- 21 - 40
- 41 - 60
- 61 - 75
- 76 - 89

Same scale as contiguous states

Approximately 44% of scale of contiguous states

1992 data

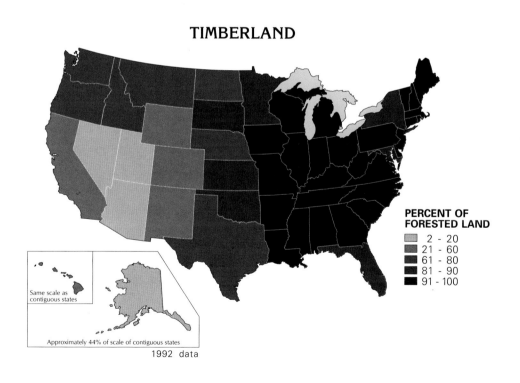

TIMBERLAND

PERCENT OF FORESTED LAND
- 2 - 20
- 21 - 60
- 61 - 80
- 81 - 90
- 91 - 100

Same scale as contiguous states

Approximately 44% of scale of contiguous states

1992 data

tained more than 100,000 acres of timberland at the time of the 1988 U.S. Forest Service Survey.

Oak and hickory forests dominate the state's timberland resources, representing just over 75 percent of all timberland area. Red oaks and white oaks are the dominant species, accounting for roughly 37 percent of the state's total growing stock. They are found in the eastern and western sections of the state. White oak is a bit more adaptable and competitive than red oak and so can be found in more environmentally diverse settings. Yellow poplar is the third most common Kentucky species after the red and white oaks. It is seldom found in the drier settings of the Bluegrass and western Kentucky, however. Soft

maple is commonly found in eastern Kentucky, while hard maple is more often encountered in the south central portions of the state. Two commercially important species that have limited ranges within Kentucky are eastern red cedar, most often found in the Bluegrass Region's shale beds, and ash, which is concentrated in the state's northern counties.

A very high percentage of Kentucky's timberland is privately owned, a situation common in much of the eastern United States. More than half of Kentucky's timberland owners—and there are nearly half a million of them—own fewer than ten acres. The average ownership is just under twenty-five acres. In general, Kentucky's timberland is more divided among private

AREA OF TIMBERLAND
(BY SIZE CLASS)

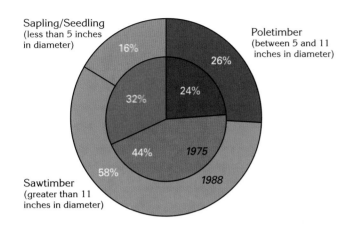

Sapling/Seedling
(less than 5 inches
in diameter)

16%

Poletimber
(between 5 and 11
inches in diameter)

26%

24%

32%

44% *1975*

58%

1988

Sawtimber
(greater than 11
inches in diameter)

TIMBERLAND

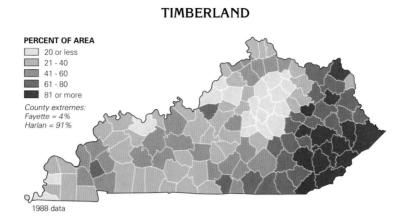

PERCENT OF AREA

- 20 or less
- 21 - 40
- 41 - 60
- 61 - 80
- 81 or more

*County extremes:
Fayette = 4%
Harlan = 91%*

1988 data

TIMBERLAND ACREAGE

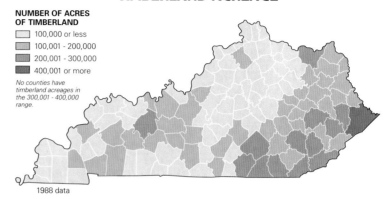

**NUMBER OF ACRES
OF TIMBERLAND**

- 100,000 or less
- 100,001 - 200,000
- 200,001 - 300,000
- 400,001 or more

*No counties have
timberland acreages in
the 300,001 - 400,000
range.*

1988 data

TIMBERLAND AND GROWING STOCK

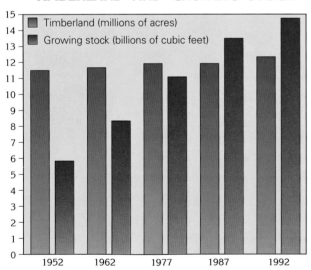

■ Timberland (millions of acres)
■ Growing stock (billions of cubic feet)

| 1952 | 1962 | 1977 | 1987 | 1992 |

*Growing stock is defined as trees of commercial species meeting
quality standards and at least 5 inches in diameter at breast height.*

owners of small tracts than is the case in other states. Large industrial timber tracts, commonly found in the Southeast and the Northwest, are essentially absent from Kentucky's rural landscape. Coordination and effective management of the valuable timberland resource are complicated by the large number of people who enjoy ownership. For instance, federal and state initiatives to educate owners in effective timberland management practices are difficult to deliver to such a large and geographically dispersed group.

Kentucky's early settlement led to especially rapid deforestation during the late 1700s and the first half of the 1800s. About half of Kentucky's forested area was cleared during these early years. Much of this land clearing was accomplished through the use of fire. The tradition of "burning off the woods" and ignoring the "fire on the mountain" lingered for decades. Commercial logging was particularly heavy during the early 1900s, an era of virtually no resource conservation. Stability has generally characterized Kentucky's timberland during the post-World War II era, however. In fact, timberland increased 8 percent between 1952 and 1992, from 11.5 million acres to 12.4 million acres. Additionally, the number and size of trees found on Kentucky's timberland have increased markedly during this forty-year period. Trees with commercial value increased in volume from 5,858 million cubic feet in 1952 to 14,781 million cubic feet in 1992, an impressive 152 percent increase. This rapid growth in tree volume was attributable both to a rapidly growing young forest that reached the mature stage and to improvement in fire prevention. The inventory of change during 1991 illustrates this point. The gross growth of Kentucky's trees was 480 million cubic feet, and 92 million cubic feet was lost to death from natural causes. Thus the net growth was 388 million cubic feet.

PRIVATELY OWNED LAND WITHIN
DANIEL BOONE NATIONAL FOREST

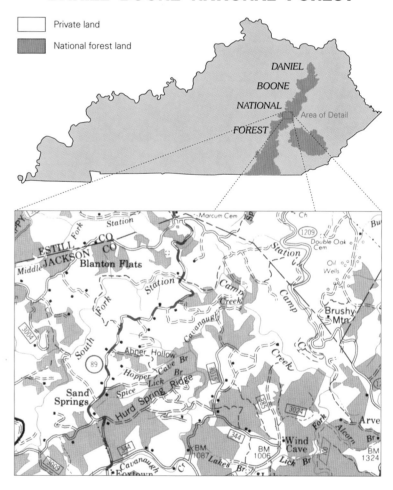

- Private land
- National forest land

PRIVATE TIMBERLAND

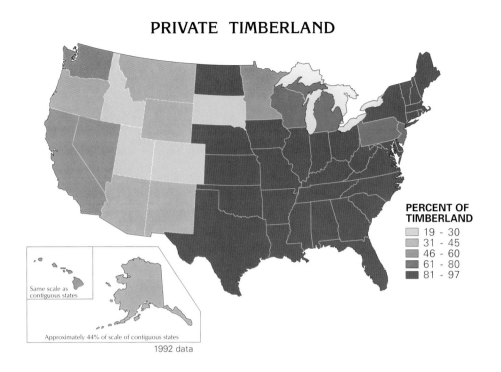

PERCENT OF
TIMBERLAND

- 19 - 30
- 31 - 45
- 46 - 60
- 61 - 80
- 81 - 97

Same scale as contiguous states

Approximately 44% of scale of contiguous states

1992 data

OWNERSHIP OF TIMBERLAND

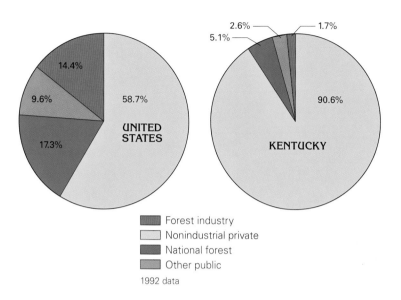

2.6%   1.7%
5.1%

14.4%

9.6%   58.7%

UNITED
STATES

17.3%

90.6%

KENTUCKY

- Forest industry
- Nonindustrial private
- National forest
- Other public

1992 data

One hundred million cubic feet was cut, for a net increase of 288 million cubic feet.

Generally, Kentucky now possesses a larger area of sawtimber than it has since the years of early settlement. The term *sawtimber* refers to trees of commercial quality that exceed eleven inches in diameter at breast height. In 1975 the U.S. Forest Service inventory of timber indicated that Kentucky had 5.3 million acres of sawtimber; by 1988 that number had grown to 7.1 million acres, representing a 34 percent increase.

The Appalachian portion of Kentucky is far more heavily forested with timberland than are the Bluegrass or western regions. In most southeastern counties, at least 80 percent of the total area is timberland. The forty-nine counties classified by the federal government as Appalachian counties comprise about 40 percent of the state's total land area but nearly 60 percent of the state's timberland. Harlan, Leslie, and Breathitt Counties are Kentucky's most forested counties; more than 90 percent of their area is timberland.

Kentucky has supplied sawn lumber to the nation and the world for about a century and a half. As early as the

1840s the state exported white oak to France for the manufacture of wine casks. By 1870, with limited tools and elementary transportation, Kentuckians were producing in excess of 200 million board feet of sawn lumber annually. Kentucky at that time was ranked in the top fifteen states in the production of all lumber and in the top three in the production of hardwood lumber. A peak in Kentucky's lumber production occurred in 1907, when 30,000 workers in 2,400 sawmills produced slightly more than 900 million board feet. Rapid decline followed, and little activity occurred during the Depression years of the early 1930s. Only 180 million board feet were produced by 104 sawmills in the record low year of 1933. Since the close of World War II, however, Kentucky's lumber production has generally increased. The U.S. Forest Service estimates that the state's lumber business reached its pinnacle in 1991, when Kentucky workers produced just over 1 billion board feet. A substantial decline immediately followed that peak, and now the industry has stabilized at about 800 million board feet annually.

Kentucky's primary users of logs are the commercial sawmills that dot the landscape from east to west. A mill that inputs logs and outputs some other product, such as

KENTUCKY SAWN LUMBER VOLUMES
(Selected Species)

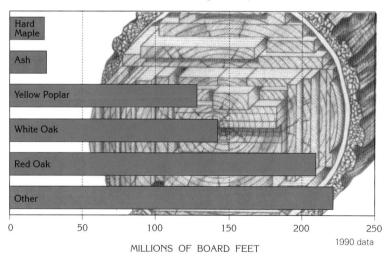

Hard Maple

Ash

Yellow Poplar

White Oak

Red Oak

Other

0   50   100   150   200   250

1990 data

MILLIONS OF BOARD FEET

KENTUCKY SAWN LUMBER PRODUCTION

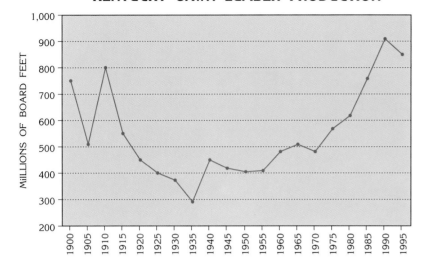

PRIMARY LUMBER INDUSTRIES

- Commercial
- Log and bolt
- Custom
- Pulp chip
- Peeled post
- Log cabin

Each ○ represents one industry

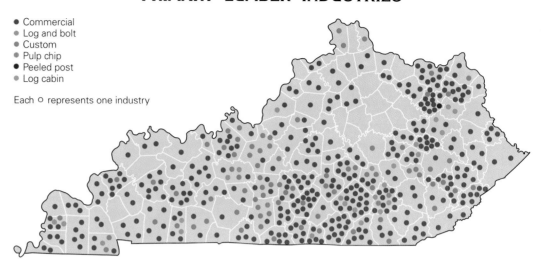

SAWN LUMBER

PERCENT OF STATE TOTAL
- None
- .01 - .25
- .26 - 1.00
- 1.01 - 2.00
- 2.01 - 3.00
- 3.01 - 5.47

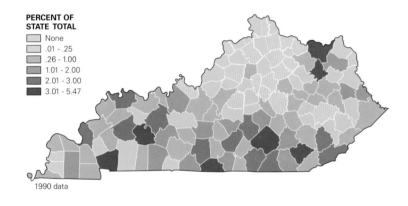

1990 data

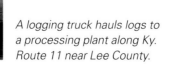

A logging truck hauls logs to a processing plant along Ky. Route 11 near Lee County.

LUMBER AND WOOD PRODUCT VALUE

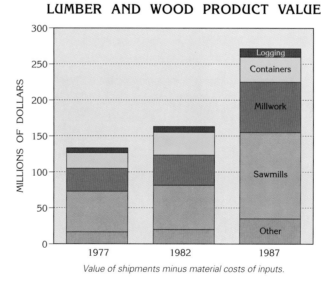

- Logging
- Containers
- Millwork
- Sawmills
- Other

Value of shipments minus material costs of inputs.

dimensioned lumber or veneer, is called a primary wood manufacturer. At least one primary wood manufacturer exists in 108 of Kentucky's 120 counties. A lower density of these mills naturally occurs in counties with less timberland, notably those in the Bluegrass. Kentucky's primary wood mills employ nearly 5,000 people in 568 mills. Red oak is the species most commonly sawn by primary producers (28 percent of the total), followed by white oak (19 percent) and yellow poplar (17 percent). Other important species include ash, maple, red cedar, hickory, and walnut. Although primary wood producers are found in most parts of the state, several of Kentucky's counties are home to larger-scale producers. Rowan County holds the distinction of being Kentucky's leading lumber producer and the leading hardwood producer in the country.

In some timber-rich counties, there is surprisingly little primary production. Primary production involves the sawing of logs into dimensioned and non-dimensioned lumber. Most of the logs provided from trees cut in the Appalachian counties of southeastern Kentucky are transported north and west to manufacturers in the state or to mills in Tennessee. Trucks laden with hardwood logs often travel more than fifty miles to deliver their loads. The value added by primary producers has grown to exceed 300 million dollars annually, with sawmills contributing nearly half of this value.

WHITE OAK

BOARD FEET
PER ACRE
- 0 - 50
- 51 - 400
- 401 - 2,000
- 2,000 or more

Average = 208

1987 data

HARD MAPLE

BOARD FEET
PER ACRE
- 0 - 12
- 13 - 100
- 101 - 700
- 701 or more

Average = 37

1987 data

ASH

BOARD FEET
PER ACRE
- 0 - 12
- 13 - 100
- 101 - 700
- 700 or more

Average = 25

1987 data

RED OAK

BOARD FEET
PER ACRE
- 50 or less
- 51 - 400
- 401 - 2,000
- 2,001 or more

Average = 219

1987 data

SOFT MAPLE

BOARD FEET
PER ACRE
- 0 - 3
- 4 - 50
- 51 - 400
- 401 or more

Average = 36

1987 data

POPLAR

BOARD FEET
PER ACRE
- 0 - 25
- 26 - 200
- 201 - 1,200
- 1,201 or more

Average = 132

1987 data

RED CEDAR

BOARD FEET
PER ACRE
- 0 - 2
- 3 - 10
- 11 - 120
- 121 or more

Average = 2

1987 data

As a strategy for economic development, Kentucky has attempted to encourage a stronger secondary wood manufacturing sector, such as manufacturers of wood furniture. To date this scheme has met with little success. In fact, the state's vast timber wealth is beginning to attract larger-scale, and externally owned, primary producers.

Pulpwood comprises a much smaller volume of wood product than sawn lumber in Kentucky. In fact, the volume of pulpwood equals only about 10 percent of the volume of sawn lumber. Construction of several pulp mills in Kentucky since 1967 has transformed the state from a net exporter of pulpwood to a net importer.

A large sawmill along US 127 in Casey County processes hardwoods from the forested Knobs region here and in surrounding counties. The state's bourbon distilling industry is dependent on cooperage manufacturing plants that use large quantities of white oak supplied, in part, by mills such as this one.

FOREST FIRES

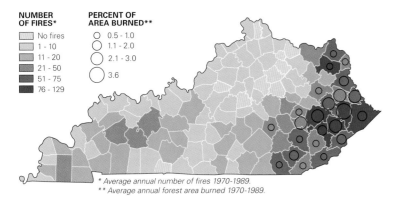

NUMBER OF FIRES*
- No fires
- 1 - 10
- 11 - 20
- 21 - 50
- 51 - 75
- 76 - 129

PERCENT OF AREA BURNED**
- ○ 0.5 - 1.0
- ○ 1.1 - 2.0
- ○ 2.1 - 3.0
- ○ 3.6

* Average annual number of fires 1970-1989.
** Average annual forest area burned 1970-1989.

Forest fires played an important role in Kentucky's early settlement, and such fires still frequent Kentucky's timberlands, although to a significantly lesser degree. Forest fires destroy notable volumes of Kentucky's hardwood resources. According to the Kentucky Division of Forestry, three-fourths of the state's forest fires can be attributed to just three factors: open burning of trash and brush that escaped control account for 31 percent of fires, arson for 28 percent, and careless smoking habits for 15 percent. Statewide, forest fires destroy 120,000 to 130,000 acres of timberland annually. The state's far eastern counties, including Floyd, Breathitt, Knott, Perry, and Pike, experience the most fires. They are also among the counties with the highest percentage of area burned. Counties that also contain parts of the Daniel Boone National Forest generally have fewer fires and a smaller area burned each year, suggesting that the U.S. Forest Service has had some success in fire prevention. The fall fire season of 1952, when severe drought conditions persisted for several months and set the stage for nearly 3,500 forest fires that burned more than 1.3 million acres of timberland, encouraged Kentuckians to organize and modernize fire prevention and response programs. Since then, Kentucky's record of forest fires is much improved. In addition to working on fire prevention and protection, Kentucky's Division of Forestry maintains tree nurseries, supplies forest information and educational programs, provides assistance to owners of forest land, and offers marketing services. These activities will prove essential in the future—Kentucky's forest wealth is now perceived to form the basis for economic development in the Commonwealth. Such sustainable economic development will require sound management now and in the future.

Each autumn, arson fires burn potentially valuable forest land in the Daniel Boone National Forest. Firefighters from Kentucky and many other states battle these blazes. Here, Indiana firefighters light a backfire in an attempt to control one of the many mountain fires.

The Agricutural Landscape

CHAPTER SEVEN

The Agricultural Landscape

Grain from western Kentucky farms is loaded on an Ohio River barge near Henderson for shipment to consumers outside the state.

Kentuckians are often proud of the state's agrarian tradition and its farming landscape, and many city residents trace their heritage to the small farms that dot the countryside. Although farm labor and agricultural products sustained the state's economy during the nineteenth century, farm earnings have declined relative to earnings from other activities. By 1992 farming ranked near the bottom of a long list of revenue-generating economic activities. For example, even though the earnings from mining dropped by about 40 percent during the decade from 1982 to 1992, mining wages still exceeded total farm earnings by 10 percent in 1992. The earnings accumulated by manufacturing activities were about 7.5 times as great as those of agriculture in the same year. And farm earnings equaled only about 3 percent of aggregate non-farm earnings for the year.

Nevertheless, the farm and the rural countryside retain a special place in the state's iconography. Though farm buildings may be relicts and fields may have long been abandoned to red cedars and hardwoods, the farm as an idea still symbolizes strong family ties to the land, as well as hard, honest work, community cohesion, and a fondly remembered past. The farm is more than a place of production, a cog in the gears of the state's economic engine. The farm is a venerable social and cultural institution, and this favored status means that farmers are not without political leverage. Such influence is no small advantage as national and state politicians struggle to accommodate increased competition from imports for the state's primary cash crop, tobacco.

The geography of productive farming in Kentucky is strikingly similar to the distribution of level to gently sloping land, to limestone bedrock, and to broad river floodplains. Land suitable for cultivation is not broadly distributed across the state but is largely confined to rather small portions of the Bluegrass, the Pennyroyal, and the Jackson Purchase Regions, as well as the Ohio Valley.

LAND USE SUITABILITY

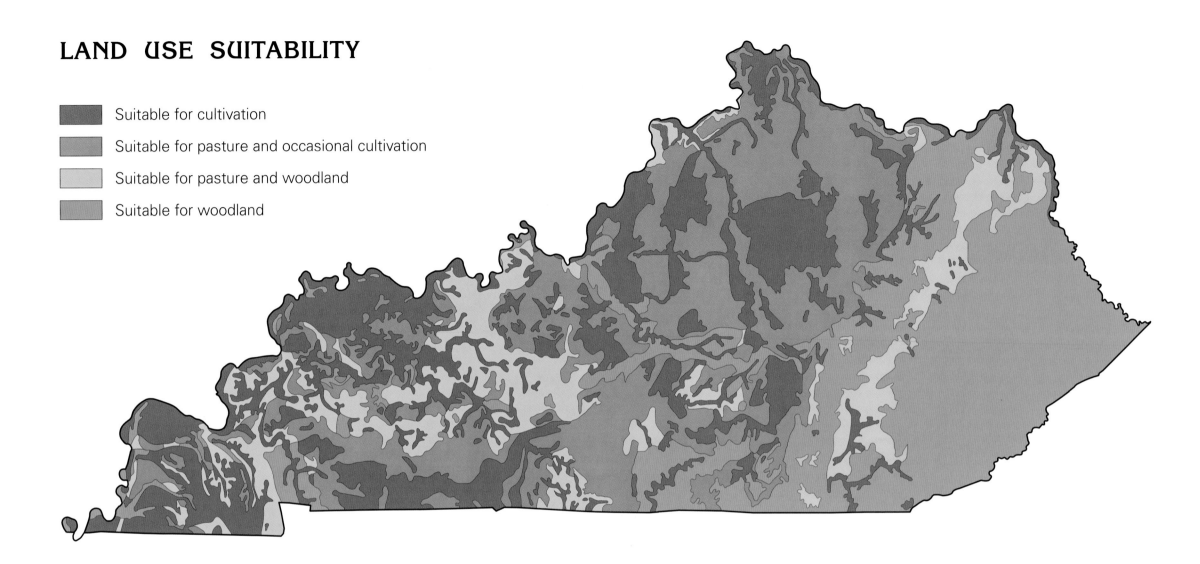

- ▮ Suitable for cultivation
- ▮ Suitable for pasture and occasional cultivation
- ▯ Suitable for pasture and woodland
- ▯ Suitable for woodland

Limestone underlies both the Inner Bluegrass and the western Pennyroyal sinkhole plain, and this foundation contributes to fertile soils. Karst sinks and underground streams readily drain these areas after heavy rainfall, and farming is not deterred by periodic flooding or standing water. Seventy-five to 80 percent of the land in these two regions is cropland. Jackson Purchase uplands, covered by a mantle of wind-deposited silt or loess, are attractive for agriculture, as are the broader stream valleys that drain west into the Mississippi. More than 80 percent of the valley bottoms and about half of the loess uplands are cropland.

The state's most productive farms—and the area where the scale of farming and crop and livestock specialties most closely resembles that of the Corn Belt to the north—can be found in the low hills and broad stream valleys that border the Ohio River in Union, Henderson, and Daviess Counties. Broad areas in these counties are fertile and nearly level, although poorly drained. The soils are stream-deposited silts and clays that accumulated during Pleistocene glacial advances. When thick ice sheets to the north—in present-day Indiana and Illinois—closed off the Ohio River and its tributaries, drainage from the Green River and adjacent streams was effectively blocked. Impounded stream waters collected across hundreds of square miles in a belt from near Sturgis and Morganfield in Union County, east to Owensboro in Daviess County, and south to Livermore in McLean County. From the still waters, silt and clay particles settled out in deep deposits of 150 feet or more. When the gla-

cial ice retreated north and the Ohio River and its Kentucky tributaries again flowed freely toward the southwest, the silts and clays became the basis for soil formation and eventually composed some of the state's most productive farmlands, although the lack of slope meant that the fields would require artificial drainage. If Kentucky has a Corn Belt, this is the place.

The state's most rugged lands have thin, acidic soils. These lands, and those areas underlain by sandstone or shale, are generally the least desirable for agriculture. Such conditions prevail throughout the eastern mountains, the eastern portion of the Pennyroyal, and the Eden Shale Hills of the Bluegrass. Only 5-7 percent of eastern Kentucky's mountain lands—primarily fertile valley floodplains—are cropland, although from the late nine-

PRIME FARMLAND

Less than 25% prime

25 - 50% prime

51 - 75% prime

76 - 100% prime

teenth century through the 1930s a much larger proportion of this land was cultivated as subsistence hillside farms. The shale and limestone lands in the Bluegrass hills and the eastern Pennyroyal are somewhat more attractive for agriculture, with 15-30 percent of the land in crops and an additional 30-75 percent in pasture, depending upon the location.

Because prime farmland is a comparatively rare but highly valued resource in the state, farmers have a vested interest in protecting and conserving it. Prime farmland is defined as an area that is flat or gently rolling, generally has a low risk of erosion, and has medium to high fertility. In the state's eastern half, prime farmland concentrates in

narrow ribbons along major stream valleys, and secondarily on the limestone plains of the Inner and Outer Bluegrass. These conditions are accentuated in the state's western half, where stream valleys become broader, loess-derived soils blanket the Jackson Purchase uplands, and Pennyroyal limestones decay into rich reddish-brown soils. Roughly 60 percent of the state's 5.8 million acres of cropland are considered "prime" by the Soil Conservation Service. Level to gently rolling land, of course, is also a primary resource for urban expansion, highways and airports, new industrial plants, and construction of dispersed residential housing. Therefore, farmers, especially those within the commuting zones of larger cities, are under substantial

Prime farmland is often associated with the fertile alluvial soils found along some of the state's rivers, as illustrated here by the Kentucky River floodplain south of Carrollton, where the Kentucky joins the Ohio River, seen in the distance.

LAND IN FARMS, 1950

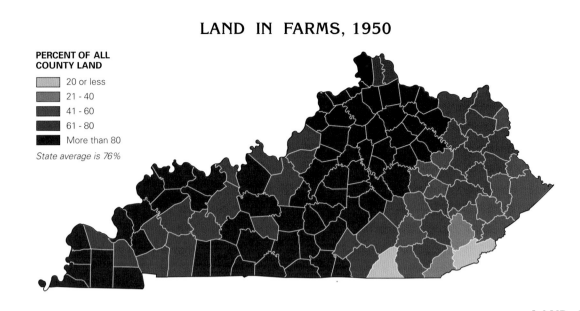

PERCENT OF ALL COUNTY LAND

- 20 or less
- 21 - 40
- 41 - 60
- 61 - 80
- More than 80

State average is 76%

LAND IN FARMS, 1992

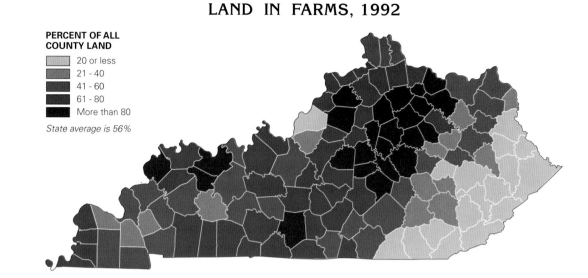

PERCENT OF ALL COUNTY LAND

- 20 or less
- 21 - 40
- 41 - 60
- 61 - 80
- More than 80

State average is 56%

AVERAGE FARM SIZE, 1950

SIZE IN ACRES

- 1 - 75
- 76 - 150
- 151 - 225
- 226 - 300
- 301 - 600

State average is 89 acres

LAND IN FARMS AND AVERAGE FARM SIZE

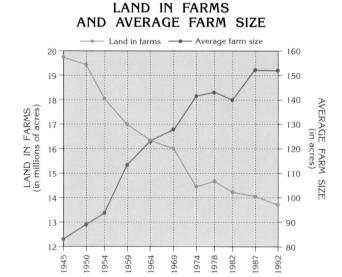

AVERAGE FARM SIZE, 1992

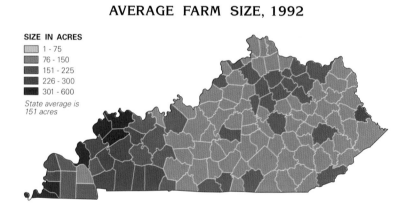

SIZE IN ACRES

- 1 - 75
- 76 - 150
- 151 - 225
- 226 - 300
- 301 - 600

State average is 151 acres

The Licking River meanders across Inner Bluegrass land near Cynthiana in Harrison County. The remnants of round hay bales leave brown circles near the barn. The bright green winter wheat fields will be ready for harvesting by June, while the open fields await corn and tobacco planting.

pressure to sell their land for nonfarm development. Between 1982 and 1987 the state lost about 350,000 acres of prime farmland when it was converted to nonfarm uses. In order to encourage farmland conservation, the state legislature passed the Agricultural District Act in 1982 that allowed farmers to protect their land from urban annexation while receiving lower property value assessments and reduced taxes. By 1991, 159 agricultural districts had been created across the state, setting aside about 167,000 acres of prime farmland.

Several districts can be found in those counties with rapidly expanding suburbs: Boone, Kenton, and Campbell Counties in northern Kentucky; Jefferson County; and Daviess and Henderson Counties in western Kentucky.

Farm Characteristics

Kentucky's farms are small, on average, when compared with the large-scale production units found

ACRES OF HARVESTED CROPLAND

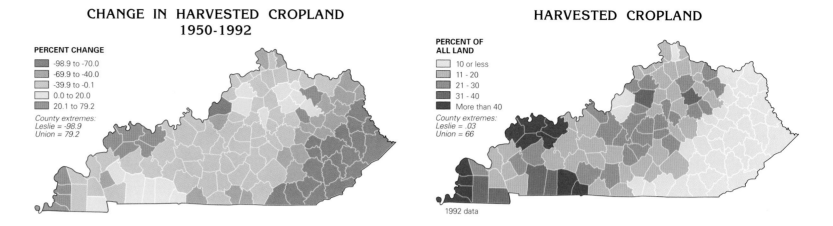

MILLIONS OF ACRES

6.0
5.5
5.0
4.5
4.0
3.5
3.0
2.5
2.0
1.5
1.0
.5
0

1959 1964 1969 1974 1978 1982 1987 1992

CHANGE IN HARVESTED CROPLAND 1950-1992

PERCENT CHANGE

-98.9 to -70.0
-69.9 to -40.0
-39.9 to -0.1
0.0 to 20.0
20.1 to 79.2

County extremes:
Leslie = -98.9
Union = 79.2

HARVESTED CROPLAND

PERCENT OF
ALL LAND

10 or less
11 - 20
21 - 30
31 - 40
More than 40

County extremes:
Leslie = .03
Union = 66

1992 data

across neighboring Corn Belt states to the north—Ohio, Indiana, Illinois, Iowa, Minnesota, and the bordering fringes of Michigan, Missouri, Nebraska, and Wisconsin. In 1992 Kentucky's average farm size was only 151 acres, of which only 55.5 acres was in harvested cropland. About 88 percent of the state's 90,281 farms claimed some harvested cropland; the remainder was primarily in pasture or woodland. Until the mid-1940s, many of Kentucky's farms were traditional general farms that produced a variety of crops and livestock, although tobacco provided the primary cash income for many. After World War II, improved transportation allowed access to broader markets, and hybrid seeds and the increasing use of fertilizers and chemicals increased production. Marginal farming operations on poor land were often consolidated or abandoned. Consequently, between 1945 and 1992 the acreage of cropland across the state declined from 10.4 to 8.88 million acres; harvested cropland declined from 5.3 to 4.4 million acres; and the number of farms dropped from 238,501 to 90,281. Generally, those farms that remained in production, or expanded their operations, were on the best land, and the economies of increased scale led to improved income; total farm income in 1992 was 145 percent of that in 1982.

In 1945 Kentucky's farms included more than 19.7 million acres of land, but by 1992 farm acreage had declined to 13.7 million acres. Farming's retreat was most pronounced on the most marginal lands. In 1950 the state average of land in farms by county exceeded 75 percent.

Between 1950 and 1992 the proportion of land in farms declined rapidly in eastern Kentucky's Appalachian counties, with a zone of counties retaining 50 percent or more of their land in farms aligning the fringing hills, the Knobs, and the Outer Bluegrass. The counties retaining the highest proportion of their land in farms by the 1990s lie in a broad diagonal swath extending south and west from Mason County on the Ohio River to the Pennyroyal, with a small but exceptionally productive outlier on the Ohio and Green Rivers in the northern part of the Western Coal Field Region.

The most precipitate drop in the number of farms from 1945 to 1992 came during the thirty-year period between 1945 and 1974, when more than 136,000 farms disappeared. The remaining farms consolidated much of the land that had been used by those that vanished. In 1945 the state's average farm size was about 83 acres, and by 1992 the average size had grown to 151 acres. Land retention meant that about 91 percent of the land that had been in farms in 1945 remained as farmland in 1992.

The proportion of farmland planted in crops, or harvested cropland, also increased from 1945 to 1992; in 1945 it was about a quarter of all farmland, and in 1992 cropland comprised more than a third of the land in farms. Harvested cropland is a good indicator of prime farmland and productive, economically viable farms that produce crops for cash sale or for livestock feed. Farms that specialize in livestock production, especially horses, often have a smaller

Many farms in the central Pennyroyal produce a variety of grains for livestock feed or cash sale. The wheat fields on these Todd County farms have been harvested recently.

PRINCIPAL CROPS

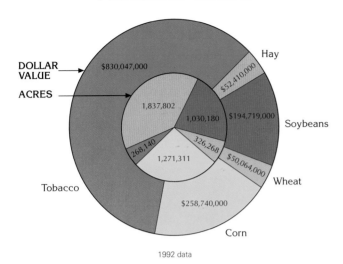

DOLLAR VALUE →

ACRES →

$830,047,000

Hay

$52,410,000

1,837,802

1,030,180

$194,719,000

Soybeans

268,140

326,268

1,271,311

$50,064,000

Wheat

Tobacco

$258,740,000

Corn

1992 data

KENTUCKY AGRICULTURAL LAND USE

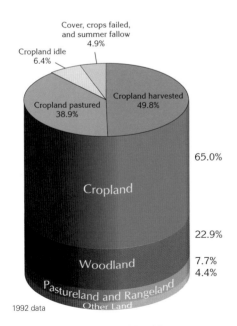

Cover, crops failed, and summer fallow
4.9%

Cropland idle
6.4%

Cropland pastured
38.9%

Cropland harvested
49.8%

65.0%

Cropland

22.9%

Woodland

7.7%
4.4%

Pastureland and Rangeland
Other Land

1992 data

Total acres = 13,665,798

A diversified farm near Monticello in Wayne County produces soybeans (the brown fields) and tobacco, and has pasture for beef cattle on irregular fields shaped by metes-and-bounds surveys.

proportion of their land in cultivated crops, while retaining substantial acreages in pasture. The highest proportion of harvested cropland remains concentrated in those regions with the most productive soils: the Corn Belt counties along the Ohio and Green Rivers—Daviess, Henderson, Union, Webster, and McLean—the Jackson Purchase, the Pennyroyal, and the Outer Bluegrass. Individual counties in these regions exceed the state average of about 17 percent of all land being harvested cropland, and some exceed 50 percent. These same areas are also the strongest retainers of cropland over time. Although total cropland declined after World War II, the trend had reversed by 1974, so that by 1992 about 83 percent of the land that had been in crops in 1945 was still being farmed as cropland.

Major Crops

Kentucky farmers raise several varieties of tobacco. Burley, used primarily in cigarette manufacturing, is grown across the entire state, although several counties in the eastern mountains have no tobac-

co growers, or fewer than ten. Tobacco is also the most important cash crop produced in the state: in 1992 state farmers produced 542,000,404 pounds of tobacco that earned $830,047,000. The crop is also an important source of cash income for those farmers who raise it, and it is often planted on the best cropland. The dependence upon tobacco as a primary income crop is suggested by the proportion of harvested cropland that it occupies. In the eastern counties, where level, fertile land is rare, a high proportion of that land is planted in tobacco. Dark tobaccos, some air-cured like burley, others fire-cured, are raised by farmers in the Jackson Purchase, the Pennyroyal, and the Green River Valley. Fire-cured tobacco is a specialty in the Purchase counties of Graves and Calloway, where the distinctive curing barns often stand sequestered from the farmstead in a grove of hardwood trees, with a pile of sawdust and sawmill wood scraps nearby. During the cure, the farmer maintains a low, smoky fire in the middle of the barn floor to raise the temperature inside the barn and to impart a distinctive color and smoke flavor to the tobacco. These tobaccos are used for blending with other tobaccos in smoking and chewing products.

TOBACCO PRODUCTION

**POUNDS PER ACRE OF
HARVESTED CROPLAND**

- [ ] Less than 1
- [ ] 1 - 100
- [ ] 101 - 200
- [ ] 201 - 300
- [ ] 301 - 500
- [ ] 501 - 640
- [ ] Missing data

County maximum:
Owsley = 640

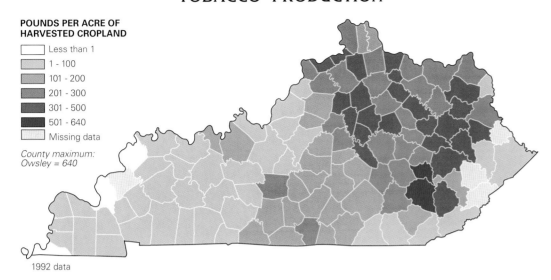

1992 data

Dark fire-cured tobacco produced in western Kentucky is cured in small barns like this one in Calloway County. When the barn has been filled with harvested tobacco, sawdust and hardwood scraps will be used to build a low fire on the barn floor. The heat and smoke will cure the tobacco over a period of several days.

Burley tobacco is often ready for harvest by mid-August. These Scott County tobacco plants stand taller than the man's head. The plants are cut off near the ground and impaled on a tobacco stick. The harvested plants—stalk and leaf together—will hang in the white tobacco barn for several months of curing.

DARK TOBACCO

POUNDS

- 5,000,000
- 2,000,000
- 1,000,000
- 500,000
- 100,000 or less

Circles are proportionally scaled according to the total number of pounds of tobacco cured.

Curing Method

AIR CURED
7,512,500 pounds

FIRE CURED
14,318,000 pounds

Kentucky Total:
21,830,500 pounds

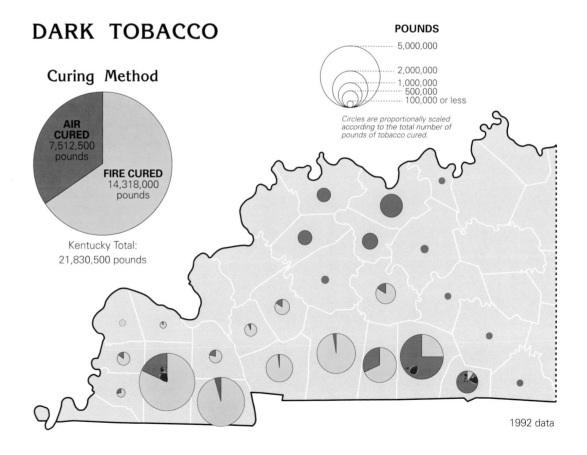

1992 data

CORN

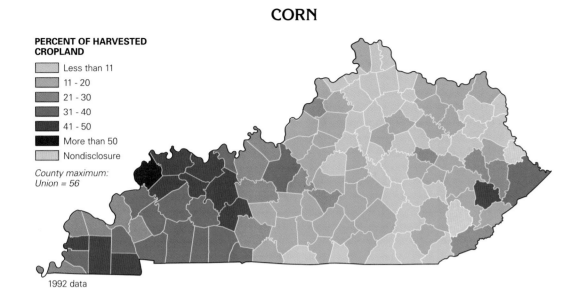

PERCENT OF HARVESTED CROPLAND

- Less than 11
- 11 - 20
- 21 - 30
- 31 - 40
- 41 - 50
- More than 50
- Nondisclosure

*County maximum:
Union = 56*

1992 data

SOYBEANS

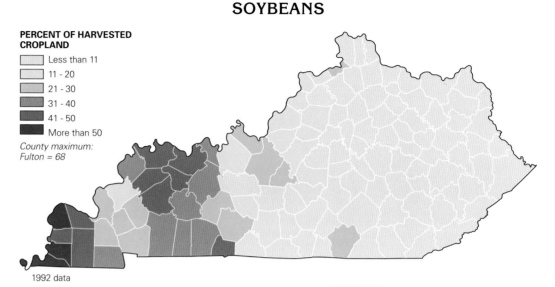

PERCENT OF HARVESTED CROPLAND

- Less than 11
- 11 - 20
- 21 - 30
- 31 - 40
- 41 - 50
- More than 50

*County maximum:
Fulton = 68*

1992 data

CORN HARVESTED FOR GRAIN

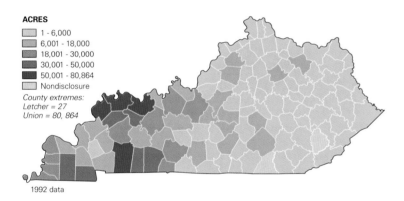

ACRES

- 1 - 6,000
- 6,001 - 18,000
- 18,001 - 30,000
- 30,001 - 50,000
- 50,001 - 80,864
- Nondisclosure

*County extremes:
Letcher = 27
Union = 80, 864*

1992 data

CORN HARVESTED FOR SILAGE

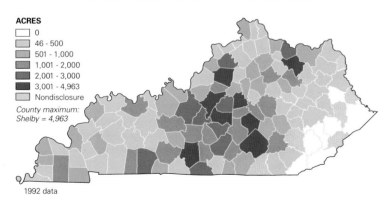

ACRES

- 0
- 46 - 500
- 501 - 1,000
- 1,001 - 2,000
- 2,001 - 3,000
- 3,001 - 4,963
- Nondisclosure

*County maximum:
Shelby = 4,963*

1992 data

WHEAT

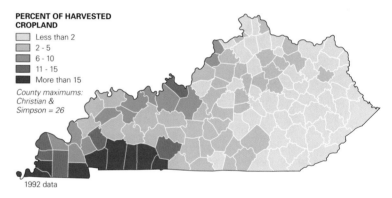

PERCENT OF HARVESTED CROPLAND

- Less than 2
- 2 - 5
- 6 - 10
- 11 - 15
- More than 15

*County maximums:
Christian &
Simpson = 26*

1992 data

Corn is a versatile commodity. As a grain it is an economical source of carbohydrate and protein if fed to livestock. When harvested green and fed to dairy cattle, it provides a nutritious silage. Or it can be manufactured into hundreds of consumer products that range from breakfast cereal and the syrup used to manufacture soft drinks to bourbon whiskey and ethanol. The latter, when used as an additive in gasoline, reduces the amount of carbon dioxide emitted by internal combustion engines. Ethanol-enriched gasoline, or gasohol, accounts for about 8 percent of the gasoline sales in the United States and 25-35 percent of sales in Kentucky. Although only a few western Kentucky counties—the Ohio and Green River counties of Daviess, Henderson, Union, Webster, and McLean—can claim a level of specialized corn, soybean, and livestock production that would roughly parallel that of the Corn Belt in the midwestern states, corn is the state's second crop in both acreage and value of agricultural products sold. Farmers planted 1,271,311 acres of corn for grain and silage in 1992, or about 29 percent of all harvested cropland. When harvested, corn sold for grain yielded a value of $258,719,000, or slightly less than 10 percent of the value of all agricultural products sold. The Western Coal Field's river valleys, the Jackson Purchase, and the Pennyroyal are the state's largest corn-producing regions, with the crop occupying 40 percent or more of the harvested cropland in many counties. These regions

CORN, SOYBEAN, AND WHEAT HARVESTS

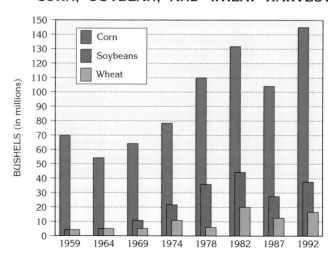

PASTURE

ACREAGE AS PERCENT OF ALL LAND IN FARMS

- 0 - 4.0
- 4.1 - 8.0
- 8.1 - 12.0
- 12.1 - 17.0
- 17.1 - 29.7

County maximum:
Martin = 29.7

1992 data

HAY

ACREAGE AS PERCENT OF ALL LAND IN FARMS

- 1.5 - 4.0
- 4.1 - 8.0
- 8.1 - 12.0
- 12.1 - 17.0
- 17.1 - 21.5

County extremes:
Fulton = 1.5
Barren = 21.5

1992 data

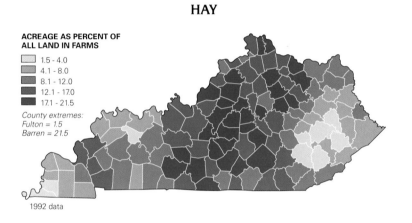

Small valley farms in the Appalachian Plateau clear land for pasture in fingerlike fields that extend upslope from the roadside farmsteads that lie along the road.

are also important hog producers. Because silage corn is used as feed for dairy cattle, the production patterns of dairy cows and silage corn are quite similar.

Soybeans are an excellent source of vegetable protein, and the crop has a wide range of uses as food for humans and feed for livestock. Beans also yield oil that can be consumed or manufactured into plastics and many other products. The soybean is a legume that fixes nitrogen on its root system and so adds to soil fertility. This makes the crop an ideal companion to corn, a high nitrogen consumer, and the two are often grown in rotation. In 1992 soybeans occupied about 23 percent of the state's harvested cropland, or 1,030,180 acres, and accounted for 7.3 percent of the value of all agricultural products sold. Although soybeans were grown in very limited quantities before World War II, soybean oil was a primary source of explosives during the war, and the price increased dramatically. In 1959 Kentucky farmers were raising 181,000 acres of soybeans that yielded just over 4 million bushels. Production increased steadily, so that by 1982 state farmers harvested more than 44.2 million bushels. Production had fallen

somewhat by 1992, to 37.8 million bushels, equal to about 26 percent of the total corn production that year.

In 1992 farmers harvested 326,268 acres of wheat, which represented 3.8 percent of the acreage of harvested cropland and about 1.9 percent of the total value of agricultural product sales. A century ago, wheat was more widely grown as a grain, and farmers often hauled their crop to a nearby mill to be ground into flour for sale or family use. While Kentucky farmers still grow some wheat, and smaller amounts of other small grains such as oats and barley, wheat is often used simply as a winter cover crop for tobacco fields. The seed is planted in late summer or early fall after the tobacco is harvested, and then it is plowed under in the spring to enhance soil quality. If allowed to mature, the crop is harvested in midsummer.

Hay is a general term for a variety of grasses and legumes—timothy and clover, for example—that are baled or chopped for livestock feed. Although hay occupied more than 1.8 million acres of Kentucky farmland in 1992, the crop's market value was less than 2 percent of the value of all agricultural commodities sold. Much of the

As farm labor has become scarce or expensive, dairy and cattle farmers have adopted the large round bale, such as these on a Mercer County farm, as a replacement for the more labor-intensive small rectangular bale. Though they are heavy, special equipment allows one person to move them easily.

CATTLE AND CALVES SOLD

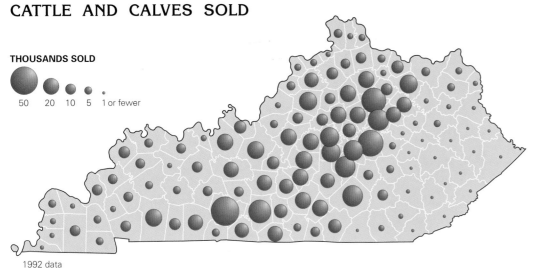

THOUSANDS SOLD

50   20   10   5   1 or fewer

1992 data

BEEF CATTLE AND DAIRY COWS

One • represents 1,000 beef cattle.
One • represents 1,000 dairy cows.

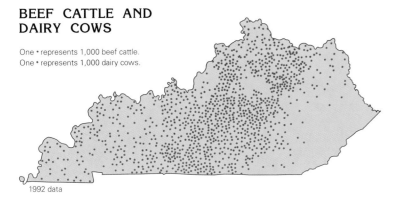

1992 data

BEEF CATTLE AND DAIRY COWS IN INVENTORY

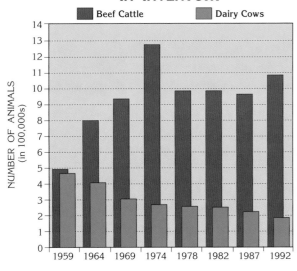

■ Beef Cattle   ■ Dairy Cows

After a heavy snow, the beef feeder cattle on this Spencer County farm track trails between shelter barns and feeding sites.

state's hay crop is not sold for cash but fed to farm cattle. Hay is a common crop across the Bluegrass and eastern Pennyroyal Regions and is associated with dairy and beef cattle production, but it is also produced in areas of marginal land. Pasture is widely distributed across the state, although farmers in the Bluegrass counties tend to keep the highest proportion of their land in pasture. Pasture is defined in different ways by the federal census, and here it includes only those lands that are not considered cropland or woodland.

Commercial Livestock

Kentucky ranked ninth in the nation in the total number of beef cattle produced in 1992 and thirteenth in the number of dairy cows. Not only do these rankings represent a change from 1974, when both beef cattle and dairy cattle resulted in a state ranking of eleventh, but Kentucky now has more beef cattle than any other state east of the Mississippi River. The number of dairy cows in inventory has been declining steadily since 1959, whereas the number of beef cattle has been increasing, the peak year being 1974. Beef cattle are an attractive option for Kentucky farmers, especially those who work part-time off the farm. Cattle can graze pastureland that might be too steep or

infertile for cropland, and a herd of beef cattle requires much less labor then does a milking dairy herd.

When the number of cattle per county is considered, a "beef belt" can be seen extending across the central Bluegrass southwest through the Pennyroyal. When the number of cattle is standardized by acres of harvested cropland, the relationship shifts further east, where farmers with limited cropland raise a few head to supplement their incomes. The total value of cattle and calves sold by state farmers in 1992 was $551,530,000, or 20.7 percent of the value of all agricultural products sold that year. Dairy products sold were valued at $266,816,000, or 10.0 percent of agricultural products sold. The state's dairy belt is more concentrated than the area of beef production. Dairy farmers in the Outer Bluegrass counties near the Louisville and Cincinnati-Northern Kentucky urban areas sell Grade A milk to bottling plants that supply the large urban markets. About 95 percent of the 2.1 billion pounds of whole milk sold in Kentucky in 1991 was Grade A milk. Manufacturing-grade milk constituted the remainder. About twenty-four processing plants in the state produce dairy products such as cheese, cottage cheese, and ice cream.

Hog production has been generally declining since 1959, with variation from one year to the next as prices rise and fall. The state ranked fourteenth in hog production in 1992 compared

HOGS AND PIGS IN INVENTORY

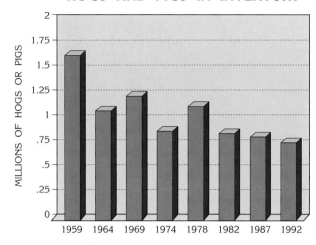

HOGS AND PIGS

One • represents 1,000 hogs or pigs.

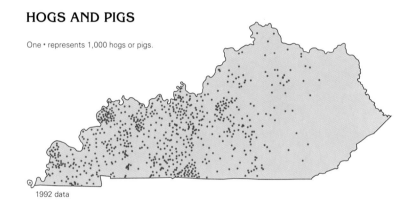

1992 data

with thirteenth in 1974. Kentucky farmers had 782,408 hogs in inventory in 1992, and they sold almost 1,465,000 as feeder pigs and for butchering. Income from hog sales was $128,774,000, or 4.8 percent of all agricultural commodity sales. Few hogs are raised in the mountains or the Bluegrass, with the notable exception of Nelson County. Instead, production is concentrated in the state's western half, especially the central Pennyroyal, the Jackson Purchase, and the Ohio and Green River counties of the Western Coal Field Region.

Chickens, especially laying hens, have long been a mainstay of the small general farm, providing self-sufficiency in eggs and meat and, if raised in sufficient number, a small cash income from eggs sold to local buyers. Although the number of laying hens declined from 5.4 million in 1959 to 2.6 million in 1992, the decline in meat-type chickens or broilers was even more dramatic, falling from 13.6 million in 1959 to 2.2 million in 1987. Since 1987, when the value of all poultry and poultry products sold by Kentucky farmers stood at $23,450,000, sales have strongly rebounded, reaching $73,194,000 in 1992. Most of this resurgence can be attributed to a rapid increase in broiler production, especially in the Jackson Purchase, where large-scale broiler facilities similar to those in Arkansas, Alabama, and Georgia have been established.

BROILERS IN INVENTORY

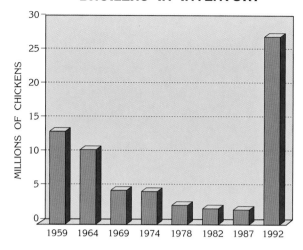

CHICKENS

One • represents 10,000 chickens.

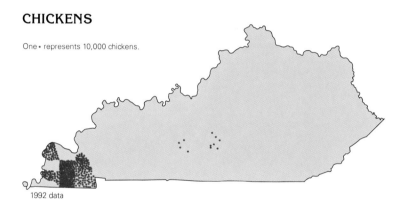

1992 data

HOGS AND PIGS SOLD

THOUSANDS SOLD

80 40 10 5 2

Indicates nondisclosure.

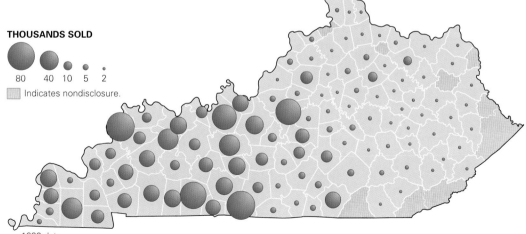

1992 data

BROILERS AND OTHER MEAT-TYPE CHICKENS SOLD

MILLIONS SOLD

13 4 2 .5 - .1

Indicates nondisclosure.

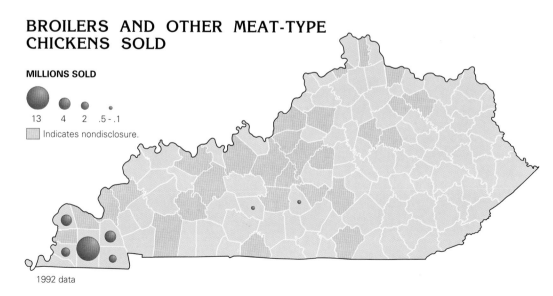

1992 data

REGISTERED THOROUGHBRED FOALS

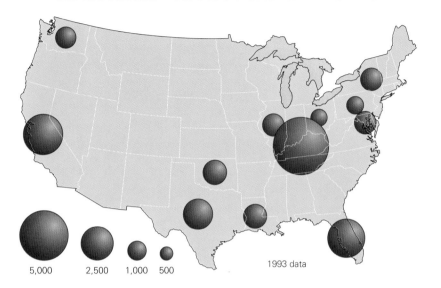

5,000   2,500   1,000   500

1993 data

Kentucky, of course, is famed for the production of Thoroughbred, Standardbred, and other recreational horse breeds. The state's total horse inventory stood at 78,083 in 1992, and 13,264 were sold for $190,413,000, or an average of more than $14,000 per animal. Income from horse sales comprised 7.1 percent of all agricultural product sales in 1992. Horses can be found in two distinct distributions. Racehorse breeding and training is concentrated in the Inner Bluegrass in Fayette, Woodford, Bourbon, and surrounding counties, with some breeding farms near Louisville in Oldham County. Thoroughbreds and Standardbreds are racehorses, and the farms that raise them often have enormous investments in land, buildings, fences, and breeding stock. Horses of various breeds are also bred and trained for recreational riding or show competitions. Often people residing in urban areas will own a horse that they board on a farm near their city of residence, and riding stables are often suburban or associated with large recreational attractions such as Lake Cumberland. This pattern is seen in the number of horses in inventory in counties that are urban or are adjacent to urban areas. The anomalous high values in Pike and adjoining counties are related to the very small amount of harvested cropland in those counties.

The largest concentration of Thoroughbred and Standardbred horse farms in the world can be found in Kentucky's Inner Bluegrass Region. The buildings on this farm between Versailles and Frankfort illustrate specialized uses; stallions, mares, and yearlings have their own separate paddocks, pastures, and barns.

HORSES

One • represents 300 horses.

HORSES TO HARVESTED CROPLAND

NUMBER OF HORSES PER THOUSAND ACRES

Fewer than 20
20 - 50
51 - 100
101 - 200
More than 200

County maximum:
Fayette = 378

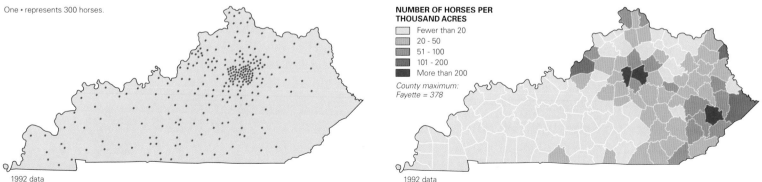

1992 data   1992 data

MARKET VALUE OF ALL AGRICULTURAL PRODUCTS SOLD

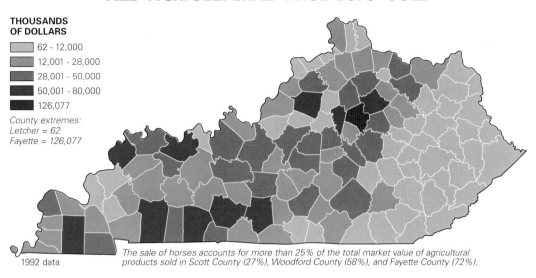

THOUSANDS OF DOLLARS

- 62 - 12,000
- 12,001 - 28,000
- 28,001 - 50,000
- 50,001 - 80,000
- 126,077

County extremes:
Letcher = 62
Fayette = 126,077

1992 data

The sale of horses accounts for more than 25% of the total market value of agricultural products sold in Scott County (27%), Woodford County (58%), and Fayette County (72%).

CROP YIELD

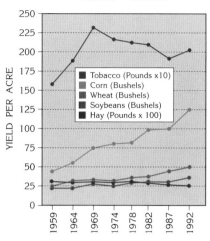

YIELD PER ACRE

- Tobacco (Pounds x10)
- Corn (Bushels)
- Wheat (Bushels)
- Soybeans (Bushels)
- Hay (Pounds x 100)

1959 1964 1969 1974 1978 1982 1987 1992

CROP VALUE

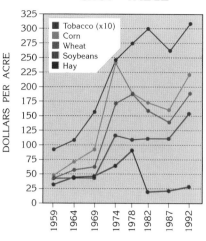

DOLLARS PER ACRE

- Tobacco (x10)
- Corn
- Wheat
- Soybeans
- Hay

1959 1964 1969 1974 1978 1982 1987 1992

Farm Income, Investment, and Ownership

The market value of all agricultural commodities sold in Kentucky has increased steadily since 1959, and most rapidly since 1974. Sales increased from $518 million in 1959 to $2.66 billion in 1992. The average value of products sold per farm, $3,431 in 1959, rose to about $29,500 in 1992. Increases are in part the result of farm consolidation and modest increases in the amount of harvested cropland. The number of farms fell from 150,986 in 1959 to 90,281 in 1992. Average farm size increased from 113 acres to 151 acres, and the state's total acreage of harvested cropland rose by more than 400,000 acres during the same period. The increase in value of products sold is also attributable to dramatic increases in crop yields brought about by increases in the application of fertilizers and chemicals, as well as by greater mechanization. For example, while corn acreage declined by more than 400,000 acres during the 1959-92 period (1,581,268 to 1,166,234), yields per acre increased almost threefold (from 44.3 bushels per acre in 1959 to 124.5 bushels in 1992). Soybean acreage increased fivefold, and yields went from 22.2 to 36.7 bushels per acre during the period. But most significant, perhaps, was the increase in tobacco acreage and yield. Total tobacco acre-

age increased from 211,692 to 268,140, and the harvest rose from 335 million to 542 million pounds, representing a pounds-per-acre increase from 1,580 in 1959 to 2,020 in 1992. Regions with the highest market value of commodities sold continued to be the Bluegrass, the Pennyroyal, the Jackson Purchase, and the Ohio and Green River counties of the Western Coal Field. With the exception of hay, other major farm crops have increased sharply in value since 1987.

VALUE OF ALL KENTUCKY AGRICULTURAL PRODUCTS SOLD

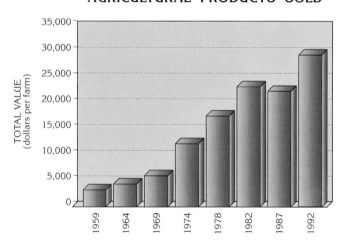

TOTAL VALUE (dollars per farm)

1959 1964 1969 1974 1978 1982 1987 1992

MARKET VALUE OF ALL KENTUCKY AGRICULTURAL PRODUCTS SOLD

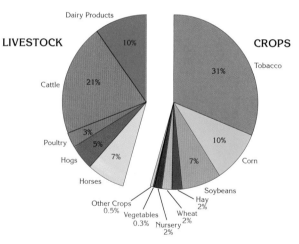

LIVESTOCK

- Dairy Products 10%
- Cattle 21%
- Poultry 3%
- Hogs 5%
- Horses 7%
- Other Crops 0.5%
- Vegetables 0.3%

CROPS

- Tobacco 31%
- Corn 10%
- Soybeans 7%
- Hay 2%
- Wheat 2%
- Nursery 2%

1992 State Total: $2,663,702,000

A dairy processing plant off Interstate 64 near Winchester, Clark County.

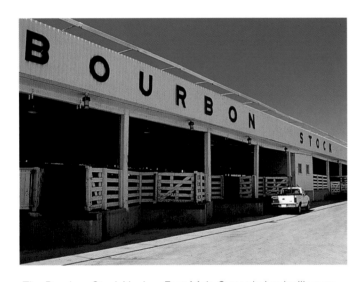

The Bourbon Stock Yard on East Main Street in Louisville was established near the historic Butchertown neighborhood on Beargrass Creek. Livestock auctioned here may be sold to farmers seeking stock to feed and fatten, or to meat processors.

DAIRY PRODUCTS SOLD

THOUSANDS
OF DOLLARS

- None
- 19 - 1,200
- 1,201 - 4,000
- 4,001 - 7,000
- 7,001 - 10,000
- 10,001 - 14,964
- Missing data

*County maximum:
Barren = 14,964*

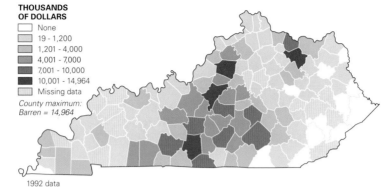

1992 data

MARKET VALUE OF DAIRY PRODUCTS SOLD

PERCENT OF ALL
AGRICULTURAL PRODUCTS

- None
- Less than 5.0
- 5.0 - 10.0
- 10.1 - 15.0
- 15.1 - 25.0
- More than 25.0

*County maximum:
Adair = 34.9*

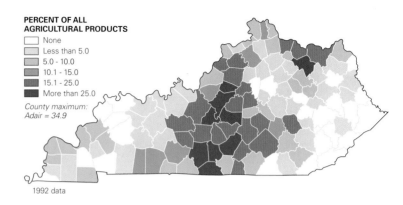

1992 data

HOGS AND PIGS SOLD

THOUSANDS
OF DOLLARS

- 2 - 1,000
- 1,001 - 2,800
- 2,800 - 5,500
- 5,501 - 8,000
- 11,091
- Nondisclosure

*County extremes:
Magoffin = 2
Fleming = 11,091*

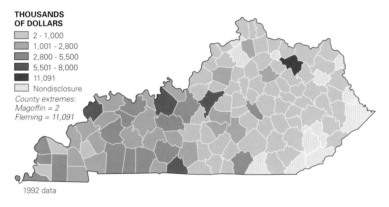

1992 data

MARKET VALUE OF HOGS AND PIGS SOLD

PERCENT OF ALL
AGRICULTURAL PRODUCTS

- Less than 2.0
- 2.0 - 6.0
- 6.1 - 11.0
- 11.1 - 18.0
- 18.1 - 32.5
- Nondisclosure

*County maximum:
Allen = 32.5*

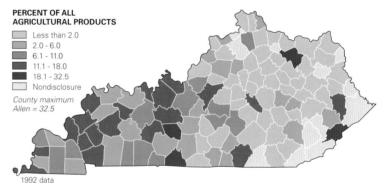

1992 data

CATTLE AND CALVES SOLD

THOUSANDS
OF DOLLARS

- 13 - 2,500
- 2,501 - 6,400
- 6,401 - 11,000
- 11,001 - 16,000
- 16,001 - 21,755

*County extremes:
Leslie = 13
Bourbon = 21,755*

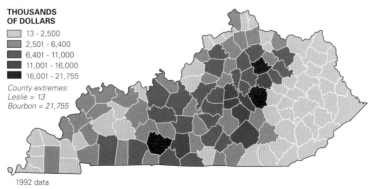

1992 data

MARKET VALUE OF CATTLE AND CALVES SOLD

PERCENT OF ALL
AGRICULTURAL PRODUCTS

- Less than 15.0
- 15.0 - 25.0
- 25.1 - 35.0
- 35.1 - 45.0
- More than 45.0

*County extremes:
Powell = 2
Bell = 79.2*

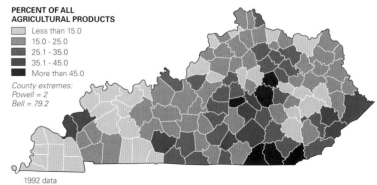

1992 data

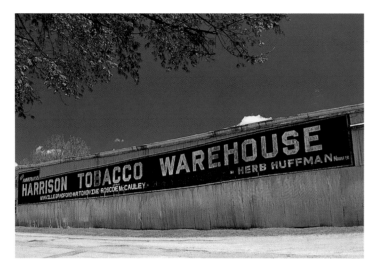

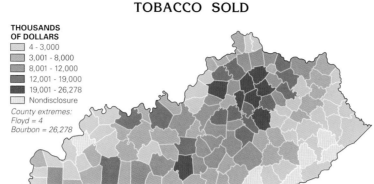

TOBACCO SOLD

THOUSANDS
OF DOLLARS
- 4 - 3,000
- 3,001 - 8,000
- 8,001 - 12,000
- 12,001 - 19,000
- 19,001 - 26,278
- Nondisclosure

County extremes:
Floyd = 4
Bourbon = 26,278

1992 data

MARKET VALUE OF TOBACCO SOLD

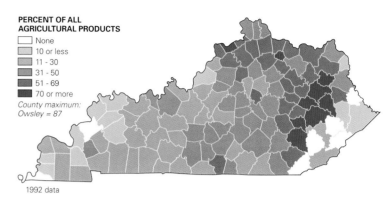

PERCENT OF ALL
AGRICULTURAL PRODUCTS
- None
- 10 or less
- 11 - 30
- 31 - 50
- 51 - 69
- 70 or more

County maximum:
Owsley = 87

1992 data

Burley tobacco farmers bring their cured crop to an auction warehouse, such as this one in Cynthiana, to be inspected and graded according to federal guidelines. Beginning after Thanksgiving, the current year's tobacco supply is auctioned to buyers across the state's Burley production belt.

Farmer's markets, such as this one in Lexington, help stabilize the fresh produce market for local farmers, a critical element in the success of small-scale horticultural farming.

VEGETABLES SOLD

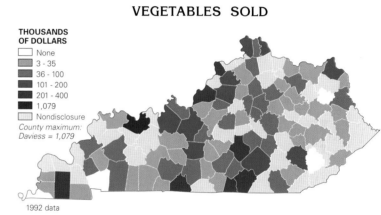

THOUSANDS
OF DOLLARS
- None
- 3 - 35
- 36 - 100
- 101 - 200
- 201 - 400
- 1,079
- Nondisclosure

County maximum:
Daviess = 1,079

1992 data

During the thirty-three-year period of 1959-92, the relative focus of Kentucky farming also shifted from livestock to crops, as reflected in the proportion of total market value derived from commodity sales. In 1959 the market value of livestock accounted for 51 percent of total farm sales, crops 48 percent. By 1992 this relationship had reversed, in part because of the labor requirements associated with raising livestock: the market value of livestock fell to 46 percent, while the value of crops rose to 55 percent.

The market value of all agricultural products sold stood at $2.66 billion in 1992. Of that total, livestock, poultry, and their products accounted for $1.2 billion and all crops $1.45 billion. The cattle and calves sold comprised 46 percent of all livestock, poultry, and product sales; dairy, 22 percent; and hogs, 11 percent. Income from crops was dominated by tobacco, which accounted for 57 percent of the market value of all crops sold (or 31 percent of all agricultural products sold). Corn and soybeans garnered 18 and 13 percent of total crop sales respectively. Vegetables are viewed by many as the logical replacement for tobacco in the Kentucky farming economy as foreign competition places the state's burley and dark tobaccos at a price disadvantage. Yet the market value of vegetables sold in 1992 accounted for only about 0.5 percent of the value of all crops sold.

Considering farming's centrality to Kentucky's economy and cultural traditions, it is surprising that the net cash returns per farm are relatively low. The highest returns accrue within the regions with the most extensive acreages of prime farmland. In more than thirty counties, the average farm has an annual net cash return of less than five thousand dollars. Government payments are a form of farm income, and the largest payments are made to the farmers on the most productive lands.

Farm income is moderated by a variety of expenditures, including purchases of feed, livestock, insurance, machinery, and numerous other items. In 1992 the purchase of feed for livestock accounted for about 13.5 percent of total state farm production expenses of $1.8 billion; purchasing livestock comprised another 13.2 percent. For those items most closely associated with increased intensification of crop and livestock production and increased crop yields, hired labor cost state farmers $202.5 million (11.1 percent); fertilizer, $176.8 million (9.7 percent); and agricultural chemicals, $70.8 million (3.9 percent). Commercial fertilizers are heavily used in tobacco and corn production. Anhydrous ammonia, a gaseous nitrogen fertilizer, cost about $200 per ton

FARM EARNINGS

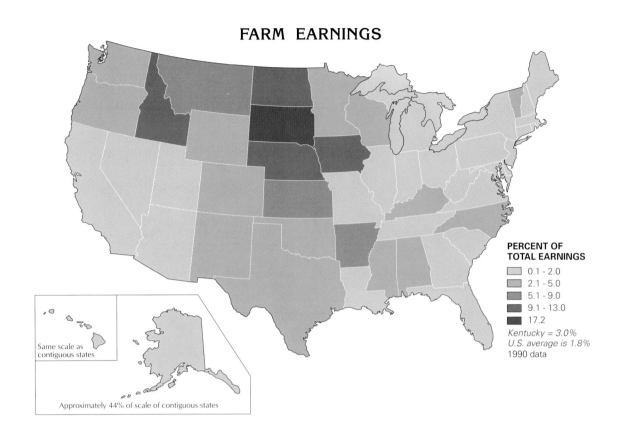

PERCENT OF TOTAL EARNINGS

- 0.1 - 2.0
- 2.1 - 5.0
- 5.1 - 9.0
- 9.1 - 13.0
- 17.2

Kentucky = 3.0%
U.S. average is 1.8%
1990 data

Same scale as contiguous states

Approximately 44% of scale of contiguous states

GOVERNMENT PAYMENTS RECEIVED

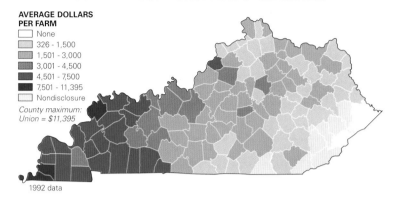

AVERAGE DOLLARS PER FARM

- None
- 326 - 1,500
- 1,501 - 3,000
- 3,001 - 4,500
- 4,501 - 7,500
- 7,501 - 11,395
- Nondisclosure

County maximum:
Union = $11,395

1992 data

NET CASH RETURNS

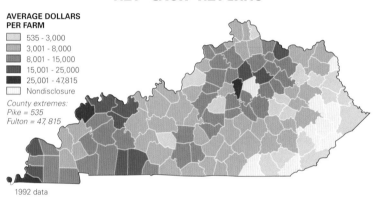

AVERAGE DOLLARS PER FARM

- 535 - 3,000
- 3,001 - 8,000
- 8,001 - 15,000
- 15,001 - 25,000
- 25,001 - 47,815
- Nondisclosure

County extremes:
Pike = 535
Fulton = 47,815

1992 data

in 1992, for example, and superphosphate was $214 per ton. Chemicals such as herbicides and insecticides are also relatively expensive. A five-gallon can of Treflan, a herbicide, cost $150 in 1992.

The most intensive use of hired labor occurs on tobacco farms and the horse farms of the Inner Bluegrass. While the ratio of labor expenses to total farm expenses is high in some mountain counties, farms there are generally small, with small total expenses. Kentucky's tobacco farmers require substantial labor support during harvest, usually August and September, when the crop is cut in the field, loaded onto wagons, and hung in air-curing barns. Labor availability has fallen in recent years, and farmers are increasingly dependent on migrant labor, especially Hispanic workers who migrate to the state each summer from Mexico to work in tobacco.

Tractors are a measure of farm mechanization. Large crop and livestock farms generally have at least two tractors: one with high horsepower to pull large tillage and planting equipment, and

a smaller one to use for towing trailers and other utility work. Generally, the larger, more profitable farms in western Kentucky have a larger number of tractors per farm, and because those farms are larger and have large-scale equipment, the tractors are higher horsepower and thus more expensive. In 1992, for example, a medium-sized 110-horsepower tractor cost forty-seven thousand dollars or more. Large tractors of 175 horsepower or more might cost more than one hundred thousand dollars. Because large tractors are so expensive, farmers tend to repair them and keep them for several years before considering replacement. In 1992, 88.8 percent of the state's tractors were more than five years old. Those farmers who purchased new tractors between 1988 and 1992 were scattered across the state, with proportionally larger numbers in those counties with a small number of farms or in urban counties. For example, Jefferson County had 564 farms averaging 79 acres each in 1992. Union County, on the Ohio River bottomlands, had 387 farms that averaged 508 acres each. Union County farmers purchased only 90 tractors between

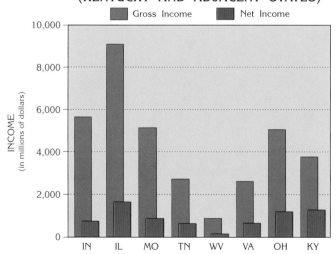

FARM INCOME, 1992
(KENTUCKY AND ADJACENT STATES)

Gross Income   Net Income

INCOME (in millions of dollars)

IN   IL   MO   TN   WV   VA   OH   KY

TOTAL FARM EXPENSES

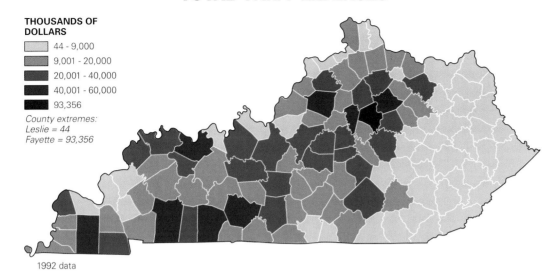

THOUSANDS OF
DOLLARS

- 44 - 9,000
- 9,001 - 20,000
- 20,001 - 40,000
- 40,001 - 60,000
- 93,356

County extremes:
Leslie = 44
Fayette = 93,356

1992 data

HISPANIC MIGRANT FARM WORKERS

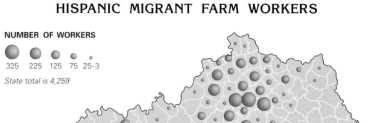

NUMBER OF WORKERS

○ ○ ○ ○ ∘
325 225 125 75 25-3

State total is 4,259

1992 data

HIRED LABOR EXPENDITURES

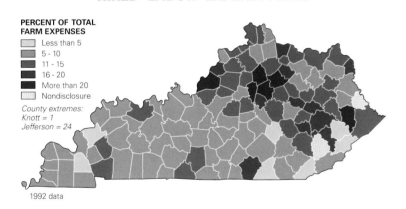

PERCENT OF TOTAL
FARM EXPENSES

- Less than 5
- 5 - 10
- 11 - 15
- 16 - 20
- More than 20
- Nondisclosure

County extremes:
Knott = 1
Jefferson = 24

1992 data

CHEMICAL EXPENDITURES

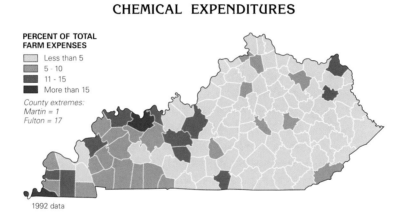

PERCENT OF TOTAL
FARM EXPENSES

- Less than 5
- 5 - 10
- 11 - 15
- More than 15

County extremes:
Martin = 1
Fulton = 17

1992 data

FARM PRODUCTION
EXPENSES

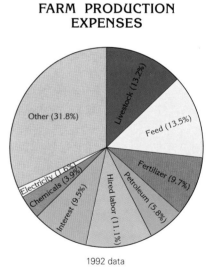

Other (31.8%)
Livestock (13.2%)
Feed (13.5%)
Fertilizer (9.7%)
Petroleum (5.8%)
Hired labor (11.1%)
Interest (9.5%)
Chemicals (3.9%)
Electricity (1.6%)

1992 data

FEED EXPENDITURES

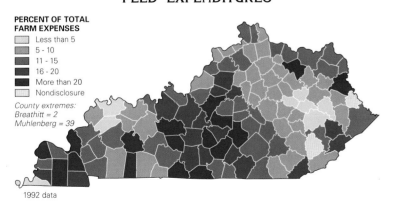

PERCENT OF TOTAL
FARM EXPENSES

- Less than 5
- 5 - 10
- 11 - 15
- 16 - 20
- More than 20
- Nondisclosure

County extremes:
Breathitt = 2
Muhlenberg = 39

1992 data

FERTILIZER EXPENDITURES

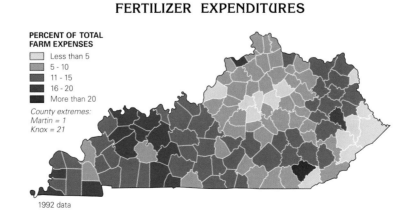

PERCENT OF TOTAL
FARM EXPENSES

- Less than 5
- 5 - 10
- 11 - 15
- 16 - 20
- More than 20

County extremes:
Martin = 1
Knox = 21

1992 data

TRACTORS

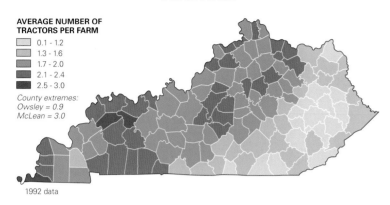

AVERAGE NUMBER OF TRACTORS PER FARM
- 0.1 - 1.2
- 1.3 - 1.6
- 1.7 - 2.0
- 2.1 - 2.4
- 2.5 - 3.0

*County extremes:
Owsley = 0.9
McLean = 3.0*

1992 data

NEW FARM TRACTORS

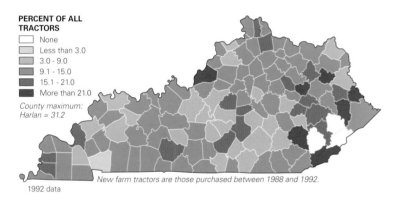

PERCENT OF ALL TRACTORS
- None
- Less than 3.0
- 3.0 - 9.0
- 9.1 - 15.0
- 15.1 - 21.0
- More than 21.0

*County maximum:
Harlan = 31.2*

New farm tractors are those purchased between 1988 and 1992.

1992 data

VALUE OF FARMLAND AND BUILDINGS

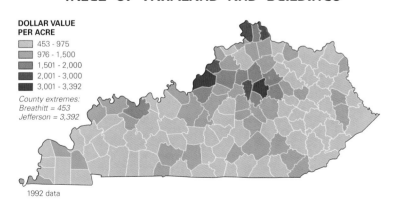

DOLLAR VALUE PER ACRE
- 453 - 975
- 976 - 1,500
- 1,501 - 2,000
- 2,001 - 3,000
- 3,001 - 3,392

*County extremes:
Breathitt = 453
Jefferson = 3,392*

1992 data

Hispanic migrant farm workers load cut tobacco in a field near Cynthiana, Harrison County. Several thousand migrant workers help bring in the tobacco crop in Kentucky each year.

VALUE OF KENTUCKY FARMLAND AND BUILDINGS

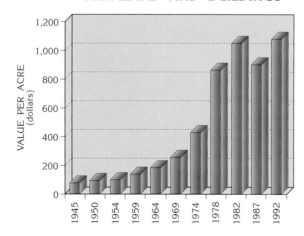

VALUE PER ACRE (dollars)

1945 1950 1954 1959 1964 1969 1974 1978 1982 1987 1992

1988 and 1992, and only 8.9 percent of these were smaller than 40 horsepower. Jefferson County farmers bought 201 tractors during the same period, and 49 percent were smaller than 40 horsepower. Small tractors are suitable for small-acreage farms that specialize in vegetables, nursery crops, tobacco, or horses.

The value of land and buildings on Kentucky farms began to accelerate in the early 1970s. Much of the United States experienced this increase; in the Corn Belt, prices for cropland exceeded three thousand dollars per acre. By 1982 Kentucky farmland cost more than one thousand dollars per acre on average, a value that it maintained for the next decade. The highest-valued land in the state centered on the Inner Bluegrass and in urbanizing counties, where developers converted farmland to subdivisions. Suburban encroachment has escalated land values from Ashland to Paducah. The highest-valued farmland in the state continues to be in Fayette County and the surrounding counties, where land for horse farms is commonly valued at eight thousand dollars per acre or more. Other factors also increase the demand for farmland, such as recreational development and the development of interstate highways and state parkways.

FARM POPULATION

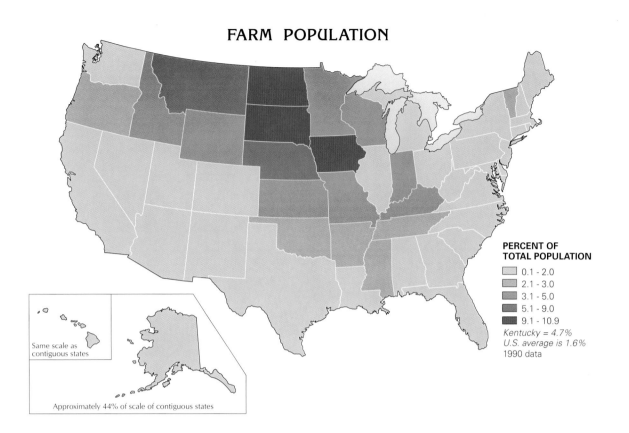

PERCENT OF TOTAL POPULATION
- 0.1 - 2.0
- 2.1 - 3.0
- 3.1 - 5.0
- 5.1 - 9.0
- 9.1 - 10.9

Kentucky = 4.7%
U.S. average is 1.6%
1990 data

Same scale as contiguous states

Approximately 44% of scale of contiguous states

FARMING AS A PRINCIPAL OCCUPATION

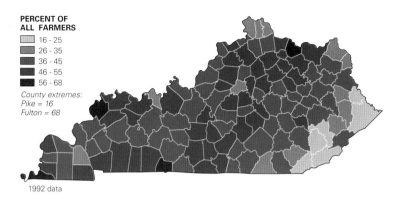

PERCENT OF ALL FARMERS
- 16 - 25
- 26 - 35
- 36 - 45
- 46 - 55
- 56 - 68

County extremes:
Pike = 16
Fulton = 68

1992 data

PART-TIME FARMERS

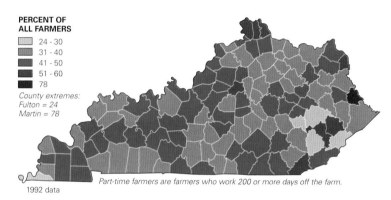

PERCENT OF ALL FARMERS
- 24 - 30
- 31 - 40
- 41 - 50
- 51 - 60
- 78

County extremes:
Fulton = 24
Martin = 78

1992 data

Part-time farmers are farmers who work 200 or more days off the farm.

A part-time farmer is one who works off the farm for a substantial number of days each year. Working off the farm allows the farmer who has a small operation with limited net cash returns to compensate by obtaining outside income. Often part-time farmers will commute to nearby towns and cities to work in manufacturing plants or service businesses. Part-time farming has increased from about 20 percent of all state farmers in 1959 to about 40 percent in 1992. The federal agricultural census reports a farmer's principal occupation. With so many part-time farmers in the state, it is not unusual for a quarter or more of the farmers in any county to consider themselves as having an occupation other than farming. Across the state, only about 43 percent of the people who lived on farms in 1992 listed their occupation as farming, with the highest proportion residing in the Bluegrass, the Pennyroyal, the Jackson Purchase, and the Ohio and Green River counties of the Western Coal Field.

In 1940, tenants operated about one-third of Kentucky's farms. By 1992 fewer than 9 percent of all farms were oper-ated by tenants. The state's highest tenancy levels tend to occur in those counties where the value of land and buildings is highest. Tenancy may approach or exceed 12 percent in sections of the Bluegrass, the Pennyroyal, and the Ohio and Green River counties of the Western Coal Field. Tenants may work for cash wages or a share of the proceeds from the sale of farm products.

If Kentucky's farmers retired when they reached sixty-five years of age, then a quarter of the state's 90,281 farms would have to be sold, rented, or shut down. A few subtle patterns can be discerned in the distribution of farmers aged sixty-five or older. The highest proportion in the retirement age bracket are in some mountain counties and the east central Bluegrass. A cluster of western Bluegrass counties with a low proportion of farmers at or above retirement age coincides with a section where dairy farms predominate and Catholicism tends to be the religious preference of the population. Generally, an influx of young people into farming reduces the average age of a county's farm population.

KENTUCKY'S PART-TIME FARMERS

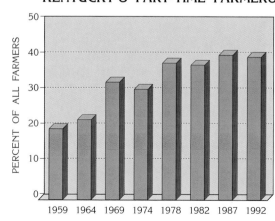

TENANT FARMS

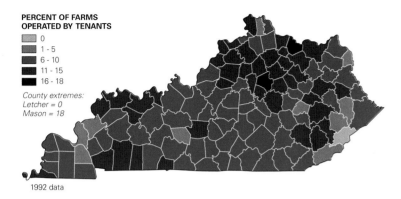

**PERCENT OF FARMS
OPERATED BY TENANTS**

- 0
- 1 - 5
- 6 - 10
- 11 - 15
- 16 - 18

*County extremes:
Letcher = 0
Mason = 18*

1992 data

FARMERS AGED 65 AND OLDER

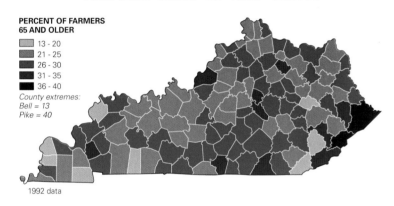

**PERCENT OF FARMERS
65 AND OLDER**

- 13 - 20
- 21 - 25
- 26 - 30
- 31 - 35
- 36 - 40

*County extremes:
Bell = 13
Pike = 40*

1992 data

KENTUCKY FARM TENURE, 1992
(principal occupation - farming)

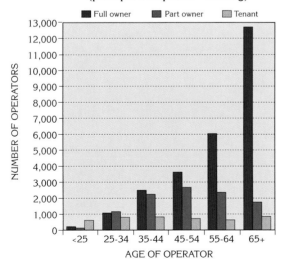

Legend: ■ Full owner ■ Part owner ▣ Tenant

Y-axis: NUMBER OF OPERATORS (0 to 13,000)
X-axis: AGE OF OPERATOR (<25, 25-34, 35-44, 45-54, 55-64, 65+)

MARIJUANA

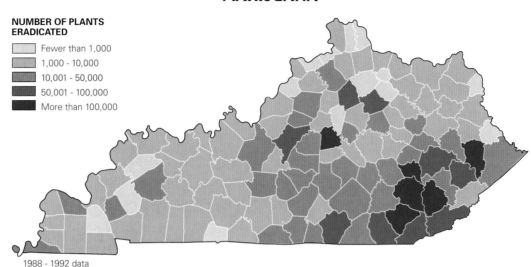

**NUMBER OF PLANTS
ERADICATED**

- Fewer than 1,000
- 1,000 - 10,000
- 10,001 - 50,000
- 50,001 - 100,000
- More than 100,000

1988 - 1992 data

Nontraditional Farm Products

During the nineteenth century, hemp was Kentucky's leading cash crop. The Bluegrass Region was the state's primary hemp producer, and Lexington was a center of hemp fiber processing. Several plants there manufactured rope, burlap sacks, and cotton bale bagging. During World War II licensed farmers grew the crop as strategic war matériel. Because hemp, or marijuana, reseeds itself, the plant grows readily in the wild. During the 1970s, federal drug enforcement authorities began to interdict marijuana shipments into the United States from Latin America and Asia. As illegal imports fell, Kentucky's rural countryside began to produce for the illicit market. In part because of favorable environmental conditions, in part because of a central location relative to large urban markets, and in part because of nagging unemployment, poverty, and a lack of income alternatives, the production of illegal marijuana has increased. Officials estimate that Kentucky is now the fourth-largest producer of marijuana in the nation. A single plant is worth up to two thousand dollars in street value. Federal and state law enforcement agencies cooperate in eradication programs each growing season. In 1992 officials located and destroyed a total of 911,000 plants.

Marijuana is Kentucky's most important illegal crop. State and federal law enforcement officers maintain a vigorous eradication program aimed at destroying the crop before it can be harvested and marketed. Here Kentucky State Police pull up plants at a remote Scott County site.

Manufacturing

Previous page: A bourbon distillery near Bardstown in Nelson County.

CHAPTER EIGHT

Manufacturing

Characteristics of Manufacturing

Manufacturing has become big business in Kentucky. In 1990 the manufacturing sector generated nearly $17 billion, representing one-quarter of the state's total gross product. More than one-fifth of the state's personal income earnings, and one-fifth of its nonagricultural employment, came from manufacturing industries in the early 1990s. In 1993, 293,696 persons were employed in manufacturing in Kentucky, and industrial production accounted for 94 percent of the value of all products from the state destined for foreign markets. These industrial exports earned more than $4.45 billion that year, representing an increase of 70 percent over the 1989 figure.

The presence of foreign ownership increased markedly in the state, especially after 1985, when Toyota Motor Manufacturing announced the construction of a major production plant in Georgetown (Scott County) to produce Camry automobiles. By 1994 more than fifty thousand workers were employed in some 240 foreign-owned plants. That is, about one of every six Kentucky manufacturing employees worked in a plant that was wholly or partly owned by a foreign company.

Kentucky's excellent location relative to markets, its skilled and productive labor force, and its excellent business climate, combined with the accessibility of the nation's transportation networks, have undoubtedly affected the increases in the number of manufacturing establishments and in manufacturing revenues. The state government has made special efforts to attract industry through a variety of economic incentive programs, including the establishment of ten state-designated enterprise zones. Companies in these zones receive exemptions from state and local taxes and other benefits to encourage investment and the creation of employment. The ten zones comprise the four largest cities—Louisville, Lexington, Owensboro, and Covington—as well as Ashland and

The Kingsford Products manufacturing plant near Burnside in Pulaski County employs about 110 people who convert hardwood sawdust into briquets that are packaged and shipped to retail outlets across the nation.

MANUFACTURING ESTABLISHMENTS

Each • represents 3 establishments

━🛡75 Interstate highway

PURCHASE PKWY Parkway

············ Connector highway

CSX Railroad

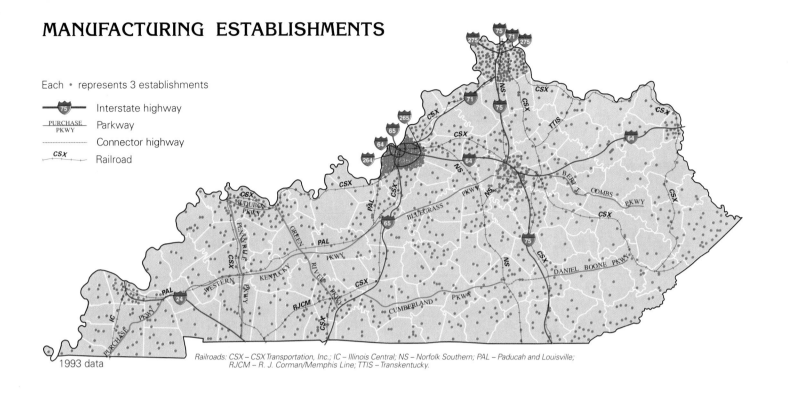

1993 data

Railroads: CSX – CSX Transportation, Inc.; IC – Illinois Central; NS – Norfolk Southern; PAL – Paducah and Louisville; RJCM – R. J. Corman/Memphis Line; TTIS – Transkentucky.

One of Louisville's largest manufacturing plants, the Ford Motor truck plant, with 3,100 employees, is the sixth-largest manufacturing plant in Kentucky. Opened in 1968, the plant has immediate access to Interstates 64, 65, and 71 and nearby markets.

Knox County in eastern Kentucky; Paducah, Hickman, and Hopkinsville in western Kentucky; and Campbell County, a large multijurisdictional zone in northern Kentucky.

Other new programs have also enhanced Kentucky's reputation as a potential location for industry. The Kentucky Rural Economic Development Authority (KREDA), established in 1989, uses corporate and personal income tax withholdings to finance projects for manufacturing companies that agree to locate in rural areas where unemployment rates have traditionally been high. KREDA projects span the rural counties of the state, but a disproportionate share of the projected employment will be located in the rural counties of western Kentucky. The Kentucky Jobs Development Authority (KJDA), established in 1992, offers tax credits to businesses in service or technology sectors that sell at least 75 percent of their output outside the state and that employ a minimum of twenty-five workers. Also, the Kentucky Industrial Development Authority (KIDA) allows eligible manufacturers to recoup up to the entire cost of their infrastructure—land, buildings, utility installation— through corporate tax credits. By the end of February 1995, incentives had been approved for some 360 businesses through these programs, and more than forty thousand new jobs were projected from these businesses. KREDA alone was projected to generate nearly two-thirds of this employment, or more than twenty-six thousand jobs.

Kentucky is within six hundred miles—or one day's drive—of nearly 70 percent of the nation's population and almost three-quarters of its manufacturing production. Five interstate highways (the north-south I-75, I-65, and I-71 routes and the east-west I-64 and I-24 routes) have encouraged the location of new manufacturing plants throughout the state and have facilitated access to the northern industrial belt, as well as to the fast-growing markets and manufacturing plants in the Southeast and Midwest. Nearly two-thirds of Kentucky's 293,696 manufacturing employees and 63.7 percent of the state's more than forty-five hundred manufacturing establishments are located in counties served by interstate highways. In addition to the interstate highway system and seven state parkways, nearly twenty-four hundred miles of mainline railroad track provide access to markets for Kentucky's industries. Sixteen national, regional, and shortline companies offer rail service over this network; the two largest companies are CSX Transportation and Norfolk Southern. The state's extensive system of rivers, lakes, and underground water sources provides

EMPLOYMENT IN MANUFACTURING

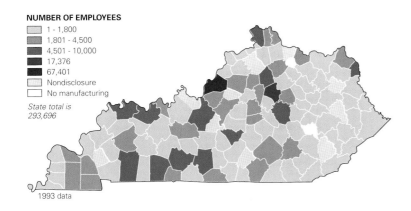

NUMBER OF EMPLOYEES
- 1 - 1,800
- 1,801 - 4,500
- 4,501 - 10,000
- 17,376
- 67,401
- Nondisclosure
- No manufacturing

State total is 293,696

1993 data

LOCATION QUOTIENT: MANUFACTURING

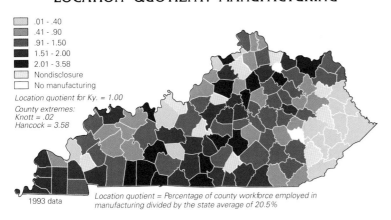

- .01 - .40
- .41 - .90
- .91 - 1.50
- 1.51 - 2.00
- 2.01 - 3.58
- Nondisclosure
- No manufacturing

Location quotient for Ky. = 1.00
County extremes:
Knott = .02
Hancock = 3.58

1993 data

Location quotient = Percentage of county workforce employed in manufacturing divided by the state average of 20.5%

CIVILIAN EMPLOYMENT IN MANUFACTURING

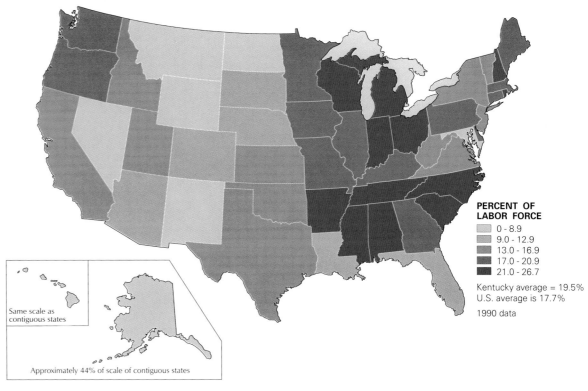

PERCENT OF LABOR FORCE
- 0 - 8.9
- 9.0 - 12.9
- 13.0 - 16.9
- 17.0 - 20.9
- 21.0 - 26.7

Kentucky average = 19.5%
U.S. average is 17.7%

1990 data

Same scale as contiguous states

Approximately 44% of scale of contiguous states

KENTUCKY'S MANUFACTURING EMPLOYMENT 1972-1993

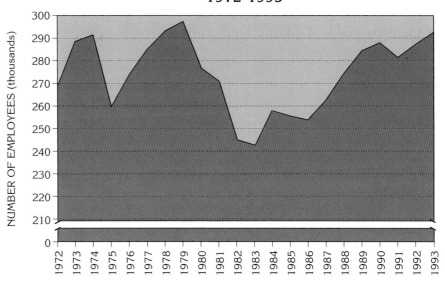

significant cost savings to manufacturers who need water for transportation, processing, and cooling. The water transportation network provides year-round barge access to major eastern industrial areas and to ports on the Gulf of Mexico. Eleven hundred miles of commercially navigable waterways include the Ohio and Mississippi Rivers, the Tennessee and Cumberland Rivers in western Kentucky (along with Kentucky Lake and Lake Barkley), and the lower portions of the Green, Kentucky, Licking, and Big Sandy Rivers.

Whereas the establishment of enterprise zones and programs such as KREDA are meant in part to encourage the location of industry in areas that have little or no manufacturing, the distribution of manufacturing in the state remains uneven. Notably, manufacturing in eastern Kentucky is virtually absent. When we consider the absolute number of manufacturing employees as well as the location of industry, we see that, with the exception of the extreme northeastern portion of the state (which has access to both I-64 and the Ohio–Big Sandy River systems), eastern Kentucky has little in the way of manufacturing. Some factors influencing this discrepancy include the terrain of the eastern part of the state, whose ruggedness offers little flat land for construction; the historic inaccessibility (both real and perceived) of the mountains; poor local markets; low levels of educational attainment among the population; and a history of labor militancy. Conversely, the metropolitan areas and those counties located at or near major transportation arteries contain a disproportionately large

MANUFACTURING JOBS AS SHARE OF NONAGRICULTURAL EMPLOYMENT

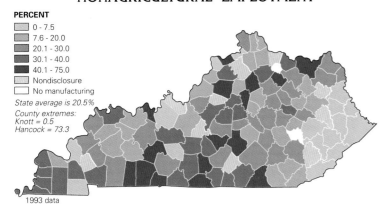

PERCENT
- ☐ 0 - 7.5
- ▨ 7.6 - 20.0
- ▨ 20.1 - 30.0
- ▨ 30.1 - 40.0
- ▨ 40.1 - 75.0
- ☐ Nondisclosure
- ☐ No manufacturing

State average is 20.5%
County extremes:
Knott = 0.5
Hancock = 73.3

1993 data

CHANGE IN MANUFACTURING EMPLOYMENT 1984-1993

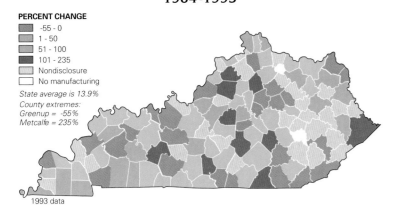

PERCENT CHANGE
- ▨ -55 - 0
- ☐ 1 - 50
- ▨ 51 - 100
- ▨ 101 - 235
- ☐ Nondisclosure
- ☐ No manufacturing

State average is 13.9%
County extremes:
Greenup = -55%
Metcalfe = 235%

1993 data

Logan Aluminum, near Russellville in Logan County, accounts for one-third of the nation's production of aluminum can stock for the soft drink and beer industry. Opened in 1983, the plant employs more than 900 workers.

share of manufacturing establishments, workers, employment, and earnings. The Golden Triangle, the area defined by the vertex cities of Louisville, Lexington, and Covington and by interstate highways I-64, I-75, and I-71, is the most important manufacturing area in the state. Together, the sixteen counties that constitute the Golden Triangle account for more than 37 percent of the state's 3.7 million residents, 44 percent of all manufacturing establishments, and 43 percent of total manufacturing value.

In 1993 the twenty-six metropolitan, or urban, counties of the state contained more than 55 percent of the state's total population and 60 percent of its manufacturing establishments and employees. Urban counties are defined here as those belonging to one of the state's seven Metropolitan Statistical Areas, designated by the U.S. Bureau of the Census, and those counties not in an MSA but containing a second-class city, that is, Warren (Bowling Green), McCracken (Paducah), Franklin (Frankfort), and Hardin (Radcliff). Thus, the state's nonmetropolitan counties contained a significant share, 40 percent, of the state's manufacturing, yet distribution of nonmetropolitan manufacturing is not uniform because the rural eastern portion of the state has very little industry.

EMPLOYMENT GENERATED BY KIDA-APPROVED PROJECTS

NUMBER OF EMPLOYEES
- · 1 - 100
- ● 101 - 300
- ● 301 - 500
- ● 501 - 1,000
- ● 1,001 - 2,390

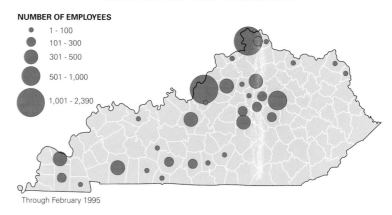

Through February 1995

EMPLOYMENT GENERATED BY KJDA-APPROVED PROJECTS

NUMBER OF EMPLOYEES
- · 1 - 100
- ● 101 - 300
- ● 301 - 500
- ● 501 - 1,000
- ● 1001 - 2,168

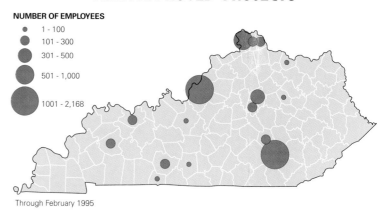

Through February 1995

EMPLOYMENT GENERATED BY KREDA-APPROVED PROJECTS

NUMBER OF EMPLOYEES
- · 1 - 100
- ● 101 - 300
- ● 301 - 500
- ● 501 - 1,000
- ● 1,001 - 2,360

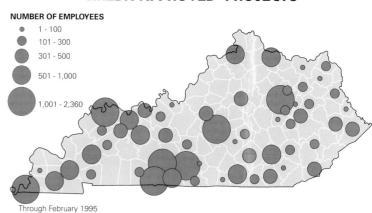

Through February 1995

KENTUCKY'S "GOLDEN TRIANGLE"

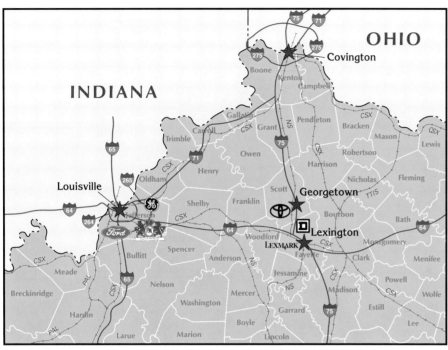

KENTUCKY'S TWENTY LARGEST PLANTS
1994

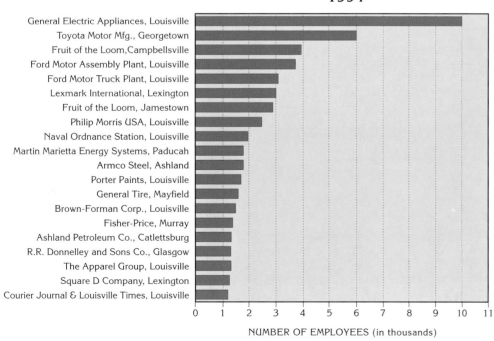

NUMBER OF EMPLOYEES (in thousands)

Japanese-owned Toyota Motor Manufacturing is the state's largest motor vehicle plant. Employing some 6,000 workers, the plant opened in 1987 and has undergone major expansion since then. This plant, which produces Camry automobiles and station wagons and the Avalon luxury automobile, has helped make Kentucky the fourth-largest motor vehicle producing state (after Michigan, Ohio, and Missouri). Positioned adjacent to Interstate 75 and the Norfolk Southern Railroad, the plant has greatly impacted Georgetown, the Scott County seat.

Average manufacturing wages for Kentucky's workers are near the national median: in 1993 hourly earnings for all manufacturing industries were more than twelve dollars in Louisville and Lexington, below the averages for cities like Detroit and Richmond, Virginia, but above those for Atlanta and Charlotte. The manufacturing industries in Kentucky that pay the highest wages are chemicals and allied products (the Standard Industrial Classification, or SIC, code for this industry is 28), petroleum refining (SIC 29), tobacco products (SIC 21), and transportation equipment (SIC 37). Manufacturing establishments producing transportation equipment include the General Motors plant in Bowling Green, where Corvettes are manufactured; two Ford Motor Company plants in Louisville, where workers build the

Ford Explorer, the Ford Ranger, and Ford trucks; the Toyota plant in Georgetown, which produces Camrys and now the Avalon; and a growing number of (mostly Japanese) plants that produce a variety of automobile accessories. Hourly earnings in these industries average more than fifteen dollars. At the other end of the spectrum, jobs that require lower skill levels, including those in industries producing textile mill products (SIC 22), apparel (SIC 23), lumber and wood products (SIC 24), and leather products (in Kentucky that means mostly shoes; SIC 31), typically pay much less. Hourly wages in these sectors average between six and eight dollars.

Whereas Kentucky, like most southern states, is not known as a strongly unionized state—16 percent of its workforce belonged to unions in 1991—

AVERAGE WEEKLY WAGES OF WORKERS
IN MANUFACTURING

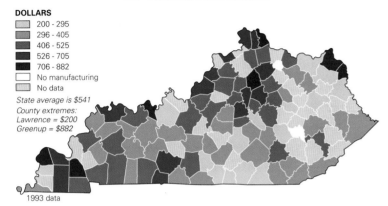

DOLLARS
- 200 - 295
- 296 - 405
- 406 - 525
- 526 - 705
- 706 - 882
- No manufacturing
- No data

State average is $541
County extremes:
Lawrence = $200
Greenup = $882

1993 data

AVERAGE WEEKLY EARNINGS
OF MANUFACTURING EMPLOYEES

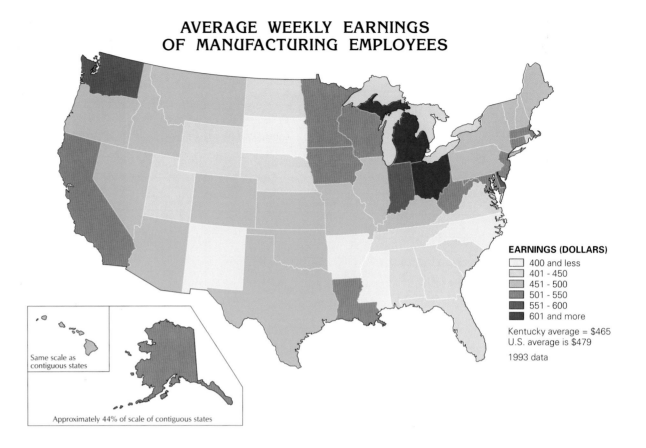

EARNINGS (DOLLARS)
- 400 and less
- 401 - 450
- 451 - 500
- 501 - 550
- 551 - 600
- 601 and more

Kentucky average = $465
U.S. average is $479

1993 data

Same scale as contiguous states

Approximately 44% of scale of contiguous states

AVERAGE EARNINGS IN MANUFACTURING
IN SELECTED CITIES, 1993

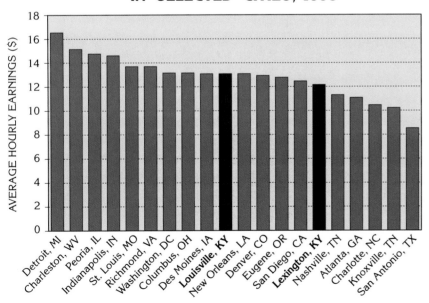

the unionization of manufacturing firms varies across the state. Those counties with industries that require skilled workers, such as the aluminum manufacturers in Henderson and Carroll Counties, chemical (including petrochemical) industries in Meade and Greenup Counties, and the steel plants in Boyd County, have a greater degree of unionization than counties in which the workforce is less skilled. Also, manufacturing firms in urban areas, such as those in Jefferson County and the northern Kentucky counties near Cincinnati, are more unionized than rural firms.

In 1994 Kentucky had twenty-nine plants each of which employed one thousand or more workers. The largest individual plant was the General Electric Appliances factory in Louisville, with ten thousand workers, that was established in 1952. General Electric Company, which produces everything from household appliances to components for aircraft engines in its seven Kentucky plants,

is the largest single manufacturing employer, with more than thirteen thousand workers in the Louisville plant and in the six other plants located across the state. The second-largest company in manufacturing employment is Fruit of the Loom, with nearly ten thousand workers in seven plants, five of them in nonmetropolitan counties. More than eight hundred thousand pairs of underwear are produced each day in Kentucky's Fruit of the Loom plants. Other nationally known consumer products manufactured in the state include Jif peanut butter, at Proctor and Gamble in Lexington; Marlboro and other cigarettes, at the Philip Morris factory in Louisville; Valvoline oil, at Ashland Petroleum near Ashland; Post-it Notes, at the 3M Company in Cynthiana; greeting cards, at American Greetings plants in Bardstown and Corbin, which employ a total of sixteen hundred workers; Louisville Slugger baseball bats, at the Hillerich and Bradsby Company, headquartered in Louisville (since 1974, however, the

NUMBER OF KENTUCKY EMPLOYEES BY SIC CODE, 1993

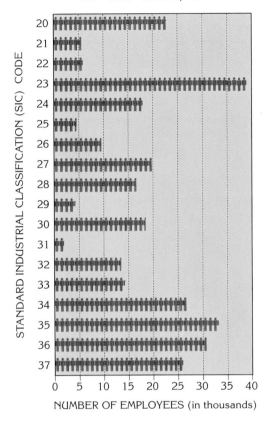

STANDARD INDUSTRIAL CLASSIFICATION (SIC) CODE

20
21
22
23
24
25
26
27
28
29
30
31
32
33
34
35
36
37

0   5   10   15   20   25   30   35   40

NUMBER OF EMPLOYEES (in thousands)

Equals 1,000 employees

KENTUCKY AND U.S. MANUFACTURING WAGES, 1992

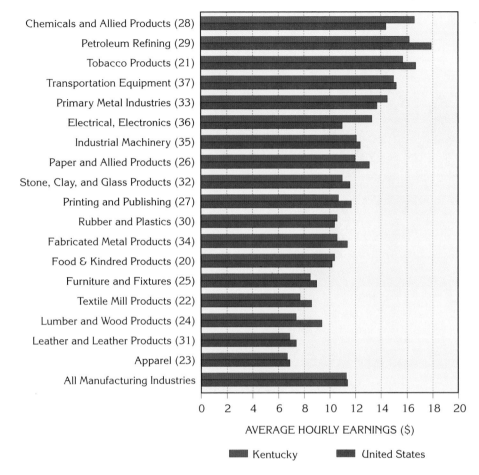

Chemicals and Allied Products (28)
Petroleum Refining (29)
Tobacco Products (21)
Transportation Equipment (37)
Primary Metal Industries (33)
Electrical, Electronics (36)
Industrial Machinery (35)
Paper and Allied Products (26)
Stone, Clay, and Glass Products (32)
Printing and Publishing (27)
Rubber and Plastics (30)
Fabricated Metal Products (34)
Food & Kindred Products (20)
Furniture and Fixtures (25)
Textile Mill Products (22)
Lumber and Wood Products (24)
Leather and Leather Products (31)
Apparel (23)
All Manufacturing Industries

0   2   4   6   8   10   12   14   16   18   20

AVERAGE HOURLY EARNINGS ($)

■ Kentucky   ■ United States

Lexmark International is a major national producer and exporter of laser printers and computer keyboards. Located in Lexington, the facility that is now home to Lexmark was an IBM plant until 1991. Employing about 3,000 workers, Lexmark went public in 1995.

MANUFACTURING FIRMS WITH UNIONS

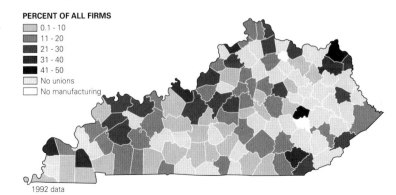

PERCENT OF ALL FIRMS
- 0.1 - 10
- 11 - 20
- 21 - 30
- 31 - 40
- 41 - 50
- No unions
- No manufacturing

1992 data

bats have been made in Jeffersonville, Indiana, still part of the Louisville metropolitan area); Oshkosh B'Gosh clothing for children, at five plants employing nearly two thousand workers; Kingsford charcoal, at Kingsford Products in Burnside; typewriters and laser printers, at Lexmark in Lexington (between 1957 and 1991 this factory was Lexington's IBM plant); two hundred types of Fisher-Price toys, at Fisher-Price in Murray (the plant opened in 1959, and today it employs one-quarter of all Fisher-Price workers); and Doritos and Fritos and other snack foods, at Frito-Lay in Louisville. Workers in Kentucky produce numerous other products as well, including, of course, automobiles and bourbon whiskey.

Automobiles have been manufactured in Kentucky since 1913, when Louisville's first Ford Motor Company assembly plant produced about a dozen Model T's daily. Since then Ford has expanded its Louisville production, and in 1985 Toyota announced its arrival in the state. In 1992 the fifth, sixth, and seventh best-selling cars and trucks in the United States were produced in Kentucky assembly plants: the Ford Explorer sport utility truck, the Toyota Camry, and the Ford Ranger pickup, respectively. The state's smallest automobile manufacturing facility is its most prestigious: the General Motors plant in Bowling Green produces the Corvette, long billed as "America's only true sports car." The first Corvette was made in 1953 at the GM plant in Flint, Michigan, and from 1954 to 1981 more than seven hundred thousand Corvettes were manufactured in the St. Louis GM plant. In late 1979 General Motors announced that it was moving its Corvette plant to Bowling Green, Kentucky, and in 1982 the first Corvettes rolled off the assembly line there. Today the eleven hundred employees at the Bowling Green plant manufacture about twenty thousand cars each year. Although the number of Corvettes produced may seem small when compared, say, with the nearly one-quarter million Camrys produced annually at the Toyota plant, in 1993 Corvettes accounted for nearly one-third of the U.S. sports car market. Automobile enthusiasts celebrated the completion of the one millionth Corvette in 1992, and in 1994 the National Corvette Museum opened in Bowling Green.

Ashland Inc.'s Catlettsburg refinery near Ashland in Boyd County employs more than 1,200 and is the state's largest producer of petroleum products and petrochemicals. Ashland Oil, Inc., which began in 1924, is today the largest Kentucky-based corporation and ranks among the one hundred largest corporations in the United States.

Types of Industry

Kentucky's manufacturing is diversified, in that industries representing about 85 percent of the Standard Industrial Classifications for manufacturing can be found within the state. The largest of these, in total employment, are apparel (SIC 23), with 38,895 workers, or more than 12 percent of all manufacturing employees in Kentucky; industrial machinery and equipment (SIC 35), with 33,105 workers; electrical and electronic machinery, equipment, and supplies (SIC 36), with 30,827 workers; fabricated metal products, except machinery and transportation equipment (SIC 34), with 26,487 workers; transportation equipment (SIC 37), with 25,899 workers; and food and kindred products (SIC 20), with 22,740 workers.

The location of manufacturing establishments and employment in the state varies by type of industry. Some industries, because of factors such as the importance of being close to a large market, the presence of a skilled labor supply, or the accessibility of transporta-

MANUFACTURING EMPLOYMENT BY TYPE OF INDUSTRY

FOOD AND KINDRED PRODUCTS
(SIC 20)

1 - 50
51 - 100
101 - 500
501 - 1,000
1,480
8,495

1993 data

TOBACCO PRODUCTS
(SIC 21)

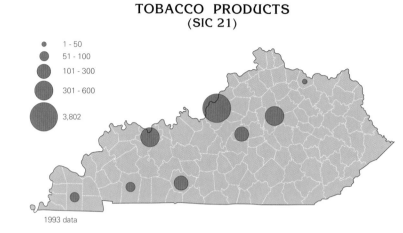

1 - 50
51 - 100
101 - 300
301 - 600
3,802

1993 data

TEXTILE MILL PRODUCTS
(SIC 22)

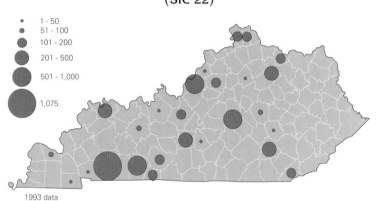

1 - 50
51 - 100
101 - 200
201 - 500
501 - 1,000
1,075

1993 data

APPAREL
(SIC 23)

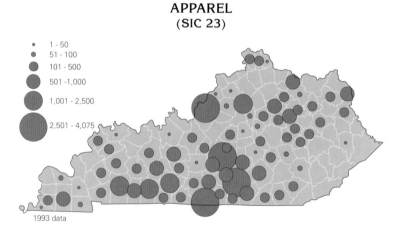

1 - 50
51 - 100
101 - 500
501 -1,000
1,001 - 2,500
2,501 - 4,075

1993 data

LUMBER AND WOOD PRODUCTS
(SIC 24)

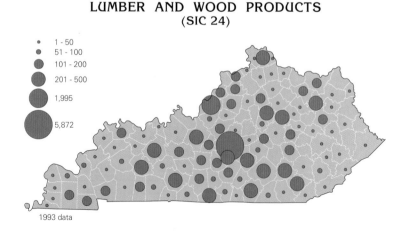

1 - 50
51 - 100
101 - 200
201 - 500
1,995
5,872

1993 data

FURNITURE AND FIXTURES
(SIC 25)

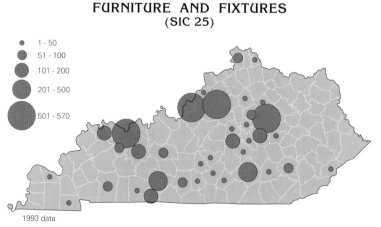

1 - 50
51 - 100
101 - 200
201 - 500
501 - 570

1993 data

MANUFACTURING EMPLOYMENT BY TYPE OF INDUSTRY

PAPER AND ALLIED PRODUCTS
(SIC 26)

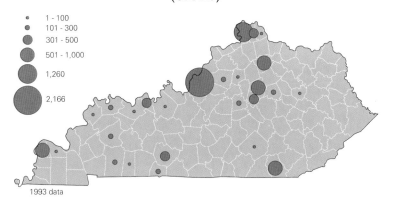

1 - 100
101 - 300
301 - 500
501 - 1,000
1,260
2,166

1993 data

PRINTING AND PUBLISHING
(SIC 27)

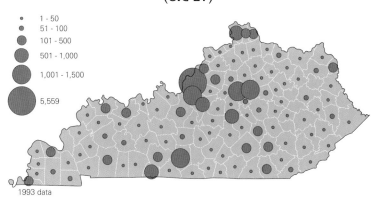

1 - 50
51 - 100
101 - 500
501 - 1,000
1,001 - 1,500
5,559

1993 data

CHEMICALS AND ALLIED PRODUCTS
(SIC 28)

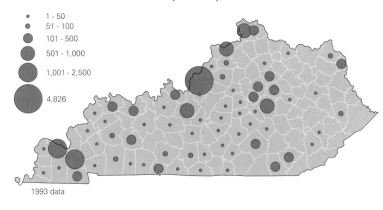

1 - 50
51 - 100
101 - 500
501 - 1,000
1,001 - 2,500
4,826

1993 data

PETROLEUM REFINING AND RELATED INDUSTRIES
(SIC 29)

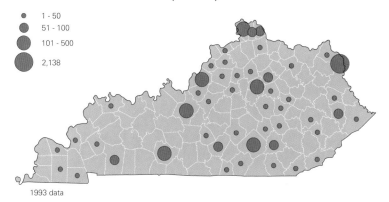

1 - 50
51 - 100
101 - 500
2,138

1993 data

RUBBER AND MISCELLANEOUS PLASTICS
(SIC 30)

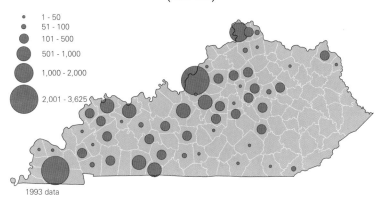

1 - 50
51 - 100
101 - 500
501 - 1,000
1,000 - 2,000
2,001 - 3,625

1993 data

LEATHER AND LEATHER PRODUCTS
(SIC 31)

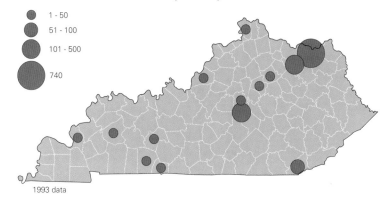

1 - 50
51 - 100
101 - 500
740

1993 data

Philip Morris, Inc., makers of the popular Marlboro brand cigarettes, employs nearly 3,000 workers in its Louisville plant. Here a quality-control employee inspects cigarettes as they pass her station.

tion networks, are found mostly in or near the larger urban areas, especially Louisville. Examples include industries related to food and kindred products, printing and publishing, fabricated metal products, transportation equipment, and machinery. Other industries are more ubiquitous and depend less on a skilled labor supply, market factors, and easy access to transportation. Workers in such industries often receive lower wages. These include employees who manufacture textile mill products, apparel, and leather and leather products. Other factors that are often critical in the locational decision include accessibility to natural resources (affecting industries manufacturing lumber and wood products, furniture, bourbon whiskey, and stone, clay, and glass products), local incentives (such as those for the Toyota plant), accessibility to related industries (such as automotive parts), and factors that might be called historical (affecting distilling and tobacco manufacturing).

Manufacturing in the state at present employs about three hundred thousand workers, although employment sectors have grown differentially. Since 1972 the number of employees in such well-known Kentucky

MANUFACTURING EMPLOYMENT BY TYPE OF INDUSTRY

STONE, CLAY, AND GLASS PRODUCTS
(SIC 32)

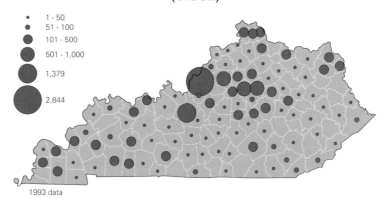

1993 data

PRIMARY METAL INDUSTRIES
(SIC 33)

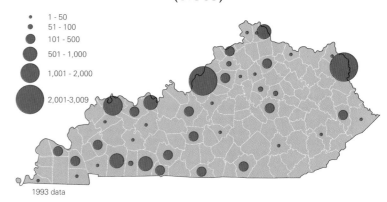

1993 data

FABRICATED METAL PRODUCTS
(SIC 34)

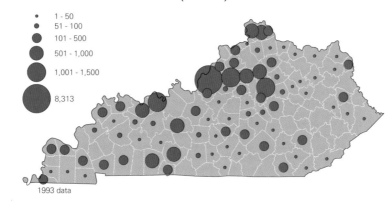

1993 data

INDUSTRIAL MACHINERY
(SIC 35)

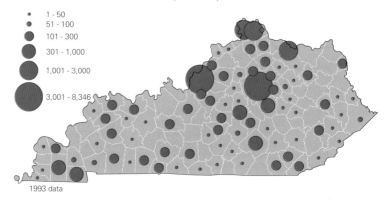

1993 data

ELECTRICAL AND ELECTRONIC MACHINERY
(SIC 36)

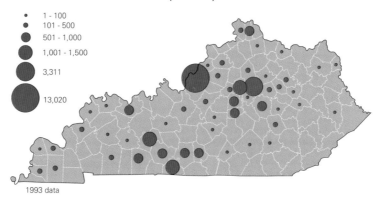

1993 data

TRANSPORTATION EQUIPMENT
(SIC 37)

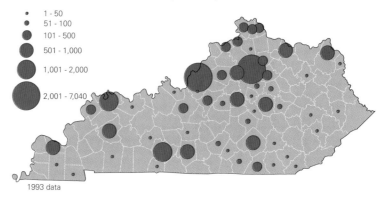

1993 data

Nearly 1,100 employees work at the General Motors Corporation's plant, situated beside Interstate 65 in Bowling Green. Since 1982, all Corvettes, which account for one-third of the American sports car market, have been manufactured at this plant.

industries as tobacco manufactures and distilled spirits has decreased rapidly, whereas employment related to textile mill products, rubber and miscellaneous plastics, and transportation equipment has increased. In fact, from 1972 to 1993 employment in tobacco manufactures declined 53 percent (from 12,400 in 1972 to 5,800 in 1993), whereas that in transportation equipment increased by 140 percent (to nearly 26,000).

CHANGE IN EMPLOYMENT IN KENTUCKY BY SELECTED INDUSTRIES, 1972-1993

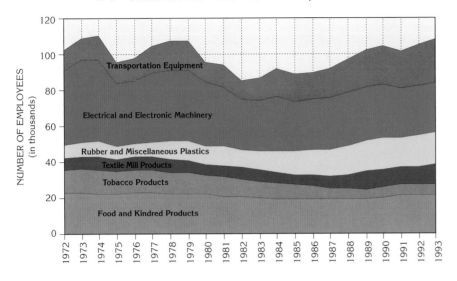

NUMBER OF EMPLOYEES (in thousands)

Transportation Equipment

Electrical and Electronic Machinery

Rubber and Miscellaneous Plastics

Textile Mill Products

Tobacco Products

Food and Kindred Products

Although it only employs a total of 3,800 workers, one of Kentucky's oldest and best-known manufacturing industries is the production of distilled spirits, especially bourbon whiskey. In 1992, the value of shipments for the industry was over $1.1 billion, or more than one-third of the national total. Kentucky, along with neighboring Tennessee, produces nearly all of the world's bourbon whiskey. Distillers produced the first whiskey in Kentucky as early as 1775; the first "bourbon" distiller may have been the Baptist minister Elijah Craig, who lived in present-day Scott County, which was then part of Bourbon County, after which the whiskey was named. Although the nation's employment in the distilled spirits industry has declined, revenues have increased. The plants that produce Kentucky's distilled spirits are not widespread; the 16 establishments are located in just 7 of the state's 120 counties. Among the state's twenty largest plants with more than 1,400 employees, Brown-Forman in Louisville, which also produces fine china and luggage, is Kentucky's largest distillery. The oldest continuously operated family distillery is the one started by Jacob Beam, a German immigrant, which has been operated by six generations of the Beam family in Clermont since 1795. The well-known brands produced by the state's distillers include Maker's Mark, Early Times, Old Forester, Ancient Age, Wild Turkey, Old Fitzgerald, Rebel Yell, Heaven Hill, I.W. Harper, and Jim Beam.

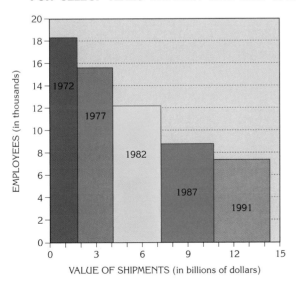

Workers hand-dip each bourbon bottle in red sealing wax at the Maker's Mark distillery in Loretto, Kentucky. Originally built in 1889, the distillery was restored in 1953 and is today a National Historic Landmark open to visitors. About 60 employees work at this distillery.

DISTILLED SPIRITS: VALUE AND EMPLOYMENT IN THE UNITED STATES
FOR SELECT YEARS BETWEEN 1972 AND 1991

EMPLOYEES (in thousands)

1972
1977
1982
1987
1991

VALUE OF SHIPMENTS (in billions of dollars)

EMPLOYMENT IN THE DISTILLED SPIRITS INDUSTRY

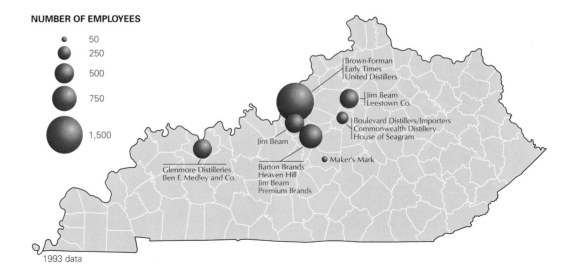

NUMBER OF EMPLOYEES

50
250
500
750
1,500

Brown-Forman
Early Times
United Distillers

Jim Beam
Leestown Co.

Boulevard Distillers/Importers
Commonwealth Distillery
House of Seagram

Jim Beam

Maker's Mark

Glenmore Distilleries
Ben F. Medley and Co.

Barton Brands
Heaven Hill
Jim Beam
Premium Brands

1993 data

EMPLOYMENT IN FOREIGN-OWNED PLANTS

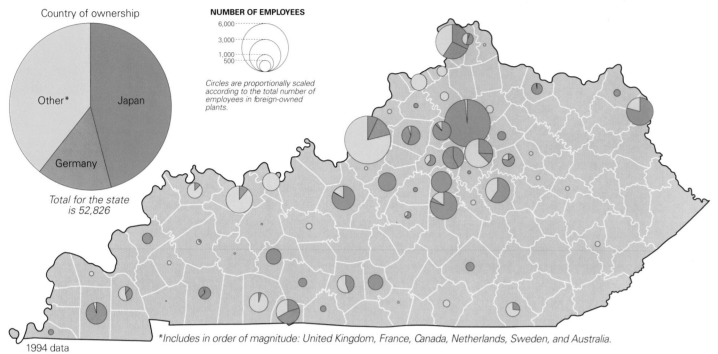

Country of ownership

NUMBER OF EMPLOYEES

6,000
3,000
1,000
500

Circles are proportionally scaled according to the total number of employees in foreign-owned plants.

Other*
Japan
Germany

Total for the state is 52,826

*Includes in order of magnitude: United Kingdom, France, Canada, Netherlands, Sweden, and Australia.

1994 data

EMPLOYMENT IN FOREIGN-OWNED INDUSTRIES IN KENTUCKY

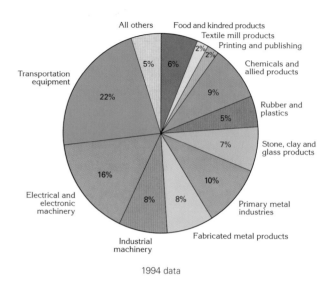

All others — 5%
Food and kindred products — 6%
Textile mill products — 2%
Printing and publishing — 2%
Chemicals and allied products — 9%
Rubber and plastics — 5%
Stone, clay and glass products — 7%
Primary metal industries — 10%
Fabricated metal products — 8%
Industrial machinery — 8%
Electrical and electronic machinery — 16%
Transportation equipment — 22%

1994 data

FOREIGN INVESTMENT IN KENTUCKY MANUFACTURING

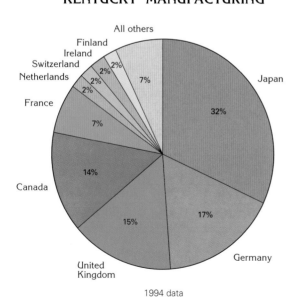

All others — 7%
Finland — 2%
Ireland — 2%
Switzerland — 2%
Netherlands — 2%
France — 7%
Canada — 14%
United Kingdom — 15%
Germany — 17%
Japan — 32%

1994 data

MANUFACTURING LABOR FORCE EMPLOYED IN FOREIGN COMPANIES

Same scale as contiguous states

Approximately 44% of scale of contiguous states

PERCENT OF ALL COMPANIES
- 0.0 - 8.9
- 9.0 - 10.9
- 11.0 - 13.9
- 14.0 - 16.9
- 17.0 - 22.7

Kentucky average = 14.3%
U.S. average is 10.4%
1990 data

FOREIGN-OWNED PRODUCERS OF AUTOMOTIVE PARTS

1992 data

One ● represents one producer

Nearly 80% of foreign-owned automotive parts producers in the U.S. are located within this circled area. Over one-fourth of that number are located in Kentucky.

EMPLOYMENT IN FOREIGN-OWNED AUTOMOTIVE PARTS PLANTS OPENED SINCE 1985

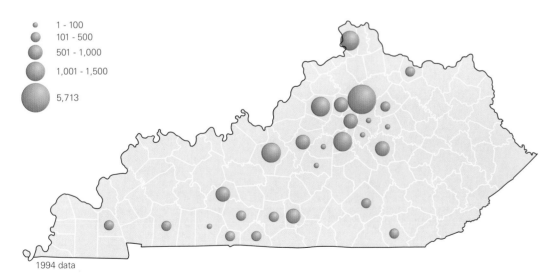

1 - 100
101 - 500
501 - 1,000
1,001 - 1,500
5,713

1994 data

EMPLOYMENT IN ALL FOREIGN-OWNED INDUSTRIES IN KENTUCKY

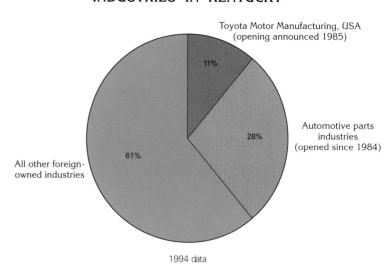

Toyota Motor Manufacturing, USA (opening announced 1985)

11%

Automotive parts industries (opened since 1984)

28%

All other foreign-owned industries

61%

1994 data

Foreign Investment and Manufacturing

Recent years have seen a flourishing of foreign investment in Kentucky manufacturing. In 1990 there were 41,100 workers in foreign-owned plants, constituting 14.3 percent of the state's manufacturing employees; by 1994 the number had grown to more than 50,000 workers. (A "foreign-owned" company is defined here as one in which at least 10 percent of the ownership is foreign. In most cases the proportion of the ownership is considerably greater than that; Toyota, for example, is wholly owned by Japanese interests.) Only six states had a greater share of their manufacturing workforce employed in foreign-owned plants in the mid-1990s: Alaska, West Virginia, South Carolina, Delaware, New Jersey, and Maryland. Plants wholly or partly foreign-owned make a wide range of products in Kentucky, but since 1985 a disproportionate share of foreign-owned companies opening plants in the state have been Japanese-owned and related to the automotive industry. Indeed, in 1994, 22 percent of all foreign-owned plants in Kentucky could be classified under SIC 37, transportation equipment. In the same year, Japan accounted for nearly one-third of the ownership of foreign-owned plants in the state, followed by Germany (17 percent), the United Kingdom (15 percent), Canada (14 percent), and France (7 percent). The Japanese impact has become so significant that the state has opened a trade office in Tokyo. The state's accessibility and central location, together with economic incentives, have

KENTUCKY EXPORTS BY INDUSTRY
(BY VALUE OF PRODUCT)

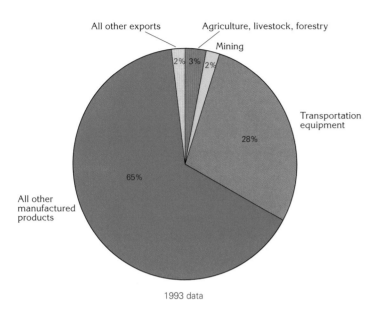

All other exports 2%
Agriculture, livestock, forestry 3%
Mining 2%
Transportation equipment 28%
All other manufactured products 65%

1993 data

KENTUCKY EXPORTS, 1989-1993

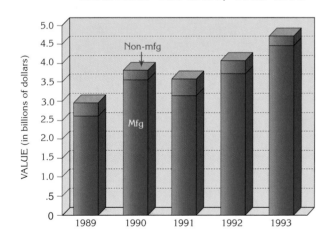

VALUE (in billions of dollars)

Non-mfg
Mfg

1989 1990 1991 1992 1993

KENTUCKY EXPORT DESTINATIONS
(BY VALUE OF PRODUCT)

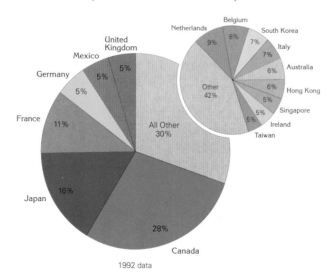

United Kingdom 5%
Mexico 5%
Germany 5%
France 11%
Japan 16%
Canada 28%
All Other 30%

Netherlands 9%
Belgium 8%
South Korea 7%
Italy 7%
Australia 6%
Hong Kong 6%
Singapore 5%
Ireland 5%
Taiwan 5%
Other 42%

1992 data

White oak from Kentucky's hardwood forests provides the raw material for bourbon barrel manufacture. Barrels from the Marion County Cooperage plant near Lebanon will be sold to distilleries and used to age bourbon whiskey, a process that normally takes eight or more years. Because bourbon barrels may not be reused (for aging bourbon whiskey), many are exported to Europe for use in the manufacture of spirits.

attracted foreign industry. Many of the foreign-owned manufacturing facilities, especially the newer plants producing automotive parts and accessories, have located in the Golden Triangle and in those counties served by interstate highways. The nearly six thousand Toyota employees, which account for 11 percent of all employment in Kentucky in foreign-owned plants, produced 240,000 Camrys in 1993. Since the plant opened, it has expanded four times, and by 1995 the factory encompassed 7.85 million square feet. More than sixty-five additional plants producing automobile parts and accessories have opened in Kentucky since 1985.

Because of the high cost of transporting relatively bulky items like seats and suspension systems to producers, the location of such automotive parts suppliers in the United States, as in Kentucky, is closely related to the location of the major automobile assembly plants. Thus, 80 percent of Japanese suppliers in the United States have located along the I-65 and I-75 corridors to be near the new assembly plants. Instead of concentrating in Michigan, the traditional home of U.S. suppliers, Japanese factories are much more widespread, with 20 percent of the total located in Ohio; 15 percent each in Kentucky, Indiana, and Michigan; 10 percent in Tennessee; and 5 percent in Illinois.

Kentucky's total exports to foreign markets increased from less than $2.97 billion in 1989 to more than $4.72 billion in 1993. The greatest share of total exports by value, 94 percent, was derived from manufactured goods. Among all exports, transportation equipment ranked first in total export value, accounting for 28 percent. Industrial machinery and computer equipment ranked next (16 percent), followed by chemicals and allied products (11 percent). Coal, Kentucky's major natural resource, accounted for only 2 percent of the state's exports in 1993. The most important export destination for Kentucky's products was Canada, which accounted for 28 percent of the total value, followed by Japan, France, and then Germany, Mexico, and the United Kingdom.

Transportation and Communications

CHAPTER NINE

Transportation and Communications

T HE IMPORTANCE OF transportation and communications to Kentucky's residents is easily demonstrated. In the early 1990s nearly one of every twenty Kentucky workers was directly employed in the transportation and communications sectors of the Commonwealth's economy. Products related to these sectors account for nearly 10 percent of the value of all goods and services produced in the state. The average state resident drives a private vehicle more than fifteen thousand miles annually, and that figure continues to increase. Kentucky has more registered vehicles (2.98 million) than licensed drivers (2.46 million). The state is home to sixty-two hundred licensed pilots, who fly out of seventy-one private and public use airports in 1,554 registered aircraft. Kentucky must maintain more than thirteen thousand bridges on a highway network that extends nearly seventy thousand miles. The state's rail system comprises nearly twenty-four hundred miles of track.

Interest in communications in Kentucky is equally illustrated by statistics: The state is served by thirty-four commercial and educational television stations, as well as twenty-three daily newspapers. Just under three hundred cable television companies serve more than one thousand Kentucky communities and eight hundred thousand subscribers. The state's 188 main and branch public library buildings house more than seven million books and serve 1.6 million registered borrowers. Additional examples could be cited, but these numbers highlight the efforts of state residents to address the need to inform and be informed, the need to visit and be visited, the need to ship and receive.

The Evolution of Waterways, Railroads, and Roads

The state's transportation history can be conveniently divided into three eras: the earliest era of river and trail, the middle era of rail dominance, and the recent auto-air era. When Kentucky gained statehood in 1792,

Construction on the Alexandria-to-Ashland Highway in Mason County.

McAlpine Locks on the Ohio River near Louisville. Surge basin is in the foreground, Sand Island is in the upper right, and New Albany, Indiana, appears beyond the railroad bridge in the distance.

NAVIGABLE WATERWAYS

◆ Lock and Dam
● Ohio River City

the dominant forms of movement involved the use of inland waterways and crude trails cut cross-country. The Ohio River provided the key outlet for the state's early exports, including tobacco, hemp, corn, lumber, cattle, hogs, and sheep. Louisville boat works manufactured river craft, most of which were simply flat-bottomed boats destined for one-way trips downriver to New Orleans. With the boats powered primarily by the river's current, trips were slow. The boats were often sold for lumber in New Orleans because, before boats were steam-powered, the return trip upriver took three months. Nonetheless, the inland river system provided the easiest method of transportation into and out of Kentucky. Distant Kentucky settlements were linked with the Ohio via internal rivers such as the Big Sandy, the Licking, the Kentucky, and the Green. These river

routes provided good accessibility north and south, but east-west connections were overland routes (see map, chapter 3), which were far less efficient during this era. Many roads in the late 1700s were little more than rutted trails that had been first imprinted by the bison that roamed freely and by Native Americans, who had a well-developed system of footpaths. Many of these early routes survive today as the basis for portions of modern highways; U.S. 25, for example, follows part of the old Wilderness Road. The Wilderness Road, which entered Kentucky through the Cumberland Gap, was the most famous of the few eastern overland connections. Trees had been cleared for the road by early 1800, and gradually the path was transformed into a log-covered "corduroy road." Workers laid logs crosswise to the direction of the road and then covered them with

Barges in the McAlpine Locks on the Ohio River.

OHIO RIVER TRANSPORTATION CORRIDOR

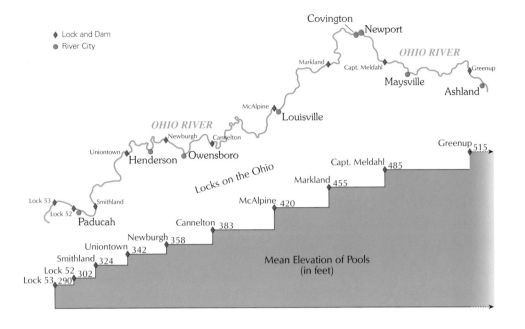

◆ Lock and Dam
● River City

OHIO RIVER

OHIO RIVER

Covington
Newport
Markland   Capt. Meldahl   Greenup
Maysville   Ashland

McAlpine
Louisville

Newburgh   Cannelton
Uniontown   Owensboro
Henderson

Locks on the Ohio

Greenup 515
Capt. Meldahl 485
Markland 455
McAlpine 420
Cannelton 383
Newburgh 358
Uniontown 342
Smithland 324
Lock 52 302
Lock 53 290

Lock 53
Lock 52   Smithland
Paducah

Mean Elevation of Pools
(in feet)

A tow with twelve barges on the Ohio River passes Ashland, Boyd County.

Coal barges on the Ohio River near Louisville. In 1994 almost 22 percent of Kentucky's coal used river barges as the primary transportation mode. Coal and coke comprise nearly 60 percent of the total commodities transported on the Ohio.

OHIO RIVER FREIGHT TRAFFIC 1920-1990

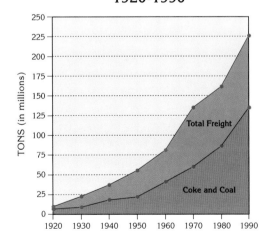

TONS (in millions)

Total Freight

Coke and Coal

1920 1930 1940 1950 1960 1970 1980 1990

OHIO RIVER WATERWAY COMMODITY DISTRIBUTION

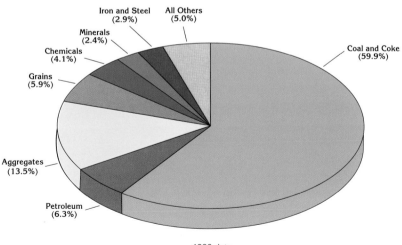

Iron and Steel (2.9%)
All Others (5.0%)
Minerals (2.4%)
Chemicals (4.1%)
Coal and Coke (59.9%)
Grains (5.9%)
Aggregates (13.5%)
Petroleum (6.3%)

1990 data

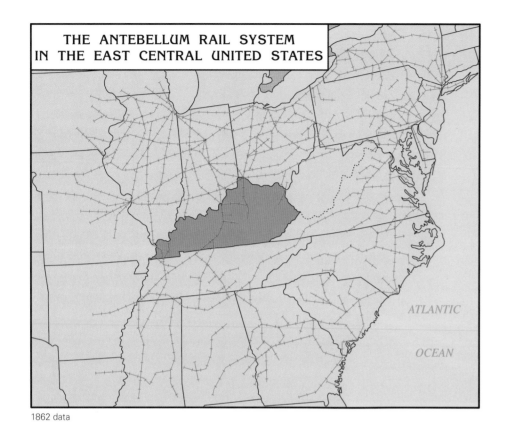

THE ANTEBELLUM RAIL SYSTEM
IN THE EAST CENTRAL UNITED STATES

ATLANTIC

OCEAN

1862 data

A railroad yard and rail cars near the Ashland Refinery in Catlettsburg, Boyd County.

dirt and later a stone surface. A basic system of toll roads evolved during the early 1800s, and stagecoaches traversed these routes as early as 1803. The use of stagecoaches, primarily for mail and passenger service, continued to grow, peaking by the 1840s. River traffic increased markedly during this era with the application of steam engines and with the improvement of river navigability through construction of dams and locks.

River transportation does not play the crucial role today that it did during the era of river and trail. The state still has eleven hundred miles of commercially navigable waterways, six public riverports, and 180 privately owned terminal facilities. Although the Ohio River is no longer the primary highway into Kentucky that it once was, it is still an essential corridor of energy-efficient transportation. Bulk shipments, especially of coal and coke, ply the Ohio's waters by way of large tows of barges (on average about fifteen), which are pushed and pulled by powerful tow-boats (on average, 4,500 horsepower). One average tow has the same capacity as 870 tractor-trailer trucks. Other Kentucky

rivers have decreased in importance for moving freight because rail and trucks provide much greater speed and are not limited to existing river channels. One exception is the lower Big Sandy River, which witnessed 15 percent annual growth in the movement of coal and petroleum products between 1982 and 1992. This growth derives from increased demand for low-sulfur bituminous coal found in eastern Kentucky and West Virginia.

Recreational use of Kentucky's river resources has increased greatly in recent years (see maps, Chapter 10). The Ohio, Cumberland, and Kentucky Rivers in particular have seen significant growth in recreational traffic. Only the first four locks on the lower Kentucky River are now operated by the U.S. Army Corps of Engineers. Locks 5-9, on the Kentucky River's upper reaches, are operated by the Commonwealth during summer months for recreational use. Locks 10-14 have been permanently closed because of deterioration. This is a source of concern to many of the river's recreational users.

After the 1840s rivers and trails gradually yielded to the railroad as the dominant mode of transportation. Rail lines were

A coal unit train transports coal on the CSX tracks near Irvine, in Estill County. Rail was the primary distribution mode for more than 100 million tons of coal, which accounted for nearly two-thirds of all coal hauled in Kentucky in 1994.

RAILROADS SERVING KENTUCKY IN 1938

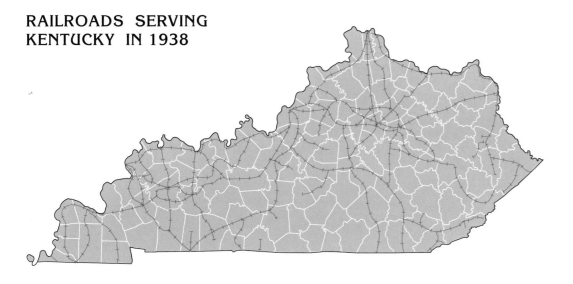

RECENT CHANGE IN KENTUCKY'S RAIL NETWORK

〰 Railroads in use in 1990

〰 Railroads abandoned during 1980-1990

1990 data

Toyota Camrys are loaded for rail shipment at the Toyota manufacturing plant in Georgetown, Scott County.

The R.J. Corman Railroad, one of nine short-line railroads that serve Kentucky.

initially designed to complement the existing river transportation system. A Lexington-Frankfort steam rail route was in place by 1835, and the trip between the two towns took about two and a half hours. It should be remembered, however, that the trip by rail, at twelve miles per hour, was at least twice as fast as the service offered by competing water transportation. One of the most important early rail connections was completed in 1859 between Louisville and Nashville, Tennessee. The Louisville & Nashville (L&N) Railroad, until it became part of the Seaboard System in 1982, provided Louisville with superior accessibility to southern markets and led to significant growth for that city. The L&N represents an early example of the growing effectiveness of the railroad in doing more than simply complementing the river system, although Kentucky's railroads generally operated at a loss during the early portion of the rail era. Additionally, the early rail system reinforced the basic north-south orientation of Kentucky's river transportation linkages. No direct east-west links existed before the end of the Civil War.

The L&N played a key role during the Civil War as a principal supply line for Union forces, and at the end of the war it was in better shape than it had been when the war began. After the war, massive investment in railroads resulted in significant increases in track mileage across the country. Kentucky's track mileage nearly tripled between 1865 and 1880, to exceed fifteen hundred miles. Railroads emerged as the fastest and most efficient method of hauling passengers and freight, and they had little competition. Rail lines put in place during the latter portion of the nineteenth century offered effective east-west connections for the first time in the Commonwealth's history. Kentucky's principal interstate rail connections, however, still ran north and south to intersect with major east-west routes that skirted the state. By 1915 rail had fully penetrated the coalfields of eastern and western Kentucky. Efficient export of coal by rail helped elevate Kentucky to a supreme position as supplier of the nation's, and the world's, most important energy resource at the time. The rail system continued to expand throughout the first third of the twentieth century. At least forty rail companies operated more than five thousand miles of track in Kentucky by 1930, when the railroad industry peaked. Railroads were the dominant mode of transporta-

EVOLUTION OF PAVED ROADS

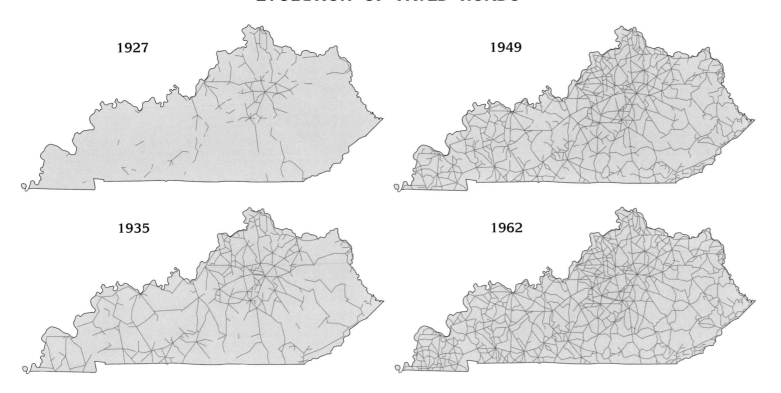

1927

1935

1949

1962

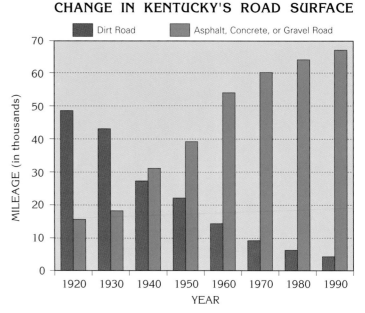

CHANGE IN KENTUCKY'S ROAD SURFACE

■ Dirt Road ■ Asphalt, Concrete, or Gravel Road

MILEAGE (in thousands)

YEAR

Cumberland Gap and the opening of the highway tunnel being constructed through Cumberland Mountain. At this gap Dr. Thomas Walker and Daniel Boone entered Kentucky in 1750 and 1773, respectively. Construction on the tunnel was completed in 1996.

tion throughout the late nineteenth century and the first half of the twentieth. The transportation environment was changing, however, as a result of technological innovations that appeared before World War II, and these would be rapidly adopted and embraced in Kentucky after the war's end. These changes, associated with highway and air transportation, ushered in the auto-air era.

Kentucky's rail system has undergone significant transformation since freight shipment and passenger traffic peaked in the early 1930s. Most notably, rail passenger service has essentially disappeared. Today, Amtrak provides minimal service, with stops in eastern Kentucky at Catlettsburg and Maysville and in western Kentucky at Fulton. Even this limited service may be curtailed or terminated as Amtrak retrenches in the face of financial loss. Private auto and air travel have captured the dominant share of local and long-distance passenger travel. The railroad industry has abandoned numerous routes: from a five-thousand-mile rail system in the 1930s, the network has shrunk so that only about twenty-

five hundred miles of track remained in the early 1990s. Although sixteen rail companies operate in the state, the two largest carriers are CSX Transportation and Norfolk Southern. Key routes on the remaining system are once again oriented in a north-south direction. High track densities remain in the eastern and western coalfields, attesting to the importance of this single commodity for the continued economic viability of railroads in Kentucky.

A recent advance in rail transportation that has generated optimism for future railroad growth is the development of intermodal traffic. Many of the consumer goods now traded between countries and across long distances within the United States travel inside containers. A container is a large metal box, twenty to forty feet long, that can be transported on a specially designed rail car, or placed on a dolly and towed behind a truck, or loaded efficiently onto a large ship in port. Containers permit the use of all modes of transportation in the most effective combination, and they move across Kentucky by a variety of methods, including rail. Other types of

1995 data

An interstate highway interchange near Louisville, Jefferson County.

intermodal movements, including the trailer on flatcar and finished autos on tiered and covered rail cars, are also common. Louisville and Georgetown are important intermodal hubs.

Kentucky's first roads date to the late 1700s, but these were important only for local movements of residents in their daily activities. It was left to rivers and then railroads to move people and goods over long distances during the 1800s and the early 1900s. Kentucky's early roads were private toll roads. These roads evolved from trails to corduroy roads, to graveled surfaces, to macadamized turnpikes, and finally to the concrete and asphalt surfaces used today. The state government generally was not directly involved in road construction. Therefore, progress in the development of a complete system of roads was slow at best, since it was left to private investors faced with the prospect of a small rate of return. By 1840 only nine hundred miles of macadamized routes were in place or under construction in Kentucky. The Covington-Lexington Turnpike, later part of the famous Dixie Highway, was probably the state's most widely used long-distance road. Road construction slowed during the 1850s, 1860s, and 1870s; many existing routes were not well maintained and rapidly deteriorated. The tolls on Kentucky's early roads fueled debate and anger, especially among farmers, who needed an inexpensive method for getting produce to local markets, and

among bicyclists, who wanted more roads and improved maintenance. Although Kentucky's fourth constitution, adopted in 1891, forbade the Commonwealth to construct highways, a vocal Good Roads movement formed in the state, as well as nationally. The adoption of a Kentucky constitutional amendment in 1909 permitted the state to aid in road construction and authorized greater road-building responsibility for the counties. The U.S. Congress finally passed the Road Aid Act in 1916, providing a mechanism to funnel appropriations to states for highway construction.

The newly created Kentucky Department of State Roads and Highways reported slightly more than five hundred miles of highways paved between 1921 and 1923. Much of this early paving radiated only short distances from urban centers to facilitate farm-to-market movements and further asserted the importance of county seats. The density of paving was greatest in the Bluegrass Region throughout the 1920s and 1930s. Intercity and interstate routes slowly appeared on Kentucky's road maps. These longer-distance routes were part of the designated national highway system. Like the river and rail routes, the earliest Kentucky long-distance highways were oriented north-south. The 1920s witnessed an explosion in the sales of vehicles, including cars, trucks, and buses. Intercity truck and bus service began to provide an important alternative to rail, especially for Kentucky's

intrastate freight and passenger movements. Public works projects activated rapid expansion of paved road mileage in the 1930s. Both growth trends, in vehicles and paved mileage, were temporarily interrupted by World War II, but the railroads could no longer claim to be the dominant mode of transportation in the state. In the 1950s Kentucky found itself in the midst of the auto and air era of transportation development.

With the close of World War II, Kentucky's highway construction accelerated, and residents increasingly looked to roads

TRAFFIC FLOWS

VEHICLES PER DAY

— Fewer than 15,000
— 15,000 - 20,000
— 20,000 - 30,000
— 30,000 - 50,000
— More than 50,000

1993 data

INTERCITY BUS ROUTES, 1975 AND 1994

—— 1994 Bus Routes
········ 1975 Routes Abandoned by 1994
● 1994 Bus Stops

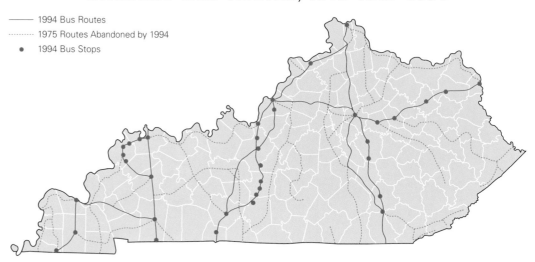

ROAD DENSITY

MILES OF PAVED ROAD PER SQUARE MILE OF AREA

▢ 1.07 - 1.50
▢ 1.51 - 1.75
▢ 1.76 - 2.00
▢ 2.01 - 2.50
▢ 2.51 - 5.74

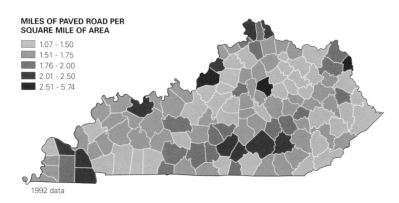

1992 data

HIGHWAY COAL TRAFFIC

PERCENT OF TON-MILES

▢ 0.1 - 0.5
▢ 0.6 - 1.0
▢ 1.1 - 5.0
▢ 5.1 - 10.0
▢ More than 10.0
▢ No coal traffic

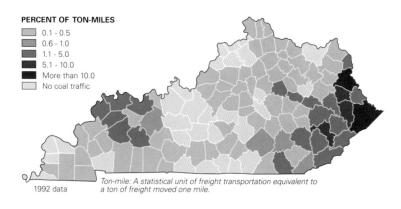

Ton-mile: A statistical unit of freight transportation equivalent to a ton of freight moved one mile.

1992 data

to meet demands for overland travel. The number of household vehicles grew rapidly, and trucking companies, with their greater speed and flexibility, started to compete with rail for freight traffic. The original forty-one-thousand-mile Interstate Highway System was adopted at the federal level in 1956, and Kentucky immediately planned its participation. A key decision made during the planning stage was to link isolated areas in Kentucky with the interstate system through a network of development parkways. Beginning with the opening of the Mountain Parkway in 1962, the interstates and parkways were systematically completed throughout the 1960s and 1970s, with the last urban section (I-265) completed around Louisville in 1987. Today, I-75, I-65, and I-71 provide north-south access, while I-64 and I-24 provide essential east-west access. Seven development parkways are designed to complement the interstate system. Although Kentucky's interstate segments and parkways (about thirteen hundred miles) constitute only 5 percent of paved mileage within the Commonwealth, they carry 30 percent of the state's highway traffic. One prospect involves the possible addition of I-66 to the existing interstate system. Only in the proposal stage, I-66 would connect the East and West Coasts of the United States and would cut across southern Kentucky, entering in the east near Pikeville and extending westward through Bowling Green.

Kentucky's highways are used by many groups for a variety of reasons. According to the U.S. Bureau of the Census,

An intermodal truck box is loaded at Burlington Northern's Louisville hub.

only 3 percent of Kentucky's labor force works at home. The remainder commutes to work on Kentucky's highways. More than three-fourths (76 percent) of Kentucky's workers drive alone to arrive at work, the average trip being nearly twenty-one minutes long. Travel times tend to be greater in congested urban settings or for commuters who reside in remote counties and must travel longer distances to get to an employment center. Because of limited job opportunities, many of Kentucky's workers find employment outside their home counties. Suburban counties witness a daily depopulation during weekday

CHANGE IN VEHICLE REGISTRATION, 1982 - 1991

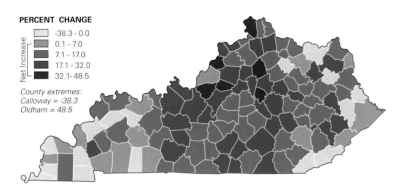

PERCENT CHANGE

Net Increase

- -38.3 - 0.0
- 0.1 - 7.0
- 7.1 - 17.0
- 17.1 - 32.0
- 32.1 - 48.5

County extremes:
Calloway = -38.3
Oldham = 48.5

VEHICLES REGISTERED

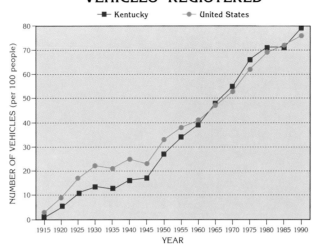

■ Kentucky ● United States

MEAN TRAVEL TIME TO WORK

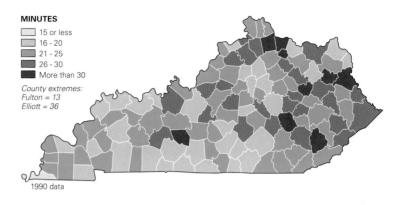

MINUTES

- 15 or less
- 16 - 20
- 21 - 25
- 26 - 30
- More than 30

County extremes:
Fulton = 13
Elliott = 36

1990 data

HOUSEHOLDS WITH NO VEHICLE

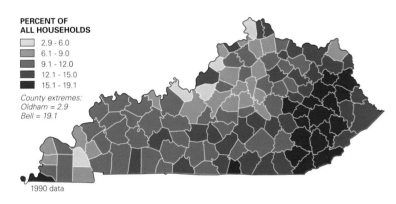

**PERCENT OF
ALL HOUSEHOLDS**

- 2.9 - 6.0
- 6.1 - 9.0
- 9.1 - 12.0
- 12.1 - 15.0
- 15.1 - 19.1

County extremes:
Oldham = 2.9
Bell = 19.1

1990 data

working hours. Almost three-fourths (71 percent) of Bullitt County's labor force heads out of the county each day, as those commuters drive to job opportunities, primarily in Jefferson County. Kentucky's highways are also shared with a growing number of tourists from within and outside the state. Automobiles carrying workers, shoppers, and tourists share the highways with trucks and buses. The volume of truck traffic is growing more rapidly than that of automobile traffic. More than fifty-one thousand freight carriers operate in Kentucky, with a combined fleet of nearly one million trucks that range from pickup trucks to tractors with twin trailers. The heaviest vehicles on the road are loaded coal trucks. Kentucky's thirty-eight hundred registered coal trucks can carry up to 120,000 pounds each, 40,000 pounds more than other trucks. In the early 1990s trucks carried about three times as much coal tonnage as did railroads. The state government collects a user fee from coal shippers to help cover the costs of added highway maintenance.

Unlike the trucking industry, Kentucky's scheduled bus service has declined drastically in the recent past. Almost all rural routes have been abandoned, and much better north-south access is provided than east-west. Owned and rented automobiles are preferred by most consumers for short and midrange trips. Longer trips are captured by the air industry. In contrast, chartered buses continue to perform valuable specialized services, and growth in that industry is anticipated as the retired and elderly component of the national population expands.

PEOPLE WHO WORK OUTSIDE COUNTY OF RESIDENCE

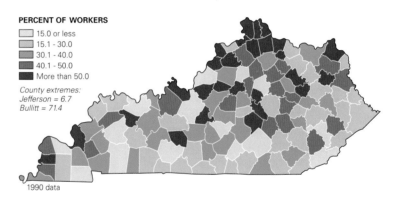

PERCENT OF WORKERS

- 15.0 or less
- 15.1 - 30.0
- 30.1 - 40.0
- 40.1 - 50.0
- More than 50.0

County extremes:
Jefferson = 6.7
Bullitt = 71.4

1990 data

TRAFFIC ACCIDENTS

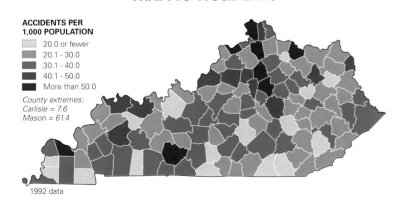

**ACCIDENTS PER
1,000 POPULATION**

- 20.0 or fewer
- 20.1 - 30.0
- 30.1 - 40.0
- 40.1 - 50.0
- More than 50.0

County extremes:
Carlisle = 7.6
Mason = 61.4

1992 data

CARPOOLING TO WORK

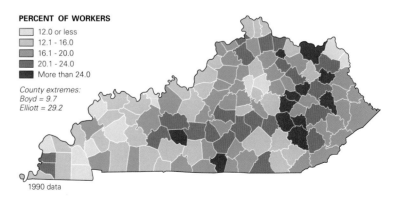

PERCENT OF WORKERS

- 12.0 or less
- 12.1 - 16.0
- 16.1 - 20.0
- 20.1 - 24.0
- More than 24.0

County extremes:
Boyd = 9.7
Elliott = 29.2

1990 data

PUBLIC USE AIRPORTS

✈ Scheduled Commercial
✈ Nonscheduled Commercial
✈ Turboprop and Small Jet
✈ Small Aircraft

1994 data

DIRECT CONNECTION COMMUTER FLIGHTS FROM KENTUCKY AIRPORTS

(Turboprop service on a daily basis)

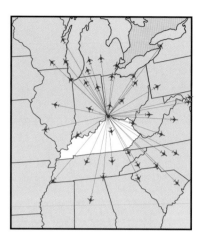

CINCINNATI
Cincinnati/Northern Kentucky
International Airport

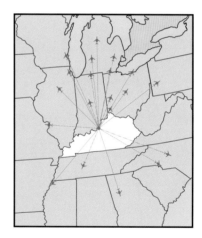

LOUISVILLE
International Airport
at Standiford Field

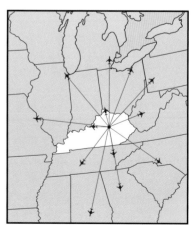

LEXINGTON
Blue Grass Airport

1994 data

AIR PASSENGER TRAFFIC

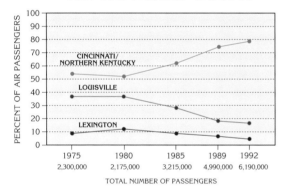

PERCENT OF AIR PASSENGERS

CINCINNATI/NORTHERN KENTUCKY
LOUISVILLE
LEXINGTON

1975	1980	1985	1989	1992
2,300,000	2,175,000	3,215,000	4,990,000	6,190,000

TOTAL NUMBER OF PASSENGERS

AIR FREIGHT TRAFFIC

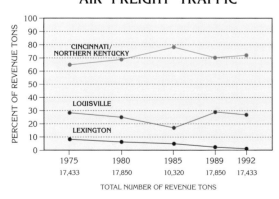

PERCENT OF REVENUE TONS

CINCINNATI/NORTHERN KENTUCKY
LOUISVILLE
LEXINGTON

1975	1980	1985	1989	1992
17,433	17,850	10,320	17,850	17,433

TOTAL NUMBER OF REVENUE TONS

Air Transportation

The other travel mode that attained commercial importance after World War II is air travel, for both passengers and freight. Louisville's Bowman Field was established as Kentucky's first civilian airfield in 1919. It was not until the 1930s, when the DC-3 entered air service, that air transportation could produce profits from basic passenger service and more lucrative U.S. mail carriage. The Greater Cincinnati Airport, in Boone County, opened in 1941 and became operational shortly after the close of World War II, in 1947. Several of Kentucky's larger airfields, developed by the military during the war years, were subsequently given to the local communities. Examples include Louisville's International Airport at Standiford Field, Lexington's Blue Grass Airport, and Paducah's Barkley Field. Air passenger and freight traffic increased rapidly after the introduction of jet aircraft in the late 1950s. Jets first landed at Kentucky airports in 1959. At least partly because of Kentucky's central location, United Parcel Service (UPS) decided in 1965 to operate its national air service hub at Standiford Field in Louisville. UPS operates nearly one hundred flights per day from Standiford and carries more than seven hundred thousand parcels daily to more than 180 countries and regions around the world.

Deregulation of air transportation during the late 1970s has had an impact on Kentucky's airports. With greater ease of route

SCHEDULED FLIGHTS FROM LOUSIVILLE INTERNATIONAL AIRPORT AT STANDIFORD FIELD, 1978
(Before Deregulation)

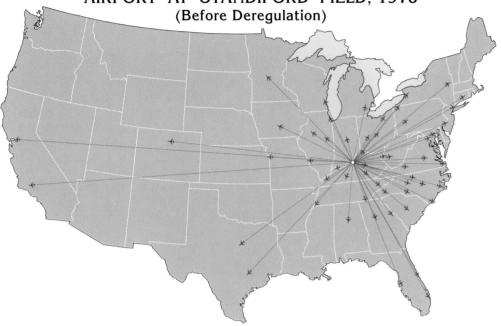

SCHEDULED FLIGHTS FROM CINCINNATI/NORTHERN KENTUCKY INTERNATIONAL AIRPORT, 1978
(Before Deregulation)

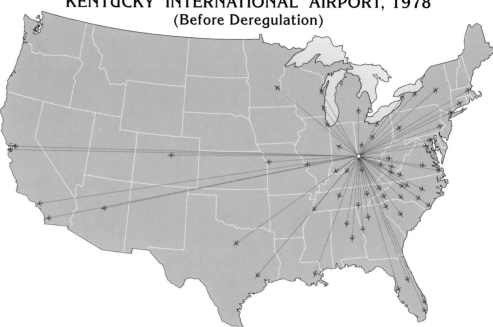

SCHEDULED FLIGHTS FROM LOUISVILLE INTERNATIONAL AIRPORT AT STANDIFORD FIELD, 1994
(After Deregulation)

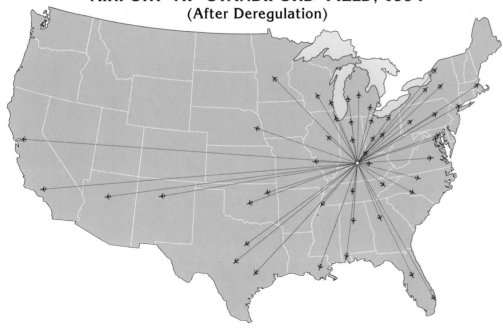

SCHEDULED FLIGHTS FROM CINCINNATI/NORTHERN KENTUCKY INTERNATIONAL AIRPORT, 1994
(After Deregulation)

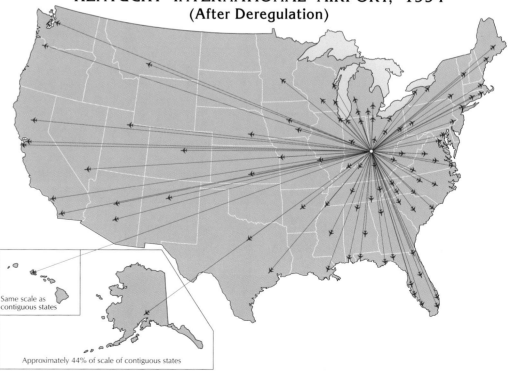

Same scale as
contiguous states

Approximately 44% of scale of contiguous states

INTERNATIONAL DIRECT FLIGHTS FROM
CINCINNATI/NORTHERN KENTUCKY INTERNATIONAL AIRPORT

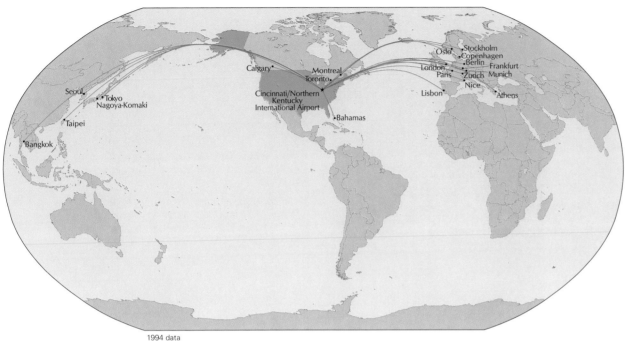

1994 data

A new passenger terminal was completed in December 1994 at the Cincinnati/Northern Kentucky International Airport to accommodate increased traffic.

entry and exit, significant changes have taken place at the Cincinnati/Northern Kentucky International Airport. This Boone County airport was selected as a hub by several airlines, most notably Delta, and its direct connections to other U.S. cities have been expanded considerably. The airport also has enjoyed a significant expansion of international connections, including European and Asian destinations. To feed this hub with passengers from surrounding airports within a radius of about five hundred miles, such as Lexington's Blue Grass Airport, a regional system of commuter flights has developed. The net result has been rapid growth in passenger traffic throughout the 1980s and 1990s at Cincinnati/Northern Kentucky International Airport. In contrast, Standiford Field in Louisville enjoys considerable freight traffic associated with UPS operations, but air passenger traffic has grown little. Standiford's direct connections with other U.S. cities have declined since deregulation, as have those from Blue Grass Airport.

United Parcel Service (UPS) aircraft are loaded at night on the tarmac of the Louisville International Airport at Standiford Field.

KENTUCKY EDUCATIONAL TELEVISION (KET) SERVICE AREAS

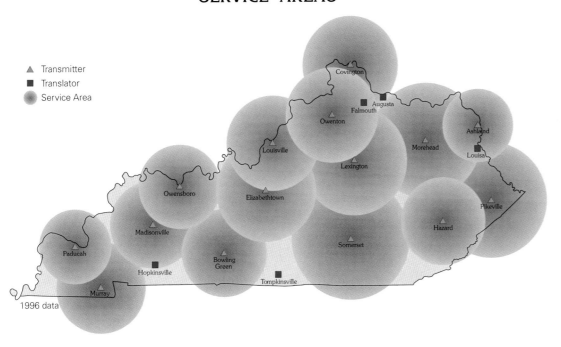

▲ Transmitter
■ Translator
⬤ Service Area

1996 data

DOMINANT TELEVISION MARKET AREAS

1994 data

TELEVISION STATIONS

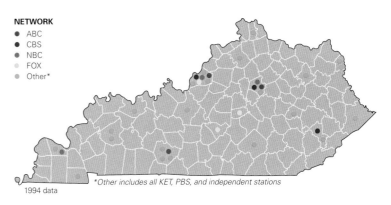

NETWORK
● ABC
● CBS
● NBC
● FOX
● Other*

1994 data   *Other includes all KET, PBS, and independent stations

Communications

The changes in communications media have also drastically affected the ability of Kentuckians to interact with one another and with the outside world. In some cases, communications can reduce the amount of transportation needed. A telephone call might substitute for a shopping trip, or an employee might complete a project at home on a personal computer and, through the use of electronic mail, transmit the work to a supervisor. Conversely, it is also apparent that the use of communications can increase the demand for travel and transportation. Effective advertising through the mass media generates shopping or tourism trips that otherwise would not have been made.

Kentucky is well served by affiliate television stations both within the state and surrounding it. Each of the major networks is represented. The state divides into ten distinctive viewing regions based on the center that captures the greatest market share. Much of the state's television market is shared with stations located outside the state. This is an important consider-

ation for any person or group, such as politicians, who must attempt to be seen by all, or at least most, Kentuckians. The Commonwealth also enjoys one of the finest public educational television systems in the country. Kentucky Educational Television (KET) went on the air in 1969 and today serves about forty thousand square miles from fifteen individual transmitters and six translators. A translator is a signal booster for use in areas of rough topography; it enables reception at a very localized level. KET also operates the highly successful STAR Channel, a statewide satellite system with which instructional programming can be beamed to any public school, state park, community college, or university in Kentucky.

Although first invented nearly a century ago, radio remains a popular communications medium in Kentucky. Nearly 290 licensed stations operate within the state, and 57 percent of these are FM. The most popular musical formats are country and country and western, which together account for 34 percent of all

The Kentucky Educational Television building in Lexington, Fayette County, houses the headquarters of Kentucky's public television network.

FM RADIO STATIONS

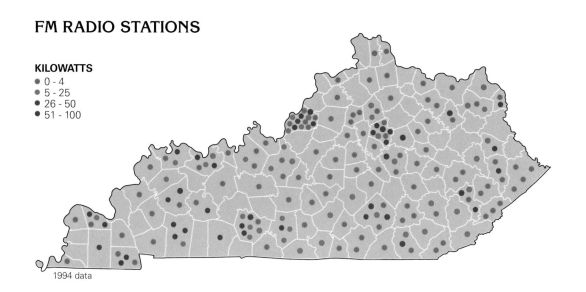

KILOWATTS
- 0 - 4
- 5 - 25
- 26 - 50
- 51 - 100

1994 data

AM RADIO STATIONS

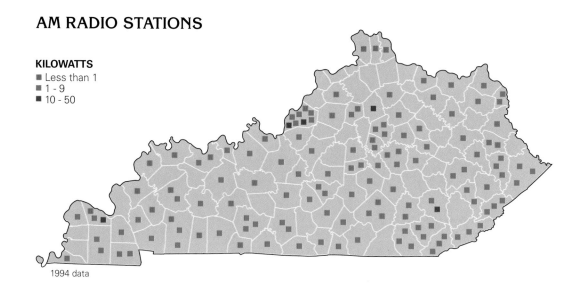

KILOWATTS
- Less than 1
- 1 - 9
- 10 - 50

1994 data

● FM STATIONS WITH 51-100 KILOWATTS

1.	WKYQ	93.3	Paducah
2.	WDDJ	96.9	Paducah
3.	WKMS	91.3	Murray
4.	WBLN	103.7	Murray
5.	WCVQ	107.9	Ft. Campbell
6.	WHOP	98.7	Hopkinsville
7.	WZZF	100.3	Hopkinsville
8.	WQXQ	101.9	Central City
9.	WKDQ	99.5	Henderson
10.	WDKM	92.5	Owensboro
11.	WSTO	96.1	Owensboro
12.	WBVR	101.1	Russellville
13.	WKYU	88.9	Bowling Green
14.	WWHR	91.7	Bowling Green
15.	WGGC	95.1	Glasgow
16.	WDCL	89.7	Somerset
17.	WFPK	91.9	Louisville
18.	WFPL	89.3	Louisville
19.	WAMZ	97.5	Louisville
20.	WKQQ	98.1	Lexington
21.	WMXL	94.5	Lexington
22.	WUKY	91.3	Lexington
23.	WVLK	92.9	Lexington
24.	WSGS	101.1	Hazard
25.	WQHY	95.5	Prestonsburg
26.	WSIP	98.9	Paintsville
27.	WRVC	93.7	Ashland

FM STATIONS FOLLOWING NATIONAL PUBLIC RADIO (NPR) FORMAT

1.	WKMS	Murray
2.	WKYU	Bowling Green
3.	WFPK	Louisville
4.	WFPL	Louisville
5.	WUOL	Louisville
6.	WUKY	Lexington
7.	WNKU	Highland Heights
8.	WEKU	Richmond
9.	WMKY	Morehead
10.	WMMT	Whitesburg

■ AM STATIONS WITH 10-50 KILOWATTS

1.	WPAD	1560	Paducah
2.	WDJX	1080	Louisville
3.	WHAS	840	Louisville
4.	WBBE	1580	Georgetown
5.	WKLB	1290	Manchester

stations. A growing number of classic rock stations attests to the radio industry's attempt to serve the post–World War II Baby Boom generation.

Twenty-three daily newspapers dominate the print media and sell just over seven hundred thousand papers each day. The *Louisville Courier-Journal* is the state's largest newspaper (circulation exceeds 240,000 daily), and the *Lexington Herald-Leader* ranks second (circulation exceeds 127,000). Nine other newspapers in Kentucky have daily circulations that exceed ten thousand. A map of circulation figures reveals zones of urban influence. The boundary between the Lexington and Louisville zones is clearly established along a north-south line that coincides with the boundary between Shelby and Franklin Counties. The *Courier-Journal* dominates west of that line, and the *Herald-Leader* dominates to the east. The success of Kentucky's larger newspapers has, at least in part, come at the expense of local newspapers. While the market areas of larger newspapers have expanded during the past decade, those for smaller city newspapers have generally contracted. This pattern coincides with enlarged retail market areas for cities that are home to Kentucky's larger shopping malls.

The Courier-Journal, *published at Fifth Street and Broadway in Louisville, is one of the nation's most respected newspapers.*

WEEKLY CIRCULATION COMPARISON OF KENTUCKY'S TWO LEADING NEWSPAPERS

- Louisville Courier-Journal
- Lexington Herald-Leader

Each • represents 500 newspapers.

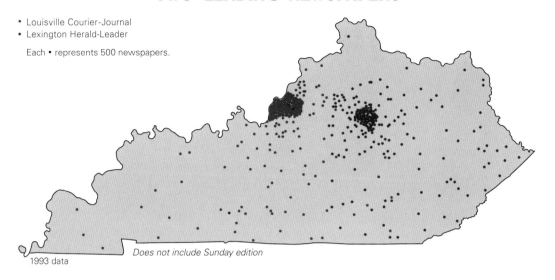

1993 data

Does not include Sunday edition

WEEKLY CIRCULATION COMPARISON OF KENTUCKY'S OTHER REGIONAL NEWSPAPERS

- Paducah Sun
- Owensboro Messenger-Inquirer
- Bowling Green News
- Cincinnati Enquirer
- Ashland Independent

Each • represents 100 newspapers.

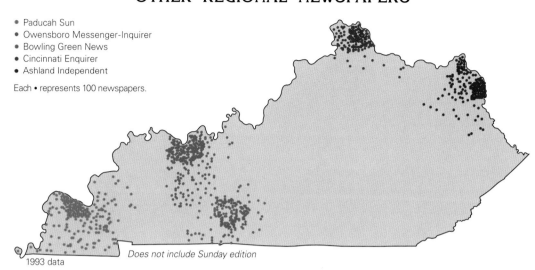

1993 data

Does not include Sunday edition

NEWSPAPER WITH THE LARGEST MARKET SHARE IN EACH COUNTY

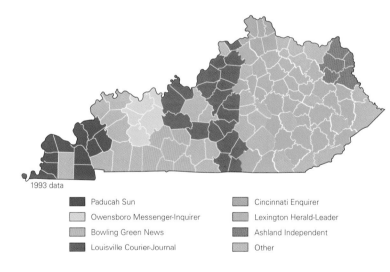

1993 data

- Paducah Sun
- Owensboro Messenger-Inquirer
- Bowling Green News
- Louisville Courier-Journal
- Cincinnati Enquirer
- Lexington Herald-Leader
- Ashland Independent
- Other

The Mitchellsburg post office and general store in Boyle County.

UNITED STATES POST OFFICES

NUMBER PER 10,000 POPULATION

- 4.0 or less
- 4.1 - 8.0
- 8.1 - 12.0
- 12.1 - 16.0
- 16.1 or more

County extremes:
Anderson, Jefferson,
Simpson, and Clark = 0.7
Magoffin = 28.3

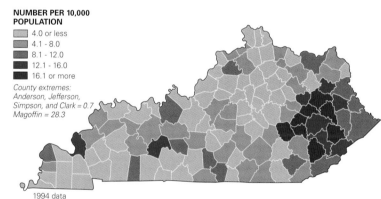

1994 data

Mail delivery, including magazine circulation, is facilitated in Kentucky through an extensive system of U.S. post offices. The U.S. Postal Service operates more than 1,360 main post offices and branches in Kentucky. The number of post offices per county ranges from just one (in Anderson, Robertson, and Simpson Counties) to sixty-two in urban Fayette County and sixty-five in rural Pike County. Eastern Kentucky has a particularly high density of post offices because of the dispersed na-

ture of the rural population, the region's rugged topography, and the importance of postal jobs to the region.

The top five magazines circulated in Kentucky are *Modern Maturity* (received by almost 20 percent of all households), *TV Guide, Reader's Digest, Better Homes and Gardens,* and *Southern Living.* This circulation ranking is identical to that of the nation, with one exception: the fifth most popular magazine in the United States is *National Geographic,* not *Southern Living.* The popularity of individual magazines varies across Kentucky. News magazines such as *Time* sell best in urban, suburban, and university counties. *Better Homes and Gardens* does well in the western and northern portions of Kentucky, but not in the east. In contrast, of all magazines circulated in eastern Kentucky, *TV Guide* ranks first. Magazines directed toward more specialized markets, like *Ebony,* sell in predictable patterns: *Ebony* circulates in largest numbers in direct proportion to the number of Kentucky's black residents. Nearly eight thousand copies are sold each month in Jefferson County, where 43 percent of the state's black residents live.

Libraries also provide access to the print and other media for 1.6 million registered borrowers. According to Kentucky's Department of Libraries and Archives, if the material circulated by Kentucky's public libraries in 1993 was stacked side by side, the line of books, magazines, and tapes would stretch from Paducah to Ashland. The largest holdings of public library books are found in Jefferson and Fayette Counties—872,000 and 600,000 volumes respectively—although six other counties also hold more than 100,000 volumes each. Two counties, McLean and Carter, do not have their own public libraries. In 1953 bookmobiles began to serve rural areas where access to libraries was limited. Libraries are of course also found in public and independent schools across the state, and by 1933 every high school had some type of library service. The single largest library in the state is that at the University of Kentucky, which houses more than 2.5 million books and receives more than twenty-seven thousand periodical and serial titles. It ranks fifty-third among the nation's 108 research libraries. The next largest libraries in the state are those at the University of Louisville (1.2 million books) and Western Kentucky University (834,000 books).

SUBSCRIBERS TO *TV GUIDE*

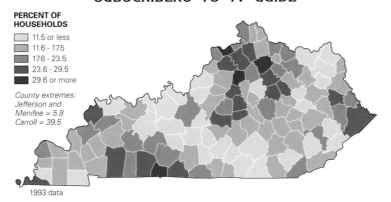

PERCENT OF HOUSEHOLDS
- 11.5 or less
- 11.6 - 17.5
- 17.6 - 23.5
- 23.6 - 29.5
- 29.6 or more

County extremes:
Jefferson and
Menifee = 5.8
Carroll = 39.5

1993 data

SUBSCRIBERS TO *TIME* MAGAZINE

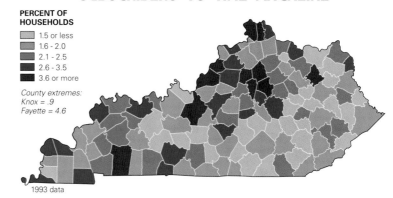

PERCENT OF HOUSEHOLDS
- 1.5 or less
- 1.6 - 2.0
- 2.1 - 2.5
- 2.6 - 3.5
- 3.6 or more

County extremes:
Knox = .9
Fayette = 4.6

1993 data

SUBSCRIBERS TO *BETTER HOMES AND GARDENS*

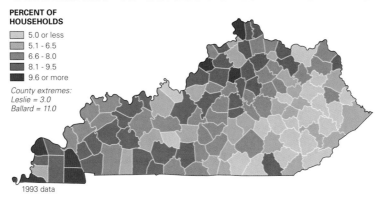

PERCENT OF HOUSEHOLDS
- 5.0 or less
- 5.1 - 6.5
- 6.6 - 8.0
- 8.1 - 9.5
- 9.6 or more

County extremes:
Leslie = 3.0
Ballard = 11.0

1993 data

SUBSCRIBERS TO *EBONY* MAGAZINE

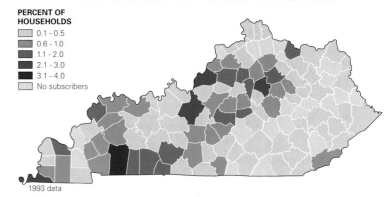

PERCENT OF HOUSEHOLDS
- 0.1 - 0.5
- 0.6 - 1.0
- 1.1 - 2.0
- 2.1 - 3.0
- 3.1 - 4.0
- No subscribers

1993 data

PUBLIC LIBRARY BOOKS

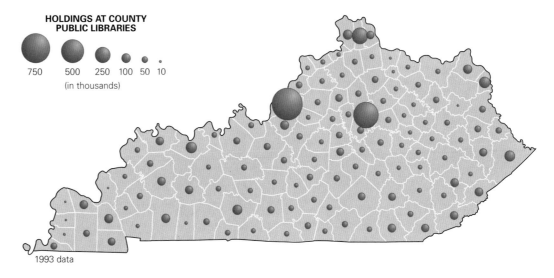

HOLDINGS AT COUNTY PUBLIC LIBRARIES

750   500   250   100   50   10
(in thousands)

1993 data

Commercial television satellite dishes line Winchester Road in Lexington.

A cellular phone tower on Palumbo Drive in Lexington.

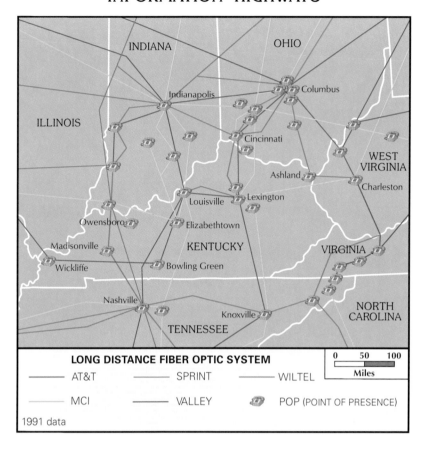

INFORMATION HIGHWAYS

LONG DISTANCE FIBER OPTIC SYSTEM

—— AT&T   —— SPRINT   —— WILTEL

—— MCI   —— VALLEY   POP (POINT OF PRESENCE)

1991 data

0   50   100
Miles

Telegraph, in the 1840s, and telephone, in the 1880s, provided the earliest forms of telecommunications in Kentucky. Transcontinental telephone calls, possible by 1915, were not commonplace until the 1920s. Not until the current digital computer era, however, would the most exciting applications of telecommunications take place. In 1992 Kentucky established a Telecommunications Research Center at the University of Louisville to assist all Kentuckians in coping with rapidly changing information technologies. Today, many state residents have grown accustomed to the benefits of the Internet, 911 emergency systems, pager systems, cellular phones, and fax services. Twenty local exchange utilities and twenty-eight cellular companies provide basic telephone service in Kentucky. Unfortunately, in some instances long-distance charges are still ascribed to telephone calls within a single county because the county has been split among local exchanges. In a deregulated industry, competition for long-distance telephone traffic is strong. More than one hundred long-distance carriers now serve Kentucky, although the high-capacity long-distance infra-

STATUS OF 911 SERVICE

TYPE OF SERVICE

No Service
Basic 911*
Enhanced 911**

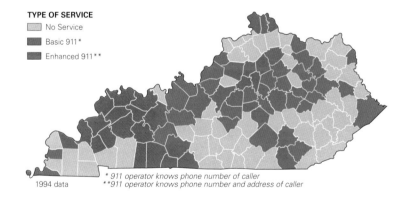

1994 data

* 911 operator knows phone number of caller
**911 operator knows phone number and address of caller

SERVICE AREAS OF TELEPHONE COMPANIES

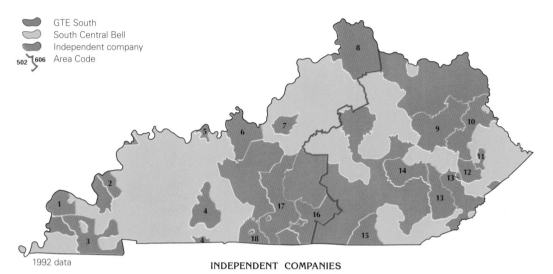

GTE South
South Central Bell
Independent company
502 606 Area Code

1992 data

INDEPENDENT COMPANIES

1. Ballard Rural Cooperative
2. Salem Telephone
3. West Kentucky Rural Telephone
4. Logan Telephone
5. Lewisport Telephone
6. Brandenburg Telephone
7. ALLTEL Kentucky
8. Cincinnati Bell
9. Mountain Rural Cooperative
10. Foothills Rural Cooperative
11. Harold Telephone
12. Thacker-Grigsby Telephone
13. Leslie County Telephone
14. Peoples Rural Telephone
15. Highland Telephone
16. Duo County Telephone
17. South Central Rural
18. North Central Cooperative

CELLULAR TELECOMMUNICATION MARKET

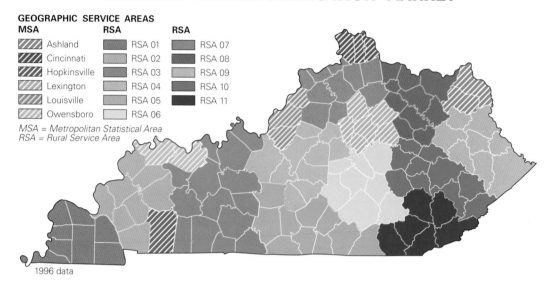

GEOGRAPHIC SERVICE AREAS

MSA	RSA	RSA
Ashland	RSA 01	RSA 07
Cincinnati	RSA 02	RSA 08
Hopkinsville	RSA 03	RSA 09
Lexington	RSA 04	RSA 10
Louisville	RSA 05	RSA 11
Owensboro	RSA 06	

MSA = Metropolitan Statistical Area
RSA = Rural Service Area

1996 data

structure (fiber optics) is owned by just a few large companies, like AT&T, MCI, and Sprint. These larger telecommunications firms lease portions of their capacity to smaller firms. Access to the long-distance fiber optic system remains problematic for some parts of Kentucky. An "on ramp" to this system is called a Point of Presence (POP). There are very few POPs in Appalachian Kentucky, and this could be an obstacle to future economic development. Another problem facing Kentucky residents as we enter the information age is household access to information highways. Many Kentucky households have no telephone (or computer) and will remain isolated from the information economy until access improves.

HOUSEHOLDS WITH NO TELEPHONE

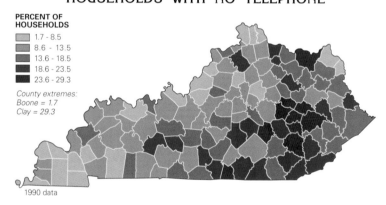

PERCENT OF HOUSEHOLDS

- 1.7 - 8.5
- 8.6 - 13.5
- 13.6 - 18.5
- 18.6 - 23.5
- 23.6 - 29.3

County extremes:
Boone = 1.7
Clay = 29.3

1990 data

A microwave DR-6 AT&T digital radio tower. This POP 135 Megabit tower system is located on Viley Road in Lexington.

Tourism and Recreation

CHAPTER TEN

Tourism and Recreation

This natural bridge, 78 feet long by 65 feet high, is the focal point of 1,900-acre Natural Bridge State Resort Park in Powell County.

ROLLING BLUEGRASS, thoroughbred horses, the Kentucky Derby, basketball, bourbon, fried chicken, caves, recreational waterways, state parks, Appalachia, poverty, tobacco, and coal mining—these are some of the images that may come to mind when one thinks about Kentucky. Such images suggest something of the natural and human diversity of the Commonwealth. Kentucky's varied landscapes, coupled with the state's situation, within a day's drive of more than two-thirds of the nation's population, help make tourism one of the state's major industries, ranking second in revenue after manufacturing. In 1993 the tourism and travel industry contributed more than $6.8 billion to the state economy; of this total, visitors to the state contributed about $4.6 billion. The Kentucky Department of Travel Development estimates that the industry generated $443.1 million in state tax revenues and $76.9 million in local government tax revenues that year. In 1994 tourism and travel supported 144,691 jobs. Of these jobs, 105,156 were the result of direct expenditures and 39,535 resulted from indirectly generated expenditures.

Tourism Employment, Facilities, and Expenditures

The Department of Travel Development divides Kentucky into nine travel and tourism regions. In revenue and employment from tourism, the regions that include the metropolitan areas of Lexington (the Bluegrass Region), Louisville (Louisville-Lincoln), and northern Kentucky easily lead the state. The Louisville-Lincoln and Bluegrass Regions together accounted for 52.5 percent of the expenditures on travel and tourism in 1994. The greatest concentration of tourist activity, however, is found in the three regions that contain Lake Cumberland (Southern Kentucky Lakes and Rivers), Mammoth Cave (Cave), and Land Between the Lakes (Western Lakes and Rivers). Mammoth Cave National Park and Land

TRAVEL AND TOURISM REGIONS

The lodge at Lake Barkley State Resort Park, one of fifteen state resort parks. Lake Barkley's 3,700 acres make it the state's largest park.

Cumberland Falls, sometimes referred to as the "Niagara of the South," plunges almost sixty feet, making it one of the largest waterfalls in the Southeast. It is the focal point for 1,657-acre Cumberland Falls State Resort Park. White-water rafting trips begin below the falls.

Between the Lakes are the state's two most popular attractions, as measured by the number of visitors in 1992: more than 2.5 and 2.3 million people visited these sites, respectively. Every year, millions of people visit Lake Cumberland, which has 63,000 surface acres and more than 1,250 miles of shoreline. Impounded in 1952 by the U.S. Army Corps of Engineers with the construction of Wolf Creek Dam, Lake Cumberland is Kentucky's largest lake and one of two major impoundments along the Cumberland River (the other is Lake Barkley in western Kentucky).

Also near this region (on the boundary between McCreary and Whitley Counties) is Cumberland Falls, which is 125 feet wide and plunges nearly 60 feet. Called the Niagara of the South, Cumberland Falls is one of the largest waterfalls in the southeastern United States. On a clear night with a full moon, the visitor can see a moonbow, created by the mist of the falls. Along another branch of the Cumberland River about twenty-five miles south of Somerset is the Big South Fork National River and Recreation Area, comprising more than 100,000 acres of land in Kentucky and Tennessee. In 1974 Congress set this land aside for preservation and recreation.

The importance of tourism to these regions can be demonstrated by the location quotient, here a measure indicating the importance of tourism employment in a county relative to the state average.

REGIONAL EMPLOYMENT IN TRAVEL AND TOURISM
1984 AND 1994

REGION	NUMBER EMPLOYED	
	1984	1994
Western Lakes and Rivers	7,110	8,945
Green River	4,614	5,281
Cave	7,064	7,990
Louisville-Lincoln	19,934	28,443
Southern Kentucky Lakes and Rivers	4,188	6,710
Northern Kentucky	7,078	12,527
Bluegrass	17,656	23,777
Eastern Highlands - North	4,253	5,222
Eastern Highlands - South	5,771	6,261

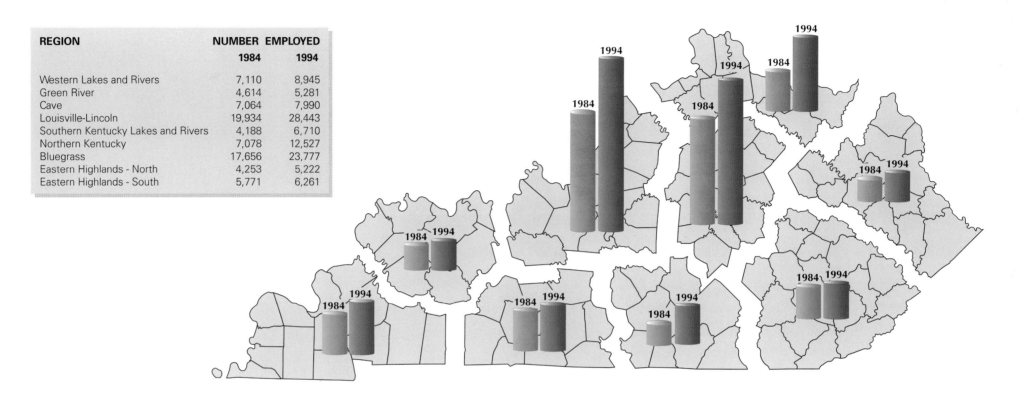

White-water rafting in the Kentucky portion of Big South Fork National River and Recreation Area in McCreary County.

LOCATION QUOTIENT: TOURISM

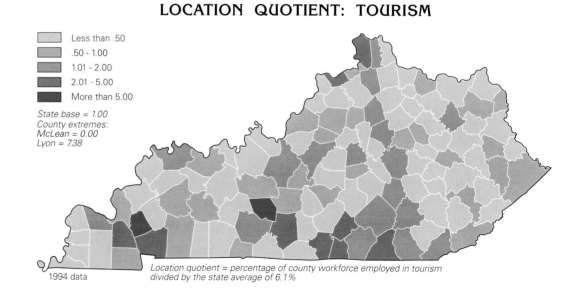

	Less than .50
	.50 - 1.00
	1.01 - 2.00
	2.01 - 5.00
	More than 5.00

State base = 1.00
County extremes:
McLean = 0.00
Lyon = 7.38

1994 data

Location quotient = percentage of county workforce employed in tourism divided by the state average of 6.1%

REGIONAL EXPENDITURES ON TRAVEL AND TOURISM, 1984 AND 1994

REGION	EXPENDITURES 1984	EXPENDITURES 1994
Western Lakes and Rivers	$154,957,674	$365,129,670
Green River	105,181,356	198,594,852
Cave	161,010,881	274,224,665
Louisville-Lincoln	499,983,938	1,271,655,256
Southern Kentucky Lakes and Rivers	95,462,285	247,519,999
Northern Kentucky	161,343,433	493,695,238
Bluegrass	379,699,259	951,011,046
Eastern Highlands - North	92,376,564	198,949,395
Eastern Highlands - South	120,143,989	235,293,188

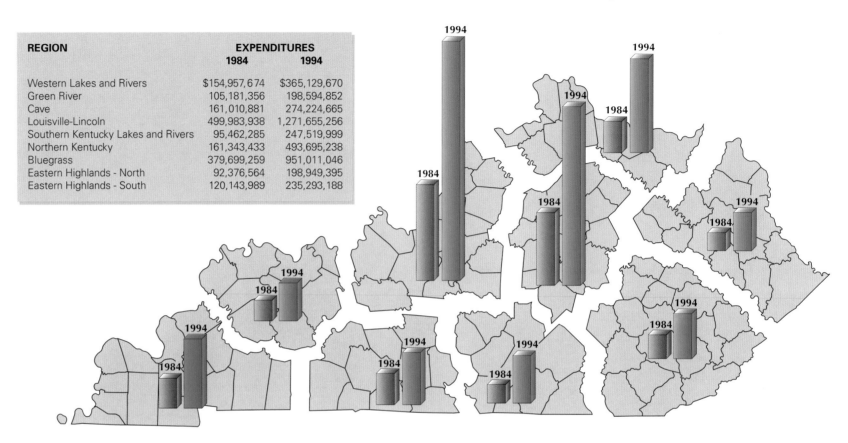

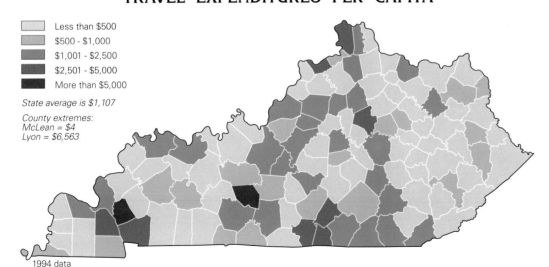

Kentucky's Red River Gorge, with its many spectacular cliffs, has become a mecca for rock climbers from across the country.

TRAVEL EXPENDITURES PER CAPITA

Less than $500
$500 - $1,000
$1,001 - $2,500
$2,501 - $5,000
More than $5,000

State average is $1,107

County extremes:
McLean = $4
Lyon = $6,563

1994 data

In 1994, 6.1 percent of the state's labor force worked in tourism. In Land Between the Lakes, however, 45 percent of the Lyon County workforce supported the tourism industry, a proportion more than seven times greater than the state average. And Edmonson County, with 33 percent of the workforce in tourism, is the home of the headquarters of Mammoth Cave National Park.

Tourist Attractions

In the 1930s private citizens donated money to preserve the beauty of the areas at Pine Mountain and Cumberland Falls, launching the state park system. Kentucky has, at present, fifteen state resort parks, all with lodging facilities. Historically, Kentucky state parks have ranked among the best in the nation. Ten of the fifteen state resort parks were ranked among the state's top twenty-five attractions in the early 1990s, as measured by number of visitors. The most visited state resort park, Kentucky Dam Village in Marshall County, had more than one million visitors in 1992. Although not a part of the state resort park system, the Breaks Interstate Park, a forty-six-hundred-acre park situated along the north-

TOP TWENTY-FIVE ATTRACTIONS

NUMBER OF VISITORS

More than 1.5 million

1-1.5 million

600,000 -1 million

400,000 - 600,000

250,000 - 400,000

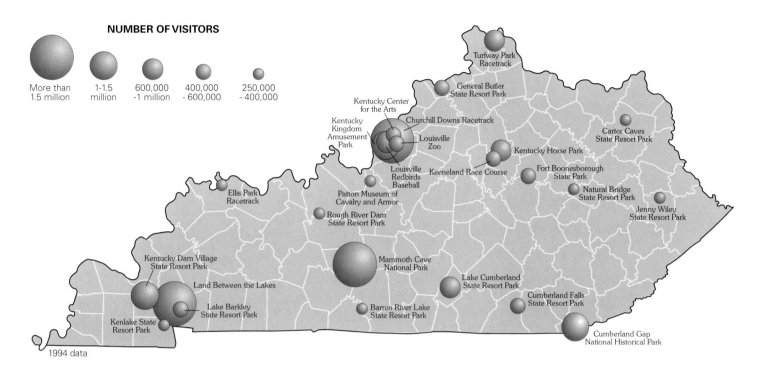

Turfway Park Racetrack

General Butler State Resort Park

Kentucky Center for the Arts

Kentucky Kingdom Amusement Park

Churchill Downs Racetrack

Louisville Zoo

Carter Caves State Resort Park

Kentucky Horse Park

Louisville Redbirds Baseball

Keeneland Race Course

Fort Boonesborough State Park

Ellis Park Racetrack

Patton Museum of Cavalry and Armor

Natural Bridge State Resort Park

Jenny Wiley State Resort Park

Rough River Dam State Resort Park

Kentucky Dam Village State Resort Park

Mammoth Cave National Park

Lake Cumberland State Resort Park

Land Between the Lakes

Lake Barkley State Resort Park

Barren River Lake State Resort Park

Cumberland Falls State Resort Park

Kenlake State Resort Park

Cumberland Gap National Historical Park

1994 data

Ashland, rebuilt in 1855-57 by the son of statesman Henry Clay on the site of his father's Lexington home, retained the original Federal floor-plan but added the Italianate ornamental details. Since 1950 it has been administered as a museum by the Henry Clay Memorial Foundation. The estate includes approximately three acres of woodlands, several dependencies, and a large formal garden.

eastern end of a ridge of Pine Mountain on the Kentucky-Virginia border, also has lodging facilities (the lodge is located on the Virginia side). This park was jointly created by the Kentucky and Virginia legislatures in 1954. It features a gorge, called the Breaks, which is more than a thousand feet deep.

In addition to the state resort parks, the Commonwealth boasts twenty-two state parks (which do not have lodging facilities) and nine state historic sites. Among the state's most visited attractions is Fort Boonesborough State Park (Madison County), which hosts more than a half million visitors annually. In April 1775 Daniel Boone and other early pioneers established Kentucky's second white settlement here, near the banks of the Kentucky River and close to a salt lick. Fort Boonesborough has been reconstructed as a working fort and includes block-houses and log cabins. Mineral Mound State Park in Lyon County is the newest addition to the state park system.

Large tracts of land throughout the state have been set aside to preserve forests, wildlife, and water and to provide recreation. These include state wildlife management areas, state and national forests, nature preserves, and trails. In 1994 the state created a new category called Watchable Wildlife viewing sites. The largest tract of land, more than 660,000 acres, is the Daniel Boone National

HOTEL, MOTEL, RESORT, AND BED AND BREAKFAST ROOMS

NUMBER OF ROOMS

1 - 250

251 - 550

501 - 1,000

1,001 - 5,000

More than 5,000

County maximum: Jefferson = 9,263 Nine counties have no lodging facilities.

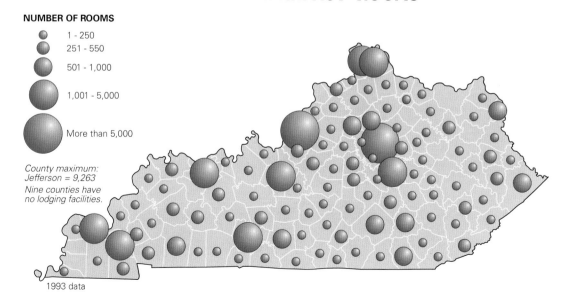

1993 data

STATE PARKS AND STATE RESORT PARKS

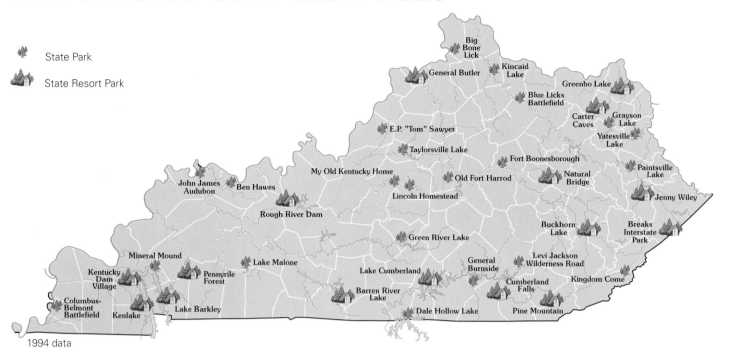

🌲 State Park

🏠 State Resort Park

1994 data

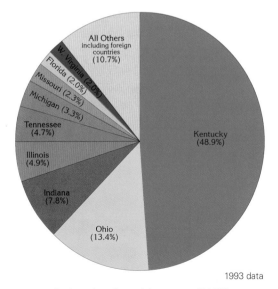

Autumn at the Breaks Interstate Park, a 4,500-acre park with lodge that was created jointly by the Kentucky and Virginia legislatures in 1954.

HOME OF OVERNIGHT GUESTS VISITING KENTUCKY'S STATE PARKS

- All Others including foreign countries (10.7%)
- W. Virginia (2.0%)
- Florida (2.0%)
- Missouri (2.3%)
- Michigan (3.3%)
- Tennessee (4.7%)
- Illinois (4.9%)
- Indiana (7.8%)
- Ohio (13.4%)
- Kentucky (48.9%)

1993 data

Total number of overnight guests = 604,020

Forest. Established in 1937 as Cumberland National Forest, the area had its name changed in 1966. The Daniel Boone National Forest contains the Red River Gorge National Geological Area, which extends more than twenty miles along the Red River, a tributary of the Kentucky River. The gorge is noted for its scenery and the diversity of its plant and animal life. Steep cliffs, rockshelters, and natural bridges are found along the gorge; sheer rock sandstone walls and hiking trails, such as the Red River Gorge National Recreation Trail, attract numerous rock climbers and hikers from throughout the Midwest and South. In December 1993 President Bill Clinton signed a bill giving this stretch of the Red River the status of Wild and Scenic River, thus ending the possibility of a dam being constructed along the river, something first proposed in the 1960s.

In the far western part of the state is another major recreation area, Land Between the Lakes. The Tennessee Valley Authority owns and manages the property; it opened its office in the area in early 1964. Almost two-thirds of the 170,000 acres that constitute LBL are located in Kentucky; the remainder are in Tennessee. Forest covers more than 90 per cent of the area. LBL is a forty-mile-long peninsula bordered by two impounded

lakes: Kentucky Lake on the west and Lake Barkley on the east. Neither a national park nor a wilderness area, LBL provides outdoor recreation and environmental education for visitors. The area is one of the most popular inland birdwatching sites in the eastern United States, and birders have observed more than 250 species there since 1963. LBL now contains no private holdings; the last private resident moved from the Kentucky portion of LBL in 1970.

Many nationally known persons have hailed from Kentucky. Opposing Civil War presidents Abraham Lincoln and Jefferson Davis were both born in the state, Lincoln near Hodgenville in 1809 and Davis at Fairview (then called Davisburg) in 1808. In 1905 writer Robert Penn Warren was born in Guthrie. National and state historic sites, including those honoring Lincoln and Davis, are found across the state. Other famous Kentucky citizens, though not necessarily born in the state, include Henry Clay, whose home, Ashland, is in Lexington; Dr. Thomas Walker, reputedly the first white man through the Cumberland Gap in 1750; the abolitionist Cassius Marcellus Clay, whose home, White Hall, is in Madison County; the naturalist and bird illustrator John James Audubon, who lived in Louisville and in

FEDERAL RECREATION LANDS

Gladie Cabin in the Red River Gorge National Geological Area, a region of cliffs, rockshelters, and natural bridges. Native American settlement of the area began as early as 8,000 B.C. Permanent settlement by whites began after 1787.

Cumberland Gap National Historical Park was created by President Franklin D. Roosevelt in 1940 and formally established in 1955, when land purchased for the park by Kentucky, Tennessee, and Virginia was turned over to the U.S. Department of the Interior. The Pinnacle Overlook rises almost 1,000 feet above the Gap.

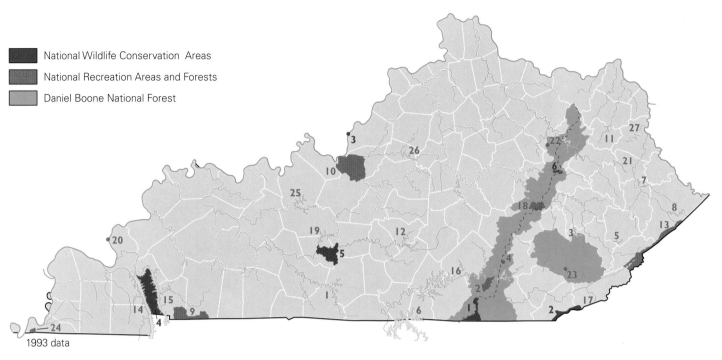

- ■ National Wildlife Conservation Areas
- ■ National Recreation Areas and Forests
- ■ Daniel Boone National Forest

1993 data

NATIONAL WILDLIFE CONSERVATION AREAS AND PARKS

1. Big South Fork National River and Recreation Area
2. Cumberland Gap National Historical Park
3. Falls of the Ohio National Wildlife Conservation Area
4. Land Between the Lakes (TVA)
5. Mammoth Cave National Park
6. Red River Gorge National Geological Area

DANIEL BOONE NATIONAL FOREST
(U.S. Forest Service)

TRAIL

Sheltowee Trace National Recreation Trail

NATIONAL RECREATION AREAS AND FORESTS
(owned by U.S. Army Corps of Engineers unless otherwise indicated)

1. Barren River Lake
2. Beaver Creek (U.S. Forest Service)
3. Buckhorn Lake
4. Cane Creek (U.S. Forest Service)
5. Carr Fork Lake
6. Dale Hollow Lake
7. Dewey Lake
8. Fishtrap Lake
9. Fort Campbell (U.S. Army)
10. Fort Knox (U.S. Army)
11. Grayson Lake
12. Green River Lake (managed by Ky. Dept. of Fish and Wildlife)
13. Jefferson National Forest (U.S. Forest Service)
14. Kentucky Lake (TVA)
15. Lake Barkley
16. Lake Cumberland
17. Martins Fork Lake
18. Mill Creek (U.S. Forest Service)
19. Nolin River Lake
20. Ohio River Islands
21. Paintsville Lake
22. Pioneer Weapons Hunting Area (U.S. Forest Service)
23. Redbird (U.S. Forest Service)
24. Reelfoot Lake National Wildlife Refuge (U.S. Fish & Wildlife Service)
25. Rough River Lake
26. Taylorsville Lake
27. Yatesville Lake

Sunset on Cave Run Lake, Rowan County. Cave Run, part of the Licking River system, was impounded by the Corps of Engineers in 1974. The lake covers more than 8,000 acres.

STATE RECREATION LANDS

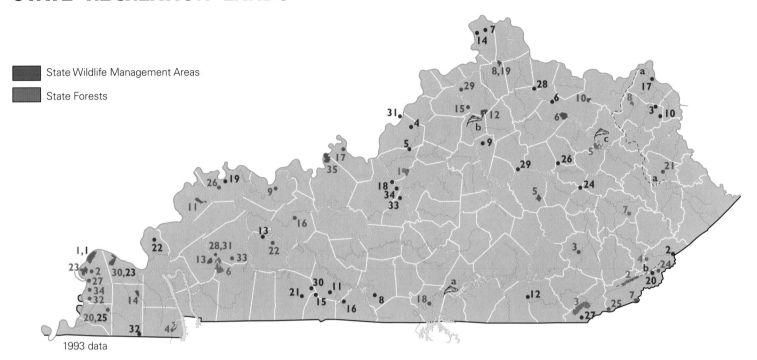

1993 data

The Shaker village at Pleasant Hill was founded in 1806 and reached a peak population of about 500 in the 1830s. Today it is an important attraction that draws more than 250,000 visitors annually.

STATE WILDLIFE MANAGEMENT AREAS
(owned by Kentucky Department of Fish and Wildlife unless otherwise indicated)

1. Ballard
2. Ballard Hunting Unit
3. Beech Creek
4. Beechy Creek
5. Central Kentucky
6. Clay
7. Cranks Creek
8. Curtis Gates Lloyd
9. Daviess Demonstration Area
10. Fleming County
11. Higginson-Henry
12. John A. Kleber
13. Jones-Keeney
14. Kaler Bottoms
15. Kentucky River
16. L.B. Davison
17. Lapland (Kimball International)
18. Mud Camp Creek
19. Mullins
20. Obion Creek
21. Paintsville Lake
22. Peabody PWA (Peabody and Beaver Dam Coal Companies and Peabody Development Company)
23. Peal
24. Pine Mountain
25. Shillalah Creek
26. Sloughs
27. Swan Lake
28. Tradewater
29. Twin Eagle

30. West Kentucky
31. West Kentucky 4-H Camp (4-H)
32. Westvaco-Columbus Bottoms (Westvaco Corp.)
33. White City
34. Winford
35. Yellowbank

STATE FORESTS
(owned by Kentucky Division of Forestry unless otherwise indicated)

1. Bernheim Arboretum and Research Forest
2. Kentenia State Forest
3. Kentucky Ridge State Forest
4. Lilley Cornett Woods
5. Olympia State Forest
6. Pennyrile State Forest
7. Robinson Forest (University of Kentucky)
8. Tygarts State Forest

STATE NATURE PRESERVES
(owned or managed by Kentucky State Nature Preserves Commission unless otherwise indicated)

1. Axe Lake Swamp
2. Bad Branch (Kentucky Nature Conservancy)
3. Bat Cave
4. Beargrass Creek
5. Blackacre
6. Blue Licks
7. Boone County Cliffs
8. Brigadoon
9. Buckley Wildlife Sanctuary (National Audubon Society)
10. Cascade Caverns

11. Chaney Lake
12. Cumberland Falls
13. Cypress Creek
14. Dinsmore Woods
15. Flat Rock Glade
16. Goodrum Cave
17. Jesse Stuart
18. Jim Scudder
19. John James Audubon
20. Kingdom Come
21. Logan County Glade
22. Mantle Rock (Kentucky Nature Conservancy)
23. Metropolis Lake
24. Natural Bridge
25. Obion Creek
26. Pilot Knob
27. Pine Mountain
28. Quiet Trails
29. Raven Run (Lexington–Fayette Urban County Government)
30. Raymond Athey Barrens
31. Six Mile Island
32. Terrapin Creek
33. Thompson Creek Glades
34. Vernon-Douglas

TRAILS

a. Jenny Wiley Trail
b. Little Shepherd Trail

FISH HATCHERIES

a. Bottom of Cumberland Lake Reservoir
b. Frankfort Fish Hatchery
c. Minor Clark State Fish Hatchery

Henderson, where a state park and museum honor him; and Daniel Boone. Boone, who died in 1820, was buried in Missouri, but in 1845 his remains and those of his wife, Rebecca, were moved to the Frankfort Cemetery in Kentucky.

Thirteen covered bridges remain in the northern and central parts of the state, of the more than four hundred that once existed. Nine working ferries persist, several of which are historic, including the Valley View Ferry on the Kentucky River between Fayette and Madison Counties. The Valley View, considered the state's oldest continuously operating business, dates from 1785. Historic cemeteries also are notable. The Lexington Cemetery, for example, known for its trees, ponds, and landscaped gardens, holds the remains of many well-known Kentuckians, including Henry Clay, who died in 1852, and basketball coach Adolph Rupp, who died in 1977.

Among the best known of Kentucky's historic attractions is Shaker Village of Pleasant Hill (Shakertown). Located between Lexington and Harrodsburg at the Kentucky River, this is the most completely restored Shaker community in the nation. Pleasant Hill was founded in 1806 and attained its largest population, about five hundred, in the 1830s. The United Society of Believers in Christ's Second Appearing —whose members were called Shakers because of the

CAMPSITES

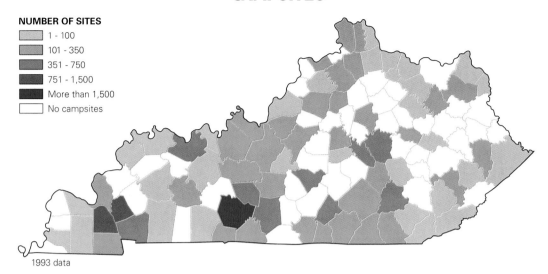

NUMBER OF SITES
- 1 - 100
- 101 - 350
- 351 - 750
- 751 - 1,500
- More than 1,500
- No campsites

1993 data

HISTORIC PLACES

1993 data

The Switzer Covered Bridge in Franklin County is one of only thirteen that remain from the more than 400 covered bridges that once existed in Kentucky.

NATIONAL HISTORIC PLACES •

Abraham Lincoln Birthplace NHS, Hodgenville
Ashland, Henry Clay Estate NHL, Lexington
Churchill Downs NHL, Louisville
Cumberland Gap NHP, Middlesboro
Fort Campbell Memorial Park, Hopkinsville
Hatfield and McCoy NHD, Pikeville
Indian Knoll NHL, near Paradise
Kentucky School for the Deaf, Danville
Kentucky Vietnam Veterans Memorial, Frankfort
Princeton NHD, Princeton
Riverside NHD, Covington
Shakertown at South Union, South Union
Shaker Village of Pleasant Hill NL, Pleasant Hill
Stearns NHD, Stearns

STATE HISTORIC SITES ★

Boone Station, Athens
Constitution Square, Danville
Dr. Thomas Walker, Barbourville
Isaac Shelby, Danville
Jefferson Davis Monument, Fairview
Old Mulkey Meeting House, Tompkinsville
Perryville Battlefield, Perryville
Waveland, Lexington
White Hall, Richmond
William Whitley House, Stanford

CEMETERIES

Camp Nelson National Cemetery, Nicholasville
Cave Hill Cemetery, Louisville
Confederate Cemetery, LaGrange
Dils Cemetery, Pikeville
Frankfort Cemetery, Frankfort
Lebanon National Cemetery, Lebanon
Lexington Cemetery, Lexington
Mill Springs Mill National Cemetery, Mill Springs
Zachary Taylor National Cemetery, Louisville

COVERED BRIDGES ⌂

1. Beech Fork Bridge, near Mooresville
2. Bennetts Mill Bridge, near Greenup
3. Cabin Creek Bridge, Vanceburg
4. Colville Bridge, Millersburg
5. Dover Bridge, Dover
6. Goddard Bridge, Goddard
7. Hillsboro Bridge, Hillsboro
8. Johnson Creek Bridge, Blue Licks
9. Oldtown Bridge, Oldtown
10. Ringos Mills Bridge, Ringos Mills
11. Switzer Bridge, near Frankfort
12. Valley Pike Bridge, Fernleaf
13. Walcott Bridge, Brooksville

FERRIES

Anderson Ferry, Covington
Augusta Ferry, Augusta
Valley View Ferry, near Richmond

ABBREVIATIONS
NHD - National Historic District
NHL - National Historic Landmark
NHP - National Historical Park
NHS - National Historic Site

trembling they did while dancing during worship—propounded pacifism, simplicity, and celibacy. The Shakers were especially noted for their artisanship and agricultural products. The last Shaker to live at Pleasant Hill died in 1923. Today, tourists can visit thirty restored original Shaker buildings on Pleasant Hill's twenty-seven hundred acres, which are crisscrossed by twenty miles of stone fences. Guests can also arrange for overnight lodging in some of the buildings. A second, smaller Shaker community in Kentucky was established in 1807 at South Union, about twelve miles southwest

of Bowling Green in western Kentucky. South Union reached its peak population, about 350 persons, in the 1840s. The property was sold in 1922, and a nonprofit organization has since restored several of the original buildings, which are open to the public. Restaurants at both of the Shaker villages feature Shaker and regional foods, including the celebrated Shaker lemon pie.

The state also boasts other attractions, such as the Louisville Zoo, planetariums, amusement parks, and museums. The varied museums cover such topics as horses and horse racing (the American Saddle Horse

WATCHABLE WILDLIFE AREAS

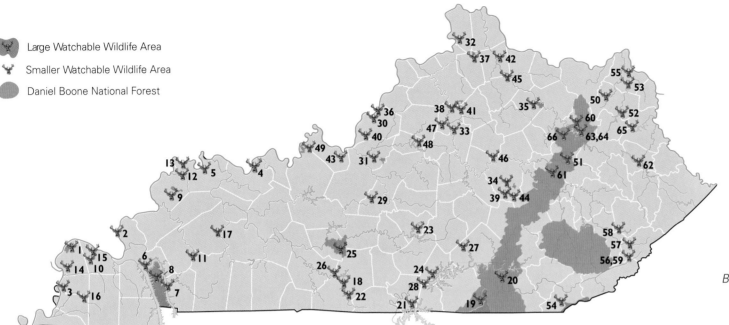

- ![] Large Watchable Wildlife Area
- ![] Smaller Watchable Wildlife Area
- ![] Daniel Boone National Forest

1994 data

Bluegrass musicians at Renfro Valley, Rockcastle County.

WESTERN REGION

1. Ballard WMA
2. Birdsville Island - Ohio River Islands WMA
3. Columbus-Belmont SP
4. Daviess Demonstration Area
5. John James Audubon SP
6. Kentucky Dam State Nongame Wildlife Natural Area
7. Lake Barkley SRP
8. Land Between the Lakes
9. Lee K. Nelson Wildlife Viewing Area
10. Metropolis Lake SNP
11. Pennyrile Forest SRP
12. Sloughs WMA, Jenny Hole-Highland Creek Unit
13. Sloughs WMA, Sauerheber Refuge Unit
14. Swan Lake WMA
15. West Kentucky WMA
16. Westvaco-Columbus Bottoms WMA
17. White City WMA

SOUTH CENTRAL REGION

18. Barren River Lake SRP
19. Big South Fork National River and Recreation Area
20. Cumberland Falls SRP
21. Dale Hollow Lake SP
22. Dry Creek Unit, Barren River Lake WMA
23. Green River Lake WMA
24. Lake Cumberland SRP
25. Mammoth Cave National Park
26. Quarry Road
27. Wesly Bend Unit, Lake Cumberland Marsh Project
28. Wolf Creek National Fish Hatchery

NORTH CENTRAL REGION

29. Abraham Lincoln Birthplace NHS
30. Beargrass Creek SNP
31. Bernheim Forest
32. Big Bone Lick SP
33. Buckley Wildlife Sanctuary
34. Central Kentucky WMA
35. Clay WMA
36. Cochran-Willig Nature Preserve
37. Curtis Gates Lloyd WMA
38. Frankfort Fish Hatchery
39. Indian Fort Mountain Trails Area, Berea College Forest
40. Jefferson County Memorial Forest
41. John A. Kleber WMA
42. Kincaid Lake SP
43. Otter Creek Park
44. Owsley Fork Reservoir, Berea College Forest
45. Quiet Trails SNP
46. Raven Run Nature Sanctuary
47. Salato Environmental Interpretive Area
48. Taylorsville Lake SP and WMA
49. Yellowbank WMA

EASTERN REGION

50. Carter Caves SRP
51. Gladie Creek Historic Site, Red River Gorge
52. Grayson Lake Nature Trail and WMA
53. Greenbo Lake SRP
54. Hensley Settlement, Cumberland Gap NHP
55. Jesse Stuart SNP
56. Kingdom Come SP
57. Lilley Cornett Woods
58. Littcarr Wildlife Viewing Area
59. Little Shepherd Trail
60. Minor Clark State Fish Hatchery
61. Natural Bridge SRP
62. Paintsville Lake SP and WMA
63. Shallow Flats Observation Area
64. Twin Knobs Campground
65. Yatesville Lake WMA Wetlands Viewing Area
66. Zilpo Recreation Area

ABBREVIATIONS	
NHP	- National Historical Park
NHS	- National Historic Site
SNP	- State Nature Preserve
SP	- State Park
SRP	- State Resort Park
WMA	- Wildlife Management Area

Museum, the International Museum of the Horse, the Kentucky Derby Museum), coal and coal mining (the Van Lear Historical Society Coal Camp Museum, the Blue Heron Coal Mining Camp, the Kentucky Coal Mining Museum), music (the Renfro Valley Museum, the Bluegrass Music Museum and Hall of Fame), and automobiles (the National Corvette Museum).

In addition to the ubiquitous county fairs (which are not included on the map of festivals and shows), numerous festivals take place across the state, mostly during the summer months. Many of the festivals celebrate those things for which Kentucky is best known: western Kentucky barbeque (the International Barbeque Festival in Owensboro), music (the Renfro Valley Bluegrass Festival, the Boone County Bluegrass Festival, the Great American Brass Band Festival in Danville, Big Singing Day in Benton), horse events (the Kentucky Derby Festival in Louisville, the Festival of the Horse in Georgetown), agricultural products (the Casey County Apple Festival, the Kentucky Beef Festival in Cynthiana, the Trigg County Ham Festival, the Washington County Sorghum and Tobacco Festival, the Kentucky Dairy Festival in Shelbyville, the World Chicken Festival in London in honor of Colonel

FESTIVALS AND SHOWS

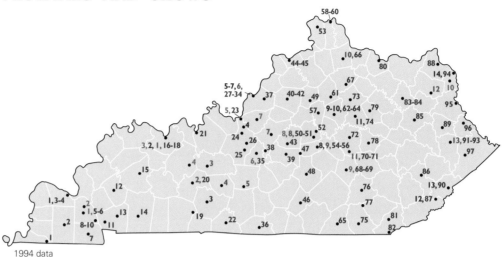

1994 data

One of many outdoor summer theater presentations is The Stephen Foster Story, performed for many years in Bardstown. The musical tells the story of one of America's foremost nineteenth-century composers whose "My Old Kentucky Home" became the state song in 1928.

Harland Sanders), and arts and crafts (the Berea Craft Festival, the Kentucky Guild of Artists and Craftsmen's fairs in Berea, the Jackson Purchase Arts and Crafts Festival in Aurora). Berea, home of Berea College, the South's first interracial college, which obtained its charter in 1859, is especially noted for arts and crafts. The Berea Craft Festival, held in July, features more than one hundred skilled artisans from throughout the Midwest and South demonstrating and selling their wares.

The performing arts are well represented in a variety of productions at Louisville's Kentucky Center for the Arts, in 1992 the thirteenth most visited attraction in the state. Another noted Louisville attraction is the Actors Theatre, which features an acting company recognized for its Humana Festival of New American Plays. Mention must also be made of the renowned center called Appalshop, located in Whitesburg. This creative media center has produced recordings, films, and performances about the history, culture, and social and economic issues of the Appalachian region.

MUSIC SHOWS

1. On Stage at D.J.'s, Benton
2. Kentucky Opry Show, Draffenville
3. Goldie's Best Little Opryhouse, Owensboro
4. Ole Barn Jamboree, Rosine
5. West Point Country Opry, West Point
6. Lincoln Jamboree, Hodgenville
7. Country Music Show, Shepherdsville
8. The Gathering Place, Harrodsburg
9. Renfro Valley, Renfro Valley
10. Coalton Country Jubilee, Cannonsburg

PERFORMING ARTS CENTERS

1. Market House Theatre, Paducah
2. RiverPark Center, Owensboro
3. Capitol Arts Center, Bowling Green
4. Alumni Performing Arts Center, Fort Knox
5. Actors Theatre, Louisville
6. Kentucky Center for the Arts, Louisville
7. Louisville Children's Theater, Louisville
8. Norton Center for the Arts, Danville
9. ArtsPlace, Lexington
10. Lexington Opera House, Lexington
11. Leeds Theater and Performing Arts Center, Winchester
12. Appalachian Cultural and Fine Arts Center, Cumberland
13. Appalshop, Whitesburg
14. Paramount Arts Center, Ashland

OUTDOOR DRAMAS AND SUMMER THEATERS

1. *Josiah*, River Park Center, Owensboro
2. *The Magic Belle*, Morgantown
3. Pine Knob Outdoor Theater, Caneyville
4. *The Death of Floyd Collins*, Brownsville
5. Horse Cave Theatre, Horse Cave
6. Kentucky Shakespeare Festival, Louisville
7. *The Stephen Foster Story*, Bardstown
8. *The Legend of Daniel Boone*, Harrodsburg
9. Pioneer Playhouse, Danville
10. Kincaid Regional Theatre, Falmouth
11. *A Point in Time*, Berea
12. *Someday* Musical, Grayson Lake SP, near Olive Hill
13. Jenny Wiley Theatre, Jenny Wiley SRP, near Prestonsburg

FESTIVALS

1. International Banana Festival, Fulton
2. Fancy Farm Picnic, Mayfield
3. Dogwood Trail Celebration, Paducah
4. Paducah Summer Festival, Paducah
5. Benton Tater Day, Benton
6. Big Singing Day, Benton
7. Freedom Fest, Murray
8. Aurora Country Festival, Aurora
9. Hardin Day, Aurora
10. Jackson Purchase Arts and Crafts Festival, Aurora
11. Four Rivers Folk Festival, Golden Pond
12. Black Patch Tobacco Festival, Princeton
13. Trigg County Ham Festival, Cadiz
14. Little River Days Festival, Hopkinsville
15. James Madison Days, Madisonville
16. IBMA's Fan Fest, Owensboro
17. International Barbeque Festival, Owensboro
18. Owensboro Summer Festival, Owensboro
19. Shaker Festival, South Union
20. Green River Catfish Festival, Morgantown
21. Sacajawea Festival, Cloverport
22. Jacksonian Days Festival, Scottsville
23. River Days Festival, West Point
24. Golden Armor Festival, Radcliff
25. Glendale Crossing Festival, Glendale
26. Kentucky Heartland Festival, Elizabethtown
27. Butchertown Oktoberfest, Louisville
28. Corn Island Storytelling Festival, Louisville
29. Kentucky Derby Festival, Louisville
30. Kentucky Music Weekend, Louisville
31. Kentucky Shakespeare Festival, Louisville
32. Kentucky State Fair, Louisville
33. Louisville CityFair, Louisville
34. Strassenfest, Louisville
35. Lincoln Days Celebration, Hodgenville
36. Monroe County Watermelon Festival, Tompkinsville
37. Jeffersontown Gaslight Festival, Jeffersontown
38. Rolling Fork Iron Horse Festival, New Haven
39. Marion County Country Ham Days, Lebanon
40. Dogwood Festival, Shelbyville
41. Kentucky Dairy Festival, Shelbyville
42. October Pork Festival, Shelbyville
43. Washington Co. Sorghum and Tobacco Festival, Springfield
44. Carroll County Tobacco Festival, Carrollton
45. Kentucky Scottish Weekend, Carrollton
46. Big Singing Day, Jamestown
47. Forkland Heritage Festival and Revue, near Gravel Switch
48. Casey County Apple Festival, Liberty
49. Capitol Expo Festival, Frankfort
50. Old Fort Harrod Festival Weekend, Harrodsburg
51. Pioneer Days Festival, Harrodsburg
52. Celebration of Autumn, Pleasant Hill
53. Boone County Bluegrass Festival, Burlington
54. Great American Brass Band Festival, Danville
55. Historic Constitution Square Festival, Danville
56. Railroad Days, Danville
57. Woodford County Days, Versailles
58. Maifest, Covington
59. Oktoberfest, Covington
60. Summer Sunfest, Covington
61. Festival of the Horse, Georgetown
62. Bluegrass State Games Festival, Lexington
63. EquiFestival, Lexington
64. Festival of the Bluegrass, Lexington
65. McCreary Festival, Whitley City
66. Kentucky Wool Festival, Falmouth
67. Kentucky Beef Festival, Cynthiana
68. Heritage Craft Festival, Renfro Valley
69. Renfro Valley Bluegrass Festival, Renfro Valley
70. Berea Craft Festival, Berea
71. Ky Guild of Artists and Craftsmen's Fairs, Berea
72. Kit Carson Days, Richmond
73. Pumpkin Festival, Paris
74. Daniel Boone Pioneer Festival, Winchester
75. Williamsburg Old Fashioned Trading Days, Williamsburg
76. World Chicken Festival, London
77. Corbin Nibroc Festival, Corbin
78. Mountain Mushroom Festival, Irvine
79. October Court Days, Mount Sterling
80. Mason County Court Day, Maysville
81. Kentucky Mountain Laurel Festival, Pineville
82. Cumberland Mountain Fall Festival, Middlesboro
83. Appalachian Celebration, Morehead
84. Kentucky Hardwood Festival, Morehead
85. Morgan County Sorghum Festival, West Liberty
86. Black Gold Festival, Hazard
87. Kingdom Come Swappin' Meetin', Cumberland
88. Greenup Old Fashioned Days, Greenup
89. Kentucky Apple Festival, Paintsville
90. Mountain Heritage Festival, Whitesburg
91. Festival of F.A.C.E.S., Prestonsburg
92. Jenny Wiley Festival, Prestonsburg
93. Kentucky Highlands Folk Festival, Prestonsburg
94. TriState Fair and Regatta, Ashland
95. Lawrence County Septemberfest, Louisa
96. Martin County Pumpkinfest, Inez
97. Hillbilly Days, Pikeville

MUSEUMS, ZOOS, PLANETARIUMS, AND AMUSEMENT PARKS

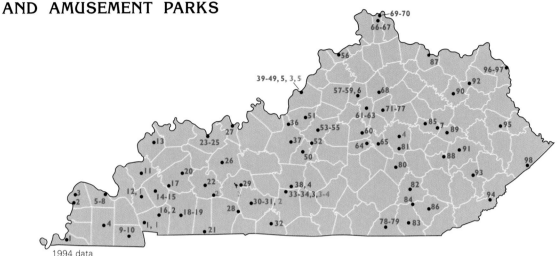

1994 data

The Berea Craft Festival is held every July and draws more than one hundred artisans from across the Southeast and Midwest. In May and October, Berea hosts the Kentucky Guild of Artists and Craftsmen's Fair. All three events are held at the Indian Fort Theater.

MUSEUMS

1. Warren Thomas Museum, Hickman
2. Wickliffe Mounds Museum and Archaeological Site, Wickliffe
3. Barlow House Museum, Barlow
4. Western Kentucky Museum, Mayfield
5. Alben W. Barkley Museum, Paducah
6. Market House Museum, Paducah
7. Museum of the American Quilter's Society, Paducah
8. Yeiser Art Center, Paducah
9. National Scouting Museum, Murray
10. Wrather West Kentucky Museum, Murray
11. Crittenden County Bob Wheeler Museum, Marion
12. Lyon County Museum, Eddyville
13. Union County Museum, Morganfield
14. Adsmore House Museum, Princeton
15. Princeton Art Guild Gallery, Princeton
16. Trigg County Museum, Cadiz
17. Dawson Springs Museum, Dawson Springs
18. Pennyroyal Area Museum, Hopkinsville
19. Trail of Tears Commemorative Park and Heritage Museum, Hopkinsville
20. Hopkins County Museum, Madisonville
21. Robert Penn Warren Birthplace Museum, Guthrie
22. Duncan House Museum and Art Gallery, Greenville
23. Bluegrass Music Museum and Hall of Fame, Owensboro
24. Owensboro Area Museum of Science and History, Owensboro
25. Owensboro Museum of Fine Art, Owensboro
26. Ohio County Museum, Hartford
27. Hancock County Museum, Hawesville
28. Shaker Museum, South Union
29. Green River Museum, Woodbury
30. Kentucky Museum and Library, Bowling Green (WKU)
31. National Corvette Museum, Bowling Green
32. Scottsville Historical Museum, Scottsville
33. Mammoth Cave Wax Museum, Cave City
34. Mammoth Cave Wildlife Museum, Cave City
35. The Wayfarer and Floyd Collins Museum, Cave City
36. Patton Museum of Cavalry and Armor, Fort Knox
37. Schmidt's Museum of Coca-Cola Memorabilia, Elizabethtown
38. American Museum of Caves and Karstlands, Horse Cave
39. Colonel Harland Sanders Museum, Louisville
40. The Filson Club, Louisville
41. J.B. Speed Art Museum, Louisville
42. John Conti Coffee Museum, Louisville
43. Joseph A. Callaway Archaeological Museum, Louisville
44. Kentucky Derby Museum, Louisville
45. Louisville Science Center, Louisville
46. Louisville Visual Art Association at the Water Tower, Louisville
47. National Society of the Sons of the Revolution Museum, Louisville
48. Portland Museum, Louisville

49. Thomas Edison House, Louisville
50. Lincoln Museum, Hodgenville
51. Jim Beam American Outpost Museum, Clermont
52. Kentucky Railway Museum, New Haven
53. America's Miniature Soldier Museum, Bardstown
54. Doll Cottage, Bardstown
55. Oscar Getz Museum of Whiskey History, Bardstown
56. Old Stone Jail Museum, Carrollton
57. Kentucky Military History Museum, Frankfort
58. Luscher's Museum of the American Farmer, Frankfort
59. Old State Capitol/Kentucky History Museum, Frankfort
60. Harrodsburg Historical Society Museum, Harrodsburg
61. Bluegrass Scenic Railroad and Museum, Versailles
62. Nostalgia Station Toy and Train Museum, Versailles
63. Woodford County Historical Society Museum, Versailles
64. Danville-Boyle County Historical Society Museum, Danville
65. Garrard County Jail Museum, Lancaster
66. American Museum of Brewing History and Arts, Fort Mitchell
67. Vent Haven Ventriloquist Museum, Fort Mitchell
68. Georgetown-Scott County Museum, Georgetown
69. Behringer-Crawford Museum, Covington
70. Railway Exposition Museum, Covington
71. American Saddle Horse Museum, Lexington
72. Headley-Whitney Museum, Lexington
73. International Museum of the Horse, Lexington
74. Lexington Children's Museum, Lexington
75. Loudoun House, Lexington
76. Museum of Anthropology, Lexington (UK)
77. University of Kentucky Art Museum, Lexington
78. Blue Heron Coal Mining Camp, Stearns
79. Stearns Museum, Stearns
80. Renfro Valley Museum, Renfro Valley
81. Appalachian Museum, Berea (Berea College)
82. Mays Log Cabin Craft Village and Museum, London
83. Cumberland Museum and Craft Center, Williamsburg
84. Harland Sanders Café and Museum, Corbin
85. Red River Historical Museum, Clay City
86. Knox Historical Museum, Barbourville
87. Mason County Historical Museum, Maysville
88. Pioneer Museum, Lerose
89. Wolfe County Historical Museum, Campton
90. Morehead State University Folk Art Museum, Morehead
91. Breathitt County Museum, Jackson
92. Northeastern Kentucky History Museum, Olive Hill
93. Bobby Davis Memorial Museum, Hazard
94. Kentucky Coal Mining Museum, Benham
95. Van Lear Historical Society Coal Camp Museum, Van Lear
96. Ashland Area Art Gallery, Ashland
97. Kentucky Highlands Museum, Ashland
98. Elkhorn City Railroad Museum, Elkhorn City

ZOOS

1. Woodlands Nature Center, near Golden Pond
2. Woods and Wetlands Wildlife Center, Cadiz
3. Bush's Zoo, Cave City
4. Kentucky Down Under, Horse Cave
5. Louisville Zoo, Louisville
6. Game Farm (Kentucky Dept. of Fish and Wildlife), Frankfort
7. Miami Valley Serpentarium, Slade

PLANETARIUMS

1. Golden Pond Planetarium, Golden Pond
2. Hardin Planetarium, Bowling Green (WKU)
3. Rauch Memorial Planetarium, Louisville (UL)
4. Hummel Planetarium and Space Theater, Richmond (EKU)

AMUSEMENT PARKS

1. Venture Action Park, Eddyville
2. Dogwood Lake Funpark, Dunmor
3. Guntown Mountain Amusement Park, Cave City
4. Kentucky Action Park, Cave City
5. Kentucky Kingdom Amusement Park, Louisville

John Jacob Niles (1892-1980), well-known ballad writer, folksinger, and storyteller, is shown performing at the Corn Island Storytelling Festival in Louisville.

Churchill Downs in Louisville, perhaps the world's most famous racetrack and twin spires, is home of the Kentucky Derby.

HORSE-RELATED ATTRACTIONS

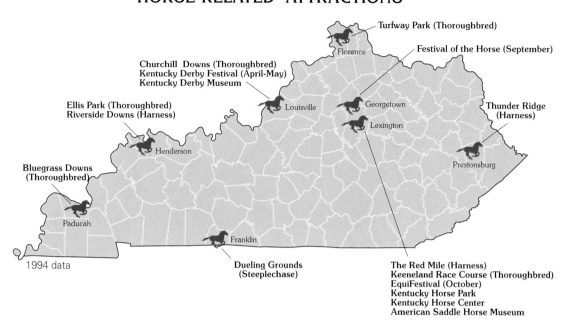

Turfway Park (Thoroughbred)

Festival of the Horse (September)

Churchill Downs (Thoroughbred)
Kentucky Derby Festival (April-May)
Kentucky Derby Museum

Ellis Park (Thoroughbred)
Riverside Downs (Harness)

Thunder Ridge
(Harness)

Bluegrass Downs
(Thoroughbred)

1994 data

Dueling Grounds
(Steeplechase)

The Red Mile (Harness)
Keeneland Race Course (Thoroughbred)
EquiFestival (October)
Kentucky Horse Park
Kentucky Horse Center
American Saddle Horse Museum

Florence
Louisville
Georgetown
Lexington
Henderson
Prestonsburg
Paducah
Franklin

Recreation

Attractions related to the horse can be found across the state. In addition to museums and festivals, nine racetracks offer thoroughbred and harness racing. The best known, of course, is Churchill Downs in Louisville, home of the Kentucky Derby. The state's newest track, Thunder Ridge in Prestonsburg, opened in 1994. Anyone interested in horses should certainly visit the Kentucky Horse Park, just north of Lexington. Located in the Bluegrass Region, amid some of world's most famous horse farms, the park spans more than a thousand acres. Visitors can enjoy a farm tour and see forty breeds of horses (including the American Saddle Horse, the only breed that originated in Kentucky), horse performances, museums, exhibits, equestrian events, and a memorial to Kentucky's best-known racehorse, Man O' War. Lexington also boasts the Kentucky Horse Center, where a visitor can tour a working thoroughbred training operation, and Keeneland Race Course, opened in 1936 and today a National Historic Landmark. The world's major horse sales event takes place at Keeneland, where the record price for a horse was set in 1985 at $13.1 million.

Befitting a state that has fifteen hundred square miles of water surface, thirteen thousand miles of streams, and fifteen hundred miles of navigable waterways (more than in any other of the contiguous states), Kentucky residents and visitors enjoy water activities. Sailing, motorboating, waterskiing, canoeing, rafting, and fishing are all popular summer sports on the recreational waterways of the Commonwealth. Kentucky has ten major lakes, each with more than five thousand surface acres of water; all are impoundments developed for flood control, the generation of hydroelectric power, and recreation, and most were built by the U.S. Army Corps of Engineers. Barren River Lake, Cave Run Lake, Dale Hollow Lake, Green River Lake, Kentucky Lake, Lake Barkley, Lake Cumberland, Laurel River Lake, Nolin River Lake, and Rough River Lake provide ample opportunity for water sports. Impounded in 1943, Dale Hollow Lake, shared by Kentucky and Tennessee, is the state's oldest artificial lake. More than 130 marinas, more than 200 campgrounds, and other facilities are located at the state's lakes and recreational lands.

The Red Mile, a track for Standardbred harness racing in Lexington, opened in 1875. Promoted as "the world's fastest mile," the track has also been known since 1982 for its Standardbred horse auctions.

MARINAS AND BOATING WATERS

CHANGE IN THE NUMBER OF BOAT REGISTRATIONS ISSUED

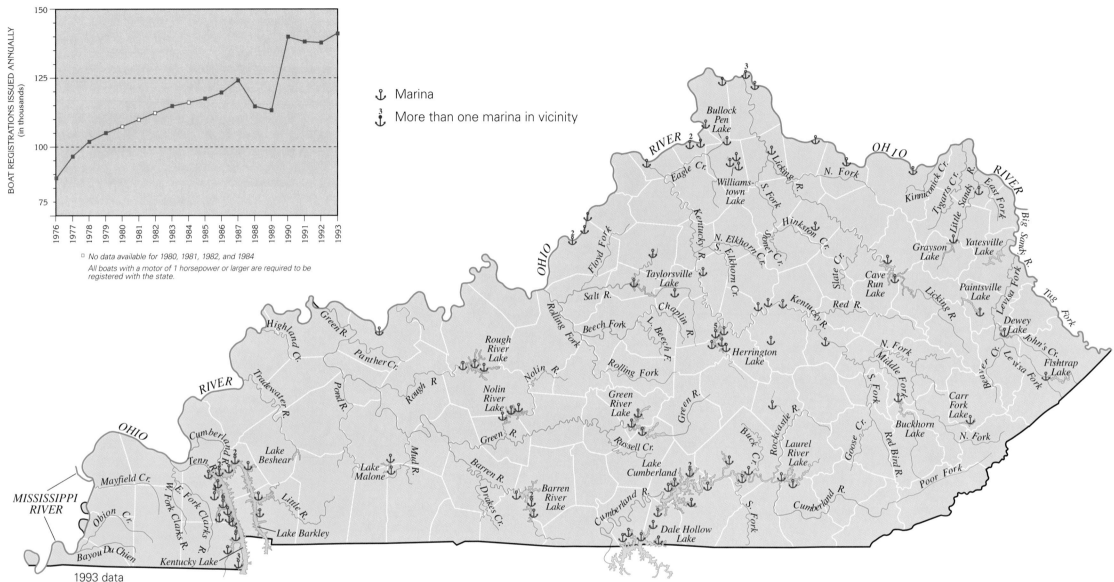

□ No data available for 1980, 1981, 1982, and 1984

All boats with a motor of 1 horsepower or larger are required to be registered with the state.

1993 data

⚓ Marina

⚓ More than one marina in vicinity

Kentucky enjoys a diversity of flora and fauna. Nearly 250 species of fish are found in the state's rivers, and about 40 are game species sought by anglers. Four major rivers that connect in the state—the Mississippi, Ohio, Tennessee, and Cumberland—together with many other smaller rivers and creeks, offer a variety of habitats for fish. Some of the game fish are found throughout the state. These include the largemouth bass, the smallmouth bass, the spotted bass (also called the Kentucky bass, as it is the state's official game fish), the bluegill, the crappie, and several species of catfish. The world-record smallmouth bass, weighing almost twelve pounds, was caught in 1955 in the Kentucky portion of Dale Hollow Lake. Nonnative species of game fish, introduced to the state, have a more limited geographic distribution and are regularly stocked in selected lakes and rivers. Such species include the brown trout, the rainbow trout, and the striped bass. Two other species, the walleye and the muskellunge, have been reintroduced to waters where they once occurred naturally. The state record for a striped bass is one

FISHING WATERS

PIKE FAMILY

⌇⌇ Muskellunge

The following fish are found virtually everywhere in Kentucky: common carp, bullhead, channel catfish, and flatheads. Muskellunge are sporadic in the rivers and creeks shown. They also occur sporadically in some minor creeks and rivers not shown.

PERCH FAMILY

⌇⌇ Sauger
⌇⌇ Walleye
⌇⌇ Sauger and Walleye

Walleye are sporadic and less common than sauger.

TEMPERATE BASS AND SUNFISH FAMILY

⌇⌇ White Bass
⌇⌇ Yellow and White Bass
⌇⌇ Striped Bass
⌇⌇ White and Striped Bass
⌇⌇ Yellow, White, and Striped Bass
▼ ▼ Smallmouth Bass
▼ ▼ Rock Bass

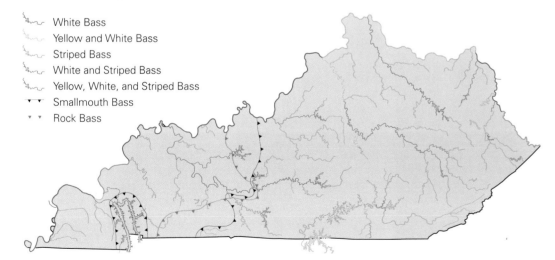

The following bass and sunfish are found virtually everywhere in Kentucky: spotted bass, largemouth bass, white and/or black crappie, sunfish (green sunfish, longear sunfish, and redear sunfish), and bluegill. Rock bass are found in the eastern 2/3 of the state as indicated on the map. Smallmouth bass are found at Land Between the Lakes and in upland streams in the eastern 2/3 of the state.

TROUT FAMILY

⌇⌇ Rainbow Trout
⌇⌇ Brook Trout
⌇⌇ Rainbow and Brook Trout

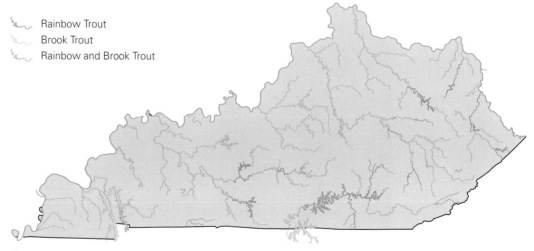

Rainbow trout — introduced, transplanted, localized, and stocked. Brown trout — introduced, exotic, localized, and stocked in tailwaters of Lakes Herrington and Cumberland as well as several lesser creeks not shown on this map. Brook trout — introduced, transplanted, established.

TOTAL NUMBER OF HUNTING AND FISHING LICENSES SOLD

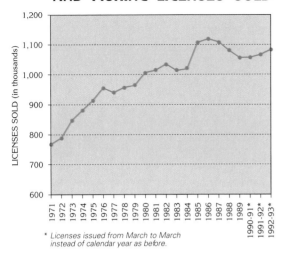

LICENSES SOLD (in thousands)

1,200
1,100
1,000
900
800
700
600

1971 1972 1973 1974 1975 1976 1977 1978 1979 1980 1981 1982 1983 1984 1985 1986 1987 1988 1989 1990-91* 1991-92* 1992-93*

* Licenses issued from March to March instead of calendar year as before.

TYPES OF HUNTING AND FISHING LICENSES SOLD

License Type	1992-1993	
	Number Sold	Revenue Received
Resident Combination Hunting and Fishing	115,800	$1,650,841
Resident Hunting	181,148	$1,295,7328
Nonresident Hunting	14,293	$794,674
Resident Fishing	375,802	$3,116,275
Nonresident Fishing	136,428	$1,512,467
Turkey Tags	12,587	$179,567
Deer Tags	187,332	$3,096,438
Trout Stamps	43,044	$119,450
Waterfowl Stamps	16,522	$83,658
Total	1,082,956	$11,849,102

HUNTING AND FISHING LICENSES SOLD

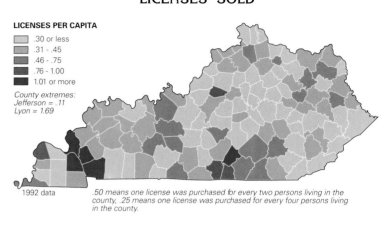

LICENSES PER CAPITA

- .30 or less
- .31 - .45
- .46 - .75
- .76 - 1.00
- 1.01 or more

County extremes:
Jefferson = .11
Lyon = 1.69

1992 data
.50 means one license was purchased for every two persons living in the county, .25 means one license was purchased for every four persons living in the county.

A bass tournament on Lake Barkley. Some 800 sanctioned bass tournaments were held in Kentucky in the mid-1980s, when their number peaked.

weighing more than fifty-eight pounds, caught in Lake Cumberland; the state record for a muskie is a forty-three-pound fish caught in Dale Hollow Lake.

Kentucky has a long tradition of hunting, and game species today include whitetail deer, squirrel, rabbit, raccoon, opossum, and bobcat. More than eighty thousand whitetail deer, found in every Kentucky county, were harvested by gun and bow hunters in 1992; the number harvested has climbed steadily each year since 1976, reflecting the increase in the population of this popular game animal. Game bird species include ruffed grouse (limited primarily to the eastern part of the state, but being restored elsewhere), bobwhites, doves, ducks, geese, and wild turkeys. Wild turkeys have made a comeback in the state. In the late 1940s and early 1950s the only known wild turkey population was located in what became Land Between the Lakes. By 1954 the statewide population was estimated at about 850. In the early 1970s a successful program was inaugurated whereby trapped birds from other states were released in various parts of Kentucky. Today, wild turkeys are found across the Commonwealth, and their population is estimated at more than twenty thousand. The popularity of hunting and fishing in Kentucky is indicated by the annual sales in recent years of more than one million hunting and fishing licenses. In 1992 these sales generated nearly twelve million dollars in income. About 15 percent of

DEER HARVEST

1977

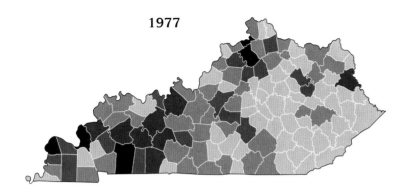

1982

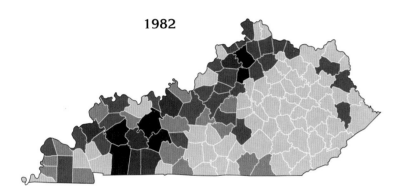

**PERCENT OF
TOTAL STATE HARVEST
FOR SELECT YEARS**

- Less than .50
- .50 - 1.00
- 1.01 - 2.00
- 2.01 - 3.00
- More than 3.00

**NUMBER OF DEER LEGALLY KILLED
FOR SELECT YEARS**

5,717

18,070

81,933

60,372

- 1977
- 1982
- 1987
- 1992

*Does not include deer killed in wildlife
management areas or special hunting
areas such as Fort Knox.*

1987

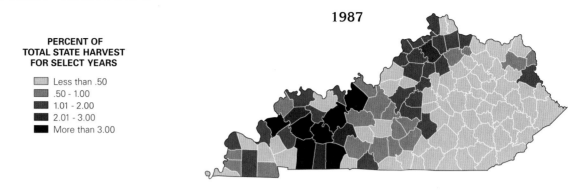

1992

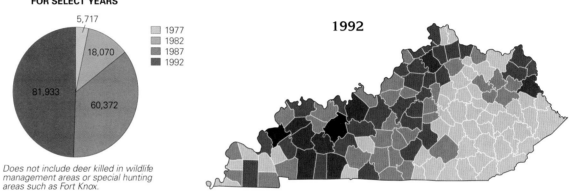

the licenses were sold to nonresidents. One former game species, black bears, were nearly eliminated from the state by 1900 but are making a comeback; since the 1980s residents of many eastern Kentucky counties have reported reliable sightings. Bears are strictly protected in Kentucky.

Golf is one of the most popular participatory sports in Kentucky. It is estimated that there are more than three hundred thousand golfers in the state, who play on more than two hundred public, semiprivate, and private courses. Ten of Kentucky's fifteen state resort parks have either nine-hole or eighteen-hole public courses. The state hosts the Foster Brooks Pro-Celebrity Tournament at Louisville's Hurstbourne Country Club in late spring, and until 1997 hosted the Senior Golf Classic in Lexington in September, which attracted top players on the senior PGA tour.

Perhaps the most popular sport in the state is basketball, which arrived in Kentucky in 1895, just four years after being invented by James Naismith in Springfield, Massachusetts. Some say that high school and college basketball almost constitutes a

GOLF COURSES

TYPE OF COURSE

- Public
- Private
- Semi-Private
- Resort

NUMBER OF HOLES

- ○ 9
- △ 18
- □ 27-54

*Example: ▲ denotes an 18-hole
public golf course*

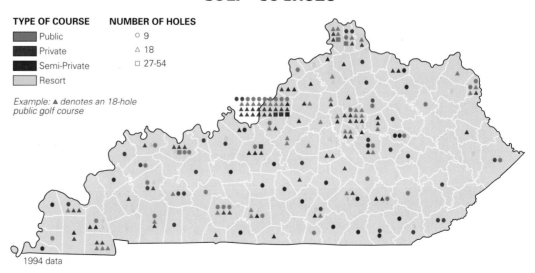

1994 data

INTERCOLLEGIATE SPORTS

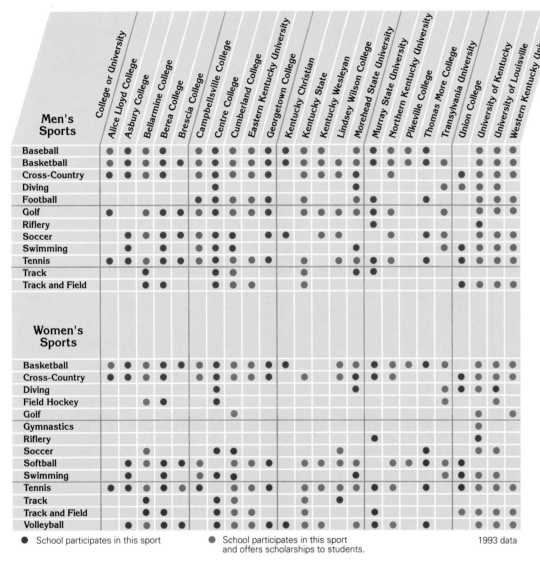

Men's Sports (1993 data)

Schools (columns): Alice Lloyd College · Asbury College · Bellarmine College · Berea College · Brescia College · Campbellsville College · Centre College · Cumberland College · Eastern Kentucky University · Georgetown College · Kentucky Christian · Kentucky State · Kentucky Wesleyan · Lindsey Wilson College · Morehead State University · Murray State University · Northern Kentucky University · Pikeville College · Thomas More College · Transylvania University · Union College · University of Kentucky · University of Louisville · Western Kentucky University

Men's Sports rows: Baseball · Basketball · Cross-Country · Diving · Football · Golf · Riflery · Soccer · Swimming · Tennis · Track · Track and Field

Women's Sports rows: Basketball · Cross-Country · Diving · Field Hockey · Golf · Gymnastics · Riflery · Soccer · Softball · Swimming · Tennis · Track · Track and Field · Volleyball

Legend:
● School participates in this sport
● School participates in this sport and offers scholarships to students.

1993 data

• One dot represents 5 season ticketholders

• One dot represents 50 season ticketholders (Fayette County only)

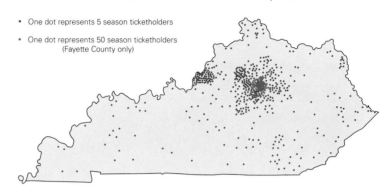

HOLDERS OF SEASON TICKETS FOR UNIVERSITY OF LOUISVILLE MEN'S BASKETBALL SEASON, 1993-94

• One dot represents 5 season ticketholders

• One dot represents 50 season ticketholders (Jefferson County only)

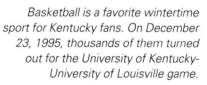

Basketball is a favorite wintertime sport for Kentucky fans. On December 23, 1995, thousands of them turned out for the University of Kentucky-University of Louisville game.

religion in the state. The state's two largest universities, the Universities of Kentucky and Louisville, together boast eight NCAA national basketball championships (Kentucky in 1948, 1949, 1951, 1958, 1978, and 1996; Louisville in 1980 and 1986). When it comes to basketball in Kentucky, everything is arguable, but probably the most famous celebrity is Adolph Rupp, who was hired to coach the University of Kentucky team in 1931. Rupp coached at Kentucky until his retirement in 1972, having recorded more wins than any other coach in any program in the nation to that time. Since 1920 the high school basketball tournament, the Sweet Sixteen, has been one of the most popular sporting events in the state. High schools from either Jefferson or Fayette Counties have won more than two-fifths of the boys' championships and one-fourth of the girls' championships. Other intercollegiate sports include football, baseball, softball, tennis, cross-country, track and field, and volleyball.

Kentucky's only professional basketball team was the Kentucky Colonels, formed in 1967 as Louisville's entry in the new American Basketball Association. Louisville lost the franchise in 1976, when the association ceased to exist. During the twentieth century the state has not had a major league baseball team, but more than thirty Kentucky cities have had minor league teams. Louisville has had a AAA minor league team for nearly all of the twentieth century. Today, the American Association's Lou-

SWEET SIXTEEN HIGH SCHOOL BASKETBALL CHAMPIONS

GIRLS

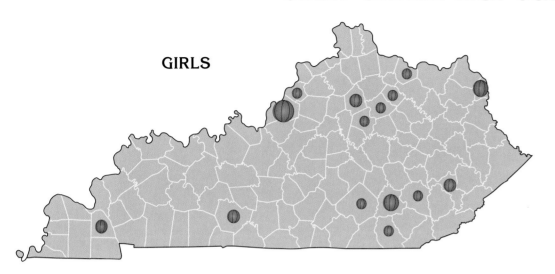

BOYS

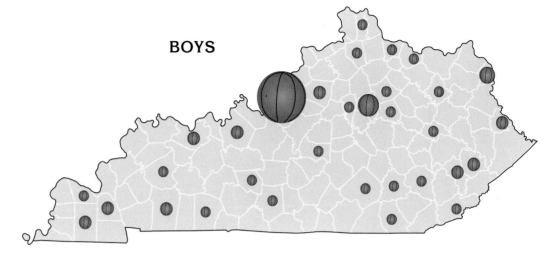

YEAR	SCHOOL	COUNTY
1920	Paris	Bourbon
1921	Ashland	Boyd
1922	Ashland	Boyd
1923	West Louisville	Jefferson
1924	Ashland	Boyd
1925	Georgetown	Scott
1926	Maysville	Mason
1927	West Louisville	Jefferson
1928	Ashland	Boyd
1929	Ashland	Boyd
1930	Hazard	Perry
1931	Woodburn	Warren
1932	Woodburn	Warren
1933-1974	No Girls Tournaments	
1975	Butler	Jefferson
1976	Sacred Heart	Jefferson
1977	Laurel County	Laurel
1978	Laurel County	Laurel

YEAR	SCHOOL	COUNTY
1979	Laurel County	Laurel
1980	Butler	Jefferson
1981	Pulaski County	Pulaski
1982	Marshall County	Marshall
1983	Warren Central	Warren
1984	Marshall County	Marshall
1985	Whitley County	Whitley
1986	Oldham County	Oldham
1987	Laurel County	Laurel
1988	Southern	Jefferson
1989	Clay County	Clay
1990	Henry Clay	Fayette
1991	Laurel County	Laurel
1992	Mercy	Jefferson
1993	Nicholas County	Nicholas
1994	M.C. Napier	Perry
1995	Scott County	Scott

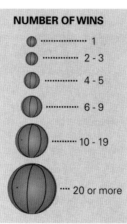

NUMBER OF WINS

- 1
- 2 - 3
- 4 - 5
- 6 - 9
- 10 - 19
- 20 or more

YEAR	SCHOOL	COUNTY
1920	Lexington	Fayette
1921	duPont Manual	Jefferson
1922	Lexington	Fayette
1923	duPont Manual	Jefferson
1924	Lexington	Fayette
1925	Manual	Jefferson
1926	St. Xavier	Jefferson
1927	M.M.I.	Bourbon
1928	Ashland	Boyd
1929	Heath	McCracken
1930	Corinth	Grant
1931	duPont Manual	Jefferson
1932	Hazard	Perry
1933	Ashland	Boyd
1934	Ashland	Boyd
1935	St. Xavier	Jefferson
1936	Corbin	Whitley
1937	Midway	Woodford
1938	Sharpe	Marshall
1939	Brooksville	Bracken
1940	Hazel Green	Wolfe
1941	Inez	Martin
1942	Lafayette	Fayette
1943	Hindman	Knott
1944	Harlan	Harlan
1945	Male	Jefferson
1946	Breckinridge Training	Rowan
1947	Maysville	Mason
1948	Brewers	Graves
1949	Owensboro	Daviess
1950	Lafayette	Fayette
1951	Clark County	Clark
1952	Cuba	Graves
1953	Lafayette	Fayette
1954	Inez	Martin
1955	Hazard	Perry
1956	Carr Creek	Knott
1957	Lafayette	Fayette

YEAR	SCHOOL	COUNTY
1958	St. Xavier	Jefferson
1959	North Marshall	Marshall
1960	Flaget	Jefferson
1961	Ashland	Boyd
1962	St. Xavier	Jefferson
1963	Seneca	Jefferson
1964	Seneca	Jefferson
1965	Breckinridge County	Breckinridge
1966	Shelby County	Shelby
1967	Earlington	Hopkins
1968	Glasgow	Barren
1969	Central	Jefferson
1970	Male	Jefferson
1971	Male	Jefferson
1972	Owensboro	Daviess
1973	Shawnee	Jefferson
1974	Central	Jefferson
1975	Male	Jefferson
1976	Edmonson County	Edmonson
1977	Ballard	Jefferson
1978	Shelby County	Shelby
1979	Lafayette	Fayette
1980	Owensboro	Daviess
1981	Simon Kenton	Kenton
1982	Laurel County	Laurel
1983	Henry Clay	Fayette
1984	Logan County	Logan
1985	Hopkinsville	Christian
1986	Pulaski County	Pulaski
1987	Clay County	Clay
1988	Ballard	Jefferson
1989	Pleasure Ridge Park	Jefferson
1990	Fairdale	Jefferson
1991	Fairdale	Jefferson
1992	University Heights	Christian
1993	Marion County	Marion
1994	Fairdale	Jefferson
1995	Breckinridge County	Breckinridge

isville Redbirds, a minor league franchise of the St. Louis Cardinals, is the tenth-most-popular tourist attraction in the state. Nearly 650,000 fans saw Redbirds games in 1992.

Kentucky has also produced its share of famous figures in sports other than basketball. A.B. ("Happy") Chandler, for example, was baseball commissioner from 1945 to 1951 and was elected to the Baseball Hall of Fame in 1982. Chandler twice served as governor of Kentucky. Born in 1898 in Corydon, he died in 1991 at his home in Versailles. Other legendary sports figures from Kentucky include boxing great Muhammad Ali and Baseball Hall of Famer Harold ("Pee Wee") Reese. Ali, born Cassius Clay in Louisville in 1942, held the world heavyweight boxing championship from 1964 to 1967, from 1974 to 1978, and from 1978 to 1979. Reese, born in Meade County in 1918, became a shortstop for the Brooklyn Dodgers and one of the "Boys of Summer".

WAGERING AT KENTUCKY'S HORSE RACING TRACKS

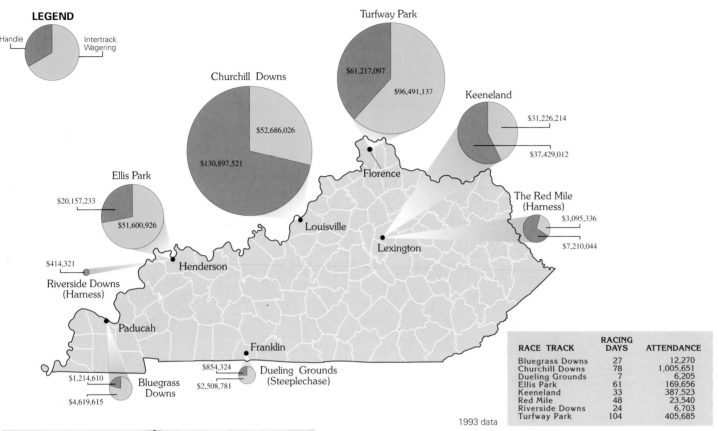

LEGEND

Handle — Intertrack Wagering

Turfway Park
$61,217,097
$96,491,137

Churchill Downs
$52,686,026
$130,897,521

Keeneland
$31,226,214
$37,429,012

Ellis Park
$20,157,233
$51,600,926

Florence

Louisville

The Red Mile (Harness)
$3,095,336
$7,210,044

Lexington

Henderson

$414,321
Riverside Downs (Harness)

Paducah

Franklin

$854,324
$2,508,781

Dueling Grounds (Steeplechase)

$1,214,610
$4,619,615

Bluegrass Downs

RACE TRACK	RACING DAYS	ATTENDANCE
Bluegrass Downs	27	12,270
Churchill Downs	78	1,005,651
Dueling Grounds	7	6,205
Ellis Park	61	169,656
Keeneland	33	387,523
Red Mile	48	23,540
Riverside Downs	24	6,703
Turfway Park	104	405,685

1993 data

LOTTERY TICKET SALES PER CAPITA
(4/4/89 - 12/29/93)

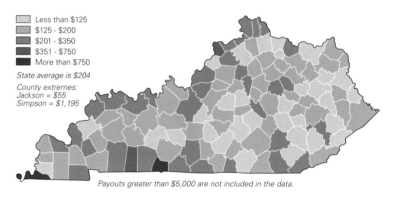

- $127 - $400
- $401 - $800
- $801 - $1,600
- $2,484
- $3,329

State average is $456
County extremes:
Jackson = $127
Simpson = $3,329

LOTTERY TICKET PAYOUTS PER CAPITA
(4/4/89 - 10/30/93)

- Less than $125
- $125 - $200
- $201 - $350
- $351 - $750
- More than $750

State average is $204
County extremes:
Jackson = $55
Simpson = $1,195

Payouts greater than $5,000 are not included in the data.

The Kentucky Lottery, designed to supplement state taxes as a source of revenue for the Commonwealth, paid proceeds to the state in fiscal 1996 that exceeded $147 million. The aggregate income from the Lottery since its inception in 1989 has exceeded $896 million.

Wagering

Kentucky is known for gambling at horse racetracks, and in 1993 the total "handle" for all races at Kentucky's eight racetracks amounted to more than five hundred million dollars. (In 1994 a ninth track for harness racing, called Thunder Ridge, opened in Prestonsburg.) About 52 percent of the total was gambled on live races, and 48 percent through intertrack wagering, or off-track betting, which began in Kentucky in the mid-1980s. The impact on Kentucky's racetracks of intertrack wagering, off-track betting outlets, riverboat casinos, and whole-card simulcasting is not possible to predict. Such new and different methods of betting on horse races through satellite communication are emerging in Kentucky and nearby states, and these will affect the state's racetracks, especially its smaller ones. Four of the state's tracks,

Churchill Downs, Turfway Park, Keeneland, and Ellis Park, ranked third, ninth, fifteenth, and twenty-first, respectively, on the list of most visited attractions in Kentucky in 1992.

Another significant form of gambling in Kentucky is the state lottery, which was approved by a majority of the electorate in a statewide vote in 1988. The previous lottery in Kentucky had been revoked in 1890, coinciding with a national trend. Ticket sales for the new state lottery began in April 1989, and as of late 1993 the lottery had taken in about $1.5 billion in receipts and paid out nearly $1.0 billion in prizes. Per capita sales and payouts varied by county. Those counties that rank high in sales and payouts tend to be near Tennessee, which does not have a state lottery, and also to be well served by the interstate highway system.

The Political Landscape

Previous page: The Capitol of the Commonwealth, Franklin County.

CHAPTER ELEVEN

The Political Landscape

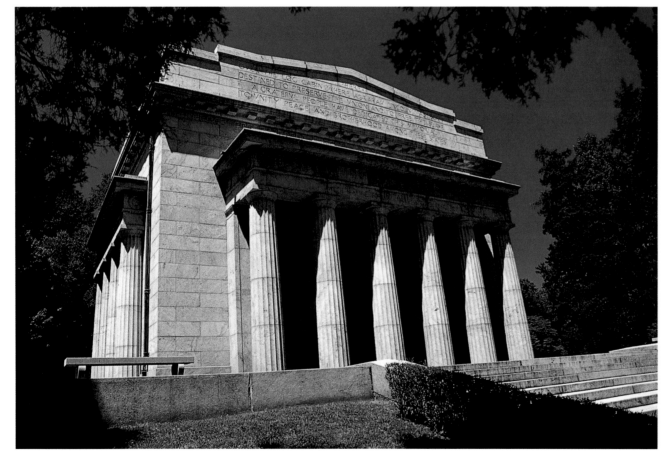

Abraham Lincoln, sixteenth president of the United States, was born in a modest log cabin near Hodgenville in 1809. The cabin is now preserved inside this grand classic temple at the Abraham Lincoln National Historic Site in Larue County.

IN THE LATE EIGHTEENTH and early nineteenth centuries, migrants poured into Kentucky, entering primarily through the Cumberland Gap or following the Ohio River and its tributaries. Kentucky's population came predominantly from Pennsylvania, Maryland, Virginia, North Carolina, and Tennessee, and the new settlers exhibited a wide range of habits, ideas, and values. Some families and individuals who settled Kentucky came from a cavalier and slaveholding tradition; others held Middle Atlantic values that focused on entrepreneurial development and religious tolerance. These diverse traditions reflected the settlers' regional origins within early America but also recalled various legacies from western Europe. Along with their knowledge of how to build houses and fences and settle forested areas and open fields, migrants also carried with them concerns about religion, education, and government. They held strong views about the roles of individuals and states in the emerging nation, and they discussed the eligibility of certain groups to participate in political processes, the meaning of representation, and the ways territory should be delimited and organized. Kentucky's unique political geography is one result of the variation and transformation in the population's traditions and beliefs. Some attributes of Kentucky politics are distinctively southern, others are northern, and still others are peculiar to the Commonwealth.

After statehood in 1792, this diversity persisted. The state's territory incorporated distinctive natural regions, and Kentuckians' responses to their physical environment resulted in a diversity of urban and rural features. Towns along the Ohio River were different from those in the Bluegrass or from the forest and mining settlements of eastern Kentucky. These in turn contrasted with settlements in the prosperous agricultural sections of the Pennyroyal or the western Ohio River lowlands. Political and social differences across the state brought forth regions distinct in their economic and political cultures. Such differences were sharp

235

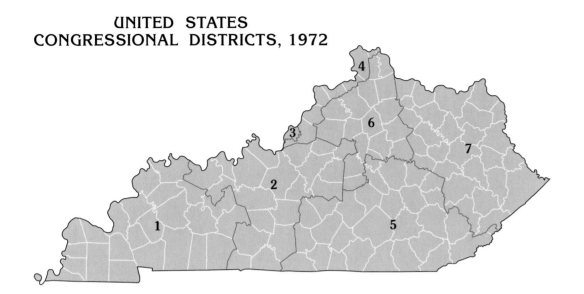

UNITED STATES
CONGRESSIONAL DISTRICTS, 1972

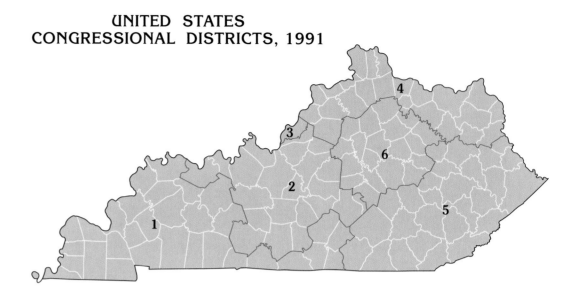

UNITED STATES
CONGRESSIONAL DISTRICTS, 1991

enough from one section of the state to another to contribute to a strong sense of regionalism. Today, regional patterns are still reflected in the retention of traditional views concerning such issues as government responsibilities, the definition of reform, appropriate qualifications for public officials, and government expenditures—whether greater emphasis should be placed on roads, education, women's issues, or environmental reform, for example. The attitudes of Kentuckians toward these and many other issues are the ingredients that permit the identification of regional political cultures across the state.

Regional Political Cultures

Voting is one of the best measures of a people's perceptions of government and political issues. Consistency of voting patterns across the Commonwealth demonstrates the stability and permanence of political regions. Kentucky has several distinct political regions within its borders. As other chapters demonstrate, historic cultural and settlement influences are still apparent in the patterns of church denominations, European ethnic heritages, extraregional and outside influences on urban economies, black migration and residence, and the influence of pre- and post- Civil War economies and government policies, including federal government programs. Often such patterns extend into adjacent states. Several Kentucky communities along the Ohio River from

Covington to Henderson, for example, share similarities with cross-border areas in Ohio and Indiana. Riverine urban and industrial growth in this area, coupled with productive agricultural hinterlands, contributed to the emergence of cities as concentrations of political strength. This development was crucial in the state's political history, as strength and power were drawn away from the traditional areas of political power, especially rural counties with economies tied to agriculture, mining, and forestry.

Tradition and political party loyalty across the state are strong, so existing social and political orders usually have sufficient inertia to resist rapid and dramatic change. The state's political history has at its core an interest in retaining an agrarian perspective and rural ways of "doing politics." Settlers who came from rural western North Carolina and eastern and central Tennessee, including the Nashville Basin, strongly influenced Kentucky's south central and southwestern counties. Western Kentucky, specifically the Jackson Purchase, was populated by settlers who came north from Tennessee. These populations were tied to rural economies and small towns, not large urban and industrial centers linked by railroads. Southern cultural traditions were important among such migrant groups and were evident in religious adherence, attitudes toward blacks, beliefs about the responsibilities of government, and the roles of elected officials, which were often custodial and intended to preserve existing order. Political scientist Daniel Elazar defines these features as characteristic of a traditionalistic political culture, quite differ-

ent from the moralistic culture he found concentrated in the northeastern and midwestern states, or the individualistic type found in portions of the central Midwest. Elazar considers Kentucky to be a traditionalistic/individualistic state, with individualistic features found only in northern Kentucky.

The state's political cultures were important during the nineteenth century in the formation and persistence of regionalism, and they remain salient today, as an examination of voting patterns will confirm. Major population increases over the past century have occurred in cities—Louisville, Lexington, and northern Kentucky near Cincinnati. These places have become more influential in statewide political leverage and voting. They have fostered candidates for political office and have contributed to the growth of the Republican Party in the state's suburban areas.

Congressional Districts

Kentucky's members of the U.S. House of Representatives are elected from districts whose number and boundaries are approved by Congress and by the state legislature.

Kentucky's representation in the U.S. House has decreased during the twentieth century because the state's population has declined relative to that of other states and the total U.S. population. From eleven members in 1910, the number had dropped to seven by 1960. As a result of the 1990 census, Kentucky lost an

KENTUCKY HOUSE OF REPRESENTATIVES DISTRICTS, 1992

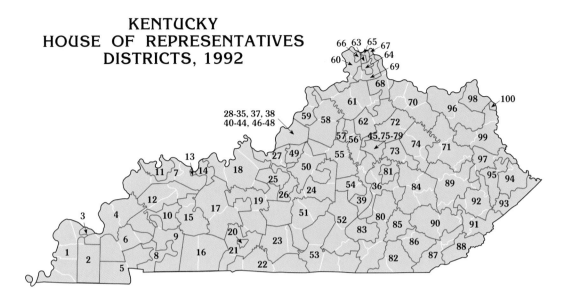

KENTUCKY SENATE DISTRICTS, 1992

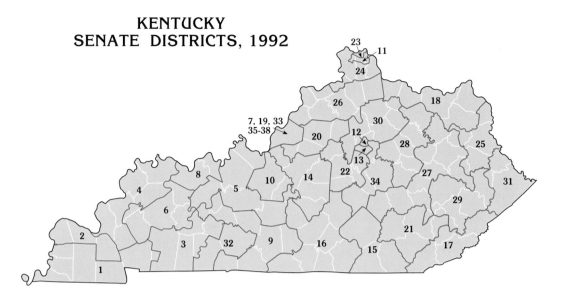

additional member, bringing the current total to six. Each reduction brought a corresponding change in the shapes and sizes of the state's congressional districts. Such adjustments are intended to maintain districts of approximately equal population size. Thus, the 1990 congressional districts are roughly balanced in population: each includes about 350,000 people.

With each change in Kentucky's population relative to that of other states, the legislature must realign existing congressional boundaries to accommodate the adjusted number of representatives. Districts that are most resistant to alteration are those whose members of Congress have served extended terms and who are members of the political party that controls the U.S. House. The "unsafe seats" are those held by members with the shortest tenures in Congress who belong to the minority party. Congressional districts where the boundaries are least likely to change, then, wield the most power.

Changes in Kentucky's congressional membership have occurred during the past several decades, not only in the number of representatives but also in the composition of the delegation. There were four Democrats and three Republicans in 1960, five Democrats and two Republicans in 1970, and four Democrats and three Republicans in 1980. In the mid-1990s there were one Democrat and five Republicans. In attempting to understand political power, one should remember that the shape and composition of districts are very important in obtaining and retaining such power. The configuration of the congressional districts is drawn

by someone or some group and is approved by others. Politicians know that maps and boundaries are critically important to their governing efforts, and some would suggest that gerrymandering plays a role in boundary adjustment. The reshaping of Kentucky's congressional districts between the 1970s and the 1990s, when one representative's seat was lost, for example, is vividly portrayed on district maps.

State Legislative and Judicial Districts

Numbers of people influence representation and political power. Population changes recorded by the decennial census necessitate alteration in Kentucky House and Senate districts, as well as districts affecting the judicial circuits and the state supreme court.

As rural areas decline in population and urban places grow, Kentucky House and Senate representation from the Louisville and Lexington areas increases. In the mid-1990s the Kentucky Senate has thirty-eight members, twelve of whom represent urban districts (seven from Jefferson County, three from northern Kentucky, and two from Fayette County). Predominantly rural districts often include portions of more than one county, which are sometimes agglomerated into peculiar shapes. Kentucky's House of Representatives has one hundred members. House districts are smaller in population and size than Senate districts, but the bias toward urban representation is similar (eighteen from

Kentucky's bicameral legislature includes the House of Representatives, shown here in special session, and the Senate.

JUDICIAL CIRCUITS, 1991

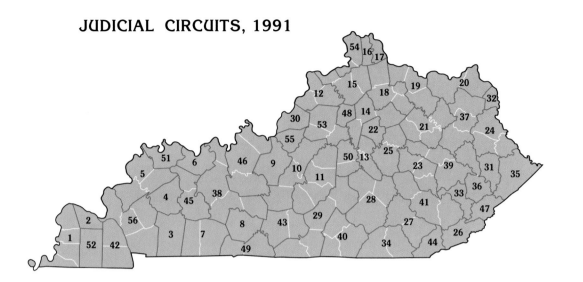

SUPREME COURT DISTRICTS, 1991

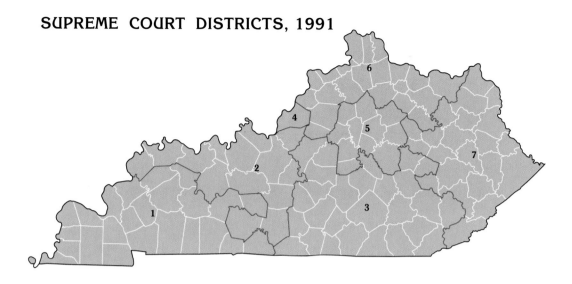

A statue of France's Louis XVI faces the Louisville City Hall across Jefferson Avenue. The city was named for the French king who aided the American colonies in their struggle with England.

Jefferson County, eight from northern Kentucky, and six from Fayette County). Achieving population parity between districts also often produces peculiar sizes and shapes. District 36, for example, looks like an appendix and extends from Pulaski County north to Madison County.

The state's fifty-six judicial circuits are based not on population size but on rough equivalence in area. Some judicial circuits are single large counties, but many are composites of two or more smaller counties. The seven state supreme court districts resemble congressional districts and also correspond roughly to the state's physiographic regions. District 5, for example, is primarily the Inner Bluegrass, and District 1 comprises the Jackson Purchase, the western Pennyroyal, and a portion of the Western Kentucky Coal Field.

Since representation translates into political power and influence, those places whose residents hold state or national office may command more influence than other places, whether measured in the politician's committee memberships or in government programs that benefit the district. Such power is not static, though. It shifts with population change. As rural-to-urban migration has accelerated since World War II, cities and their suburbs have gained representation at the expense of rural areas. Fewer people means less power in the halls of government and less leverage to promote or repress legislative initiatives.

Electoral Geography

Voting patterns suggest that Kentucky's political culture has several facets. Voting can vary from place to place and from one decade to another, and explaining why a county or district voted as it did can be a murky business.

KENTUCKY COURT SYSTEM

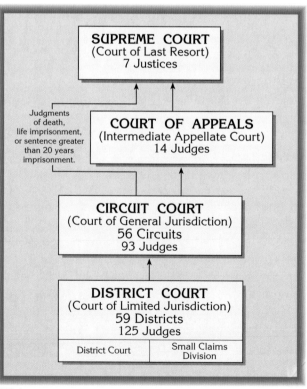

1996 data

VOTER TURNOUT
1992 PRESIDENTIAL ELECTION

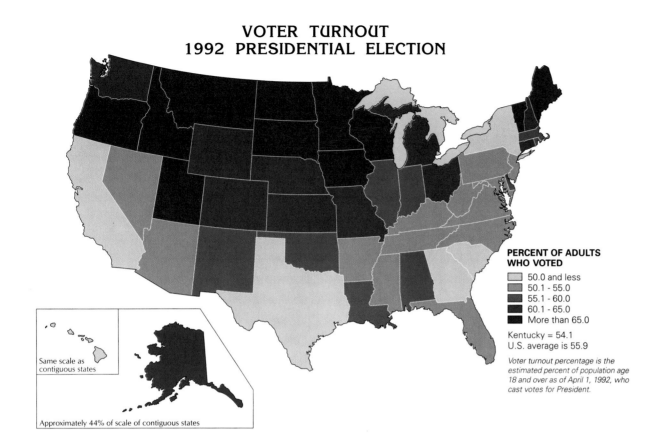

PERCENT OF ADULTS WHO VOTED

- 50.0 and less
- 50.1 - 55.0
- 55.1 - 60.0
- 60.1 - 65.0
- More than 65.0

Kentucky = 54.1
U.S. average is 55.9

Voter turnout percentage is the estimated percent of population age 18 and over as of April 1, 1992, who cast votes for President.

Same scale as contiguous states

Approximately 44% of scale of contiguous states

VOTER REGISTRATION

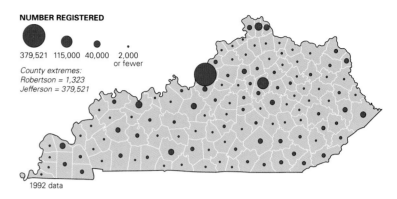

NUMBER REGISTERED

379,521   115,000   40,000   2,000 or fewer

County extremes:
Robertson = 1,323
Jefferson = 379,521

1992 data

VOTER TURNOUT
MAY 26, 1992, PRIMARY ELECTION

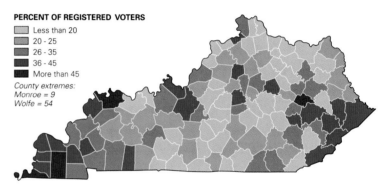

PERCENT OF REGISTERED VOTERS

- Less than 20
- 20 - 25
- 26 - 35
- 36 - 45
- More than 45

County extremes:
Monroe = 9
Wolfe = 54

VOTER TURNOUT
NOVEMBER 3, 1992, GENERAL ELECTION

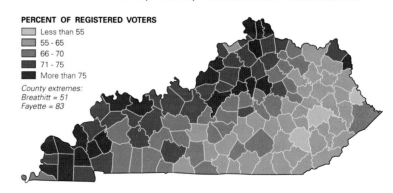

PERCENT OF REGISTERED VOTERS

- Less than 55
- 55 - 65
- 66 - 70
- 71 - 75
- More than 75

County extremes:
Breathitt = 51
Fayette = 83

Nevertheless, some correlations to voting patterns can be identified. Regional consistency in voting may relate to political party registration, party loyalty, previous voting history, voter demographics (e.g., age, occupation, income, employment level), ethnic or racial backgrounds, religious affiliations, population change, and perceptions of candidates and issues.

The five sections below address various aspects of Kentucky's electoral geography. Party registration and turnout, two important influences on election results, are considered first. Second, state Republican and Democratic votes in presidential elections are compared with national trends from 1920 to 1996, and state voting patterns in presidential elections since 1968 are explored. The next two sections examine selected recent senatorial and gubernatorial races. The final section reviews the votes of Kentuckians on selected amendments to the state constitution.

Registration and Turnout.   Voter registration is an important element in a region's political culture. Each election year, politi-

cal parties conduct registration drives to identify new voters or to reregister those who failed to participate in recent elections. Political party officials promote registration efforts to ensure that their party remains in office or wins positions from the opposition.

Total voter registration is directly related to population size. The number of registered voters in 1992, therefore, mirrored the number of inhabitants. Although numbers of registered voters change somewhat from one election year to another as new voters are added and the deceased are removed from the rolls, the overall pattern remains stable.

Voter turnout means those who actually vote. If all those who are registered voted, the turnout would be 100 percent, a figure rarely achieved in any election. Yet the proportion of people who vote is another element of regional political culture, and turnout varies from one region and one election to another. This fact is illustrated by differences in turnout for the May 1992 primary election and the November 1992 general election. In the primary election for senator and representatives, the average county turnout was only 25 percent. Not only was total turnout very low, but it varied markedly, from less than 20 percent in areas of south central and northeastern Kentucky to more than 50 percent in a few western counties and in Wolfe County in the east. Turnout for the general election in November did not mirror that for the primary. The average county turnout, at 69 percent, was much higher, perhaps because voters wished to participate in a presidential election in addition to voting for a U.S. senator, members of Congress, state representatives, and local officials, and to register their opinions on constitutional amendments and local issues. The turnout exceeded 50 percent in all counties, with the largest urban and suburban counties recording the highest levels, at above 75 percent. In general, the more populous and more industrial and service-oriented counties exhibited the highest turnouts.

Presidential Elections. Across the twenty presidential elections from 1920 to 1996, the state's voting returns were generally similar to the national aggregate vote, with a few notable exceptions. Democratic presidential candidates carried the state eleven times, Republicans nine times. The highest percentage of Kentucky's vote that any Republican received was Richard M. Nixon's 1972 total; the second-highest was the vote for Ronald Reagan's second term in 1984. In the three-person race of 1968, Nixon won easily over Hubert H. Humphrey; third-party candidate George C. Wallace received about 20 percent of the vote. Bill Clinton narrowly defeated incumbent president George Bush in the state in 1992; H. Ross Perot, an independent candidate, received nearly 15 percent of the final state vote. In 1996 Clinton narrowly defeated Bob Dole in the state's presidential election.

During the same 1920-96 period, Republicans and Democrats evenly split the twenty elections. When a Republican president was elected nationwide, Kentucky voted for that winner eight times out of ten. Republi-

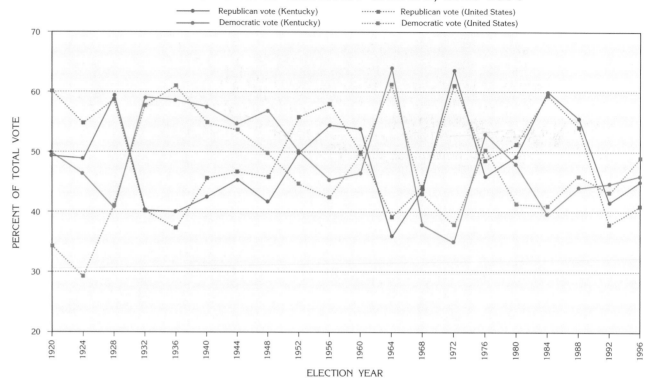

PRESIDENTIAL ELECTION TRENDS, 1920-1996

Republican vote (Kentucky) · Republican vote (United States)
Democratic vote (Kentucky) · Democratic vote (United States)

PRESIDENTIAL ELECTION POPULAR VOTE, 1920-1996

Year	Kentucky		Candidate		United States	
	% Republican	% Democrat	Republican	Democrat	% Republican	% Democrat
1920	49.3	49.7	Harding	Cox	60.3	34.1
1924	48.8	46.2	Coolidge	Davis	54.0	28.8
1928	59.3	40.5	Hoover	Smith	58.2	40.8
1932	40.2	59.1	Hoover	Roosevelt	39.6	57.4
1936	39.9	58.5	Landon	Roosevelt	36.5	60.8
1940	42.3	57.4	Wilkie	Roosevelt	44.8	54.7
1944	45.2	54.5	Dewey	Roosevelt	45.9	53.4
1948	41.5	56.7	Dewey	Truman	45.1	49.6
1952	49.8	49.9	Eisenhower	Stevenson	55.1	44.4
1956	54.3	45.2	Eisenhower	Stevenson	57.4	42.0
1960	53.6	46.4	Nixon	Kennedy	49.5	49.7
1964	35.7	64.0	Goldwater	Johnson	38.5	61.1
1968	43.8	37.6	Nixon	Humphrey	43.4	42.7
1972	63.4	34.8	Nixon	McGovern	60.7	37.5
1976	45.6	52.8	Ford	Carter	48.0	50.1
1980	49.1	47.6	Reagan	Carter	50.7	41.0
1984	60.0	39.4	Reagan	Mondale	58.8	40.6
1988	55.5	43.9	Bush	Dukakis	53.4	45.6
1992	41.3	44.6	Bush	Clinton	37.4	43.0
1996	45.1	46.1	Dole	Clinton	40.7	49.2

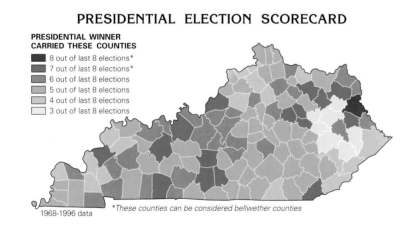

PRESIDENTIAL ELECTION SCORECARD

PRESIDENTIAL WINNER CARRIED THESE COUNTIES

■ 8 out of last 8 elections*
■ 7 out of last 8 elections*
■ 6 out of last 8 elections
□ 5 out of last 8 elections
□ 4 out of last 8 elections
□ 3 out of last 8 elections

1968-1996 data   *These counties can be considered bellwether counties

cans have won six of the last eleven presidential elections, and a majority of Kentucky voters chose the Republican candidate in all six of these elections. Democrats won the White House in 1960, 1964, 1976, 1992, and 1996, and Kentuckians voted for the winning Democrat in all but the 1960 election.

The returns in Kentucky for the eight presidential elections from 1968 through 1996 illustrate changing regional concerns during and following the Vietnam War and the last two decades of the Cold War. George Wallace and Ross Perot mustered strong third-party efforts in 1968 and 1992, respectively. Nationally, Republicans won five of these eight elections, Democrats three. Kentuckians voted for the winning candidate each time.

Because of Kentucky's record of choosing the winner, the major political parties closely observe voter attitudes and preferences in the state. One Kentucky county, Lawrence, has supported the winner in all eight presidential elections since 1968. Another sixteen counties voted for the winner in seven of those elections. These counties are scattered throughout the state, with one concentration in the Pennyroyal and another in northeastern Kentucky. At the other extreme, a pocket of counties in eastern Kentucky has voted for the winner only three times. These counties usually vote for the Democratic candidate, and only two members of that party—Jimmy Carter and Bill Clinton— have been elected to the White House since 1968.

In the presidential election of 1968, the three candidates were Richard Nixon (Republican), Hubert Humphrey (Democrat), and George Wallace (American Independent). Nixon won Kentucky, with 44 percent of the vote. Nixon's vote was highest in the state's south central counties, where it ranged from 64 to 84 percent. He also won the five counties with the highest vote totals: Jefferson, Fayette, Kenton, Campbell, and Daviess. Humphrey's strength was in southeastern Kentucky, the northern Bluegrass counties, and western Kentucky. Segregationist George Wallace won a plurality of the vote in Bullitt County, south of Louisville; in Christian and Todd Counties; and in the southwestern Jackson Purchase counties of Hickman and Fulton.

The 1972 election pitted Republican incumbent Richard Nixon against Democrat George S. McGovern. With 63 percent of the state's vote, Nixon easily defeated McGovern, winning all but eight counties. A cluster of seven rural counties in east central Kentucky was the only area of Democratic strength. Nixon's popularity was especially high (between 75 and 92 percent) in the counties of south central Kentucky, a region with a strong tradition of voting Republican for much of the twentieth century.

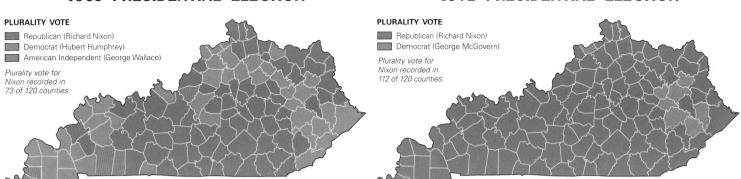

1968 PRESIDENTIAL ELECTION

PLURALITY VOTE

- Republican (Richard Nixon)
- Democrat (Hubert Humphrey)
- American Independent (George Wallace)

Plurality vote for Nixon recorded in 73 of 120 counties.

1972 PRESIDENTIAL ELECTION

PLURALITY VOTE

- Republican (Richard Nixon)
- Democrat (George McGovern)

Plurality vote for Nixon recorded in 112 of 120 counties.

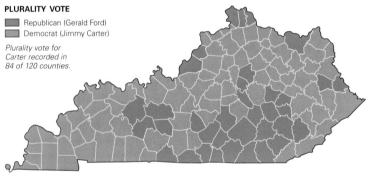

1976 PRESIDENTIAL ELECTION

PLURALITY VOTE

- Republican (Gerald Ford)
- Democrat (Jimmy Carter)

Plurality vote for Carter recorded in 84 of 120 counties.

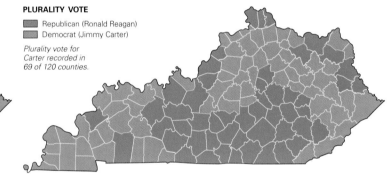

1980 PRESIDENTIAL ELECTION

PLURALITY VOTE

- Republican (Ronald Reagan)
- Democrat (Jimmy Carter)

Plurality vote for Carter recorded in 69 of 120 counties.

As in 1968, Nixon in 1972 won the five counties with the largest vote totals.

In 1976 Democrat Jimmy Carter faced Republican Gerald R. Ford. Carter was the first Democratic candidate to carry Kentucky since Lyndon Johnson defeated Barry Goldwater in 1964. Carter carried eighty-four counties to Ford's thirty-six. Support for Carter was especially strong in the Jackson Purchase counties (between 68 and 82 percent), and statewide the former Georgia governor won many counties that Nixon had carried in 1972. Ford won Jefferson, Fayette, Kenton, and Campbell Counties as well as a large cluster of traditionally Republican counties in south central Kentucky.

In 1980 Republican Ronald Reagan challenged incumbent Jimmy Carter. John Anderson, an independent, also sought the presidency. Reagan won Kentucky by less than 2 percent of the

vote; he carried Jefferson and Fayette Counties, the three suburban counties of northern Kentucky, and the traditionally Republican south central counties. Reagan received his lowest percentages in eastern Kentucky and the rural Bluegrass counties within the Louisville-Lexington-Covington triangle.

In 1984 Reagan ran for a second term, challenged by Democrat Walter F. Mondale. Reagan's margin of victory in the state widened substantially this time, and the incumbent president defeated Mondale by more than 20 percent of the vote. Reagan won 99 of the state's 120 counties, including the 5 counties with the highest vote totals. The 1984 election pattern resembles that of the 1972 Nixon-McGovern contest. Mondale's only significant following was in counties of southeastern Kentucky, where the Minnesotan received between 58 and 90 percent of the vote. South central Kentucky continued to be a Reagan bastion—he

1984 PRESIDENTIAL ELECTION

PLURALITY VOTE

■ Republican (Ronald Reagan)
■ Democrat (Walter Mondale)

Plurality vote for Reagan recorded in 99 of 120 counties.

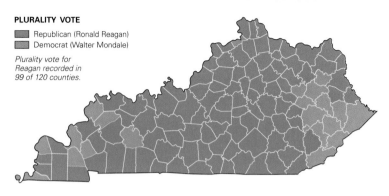

1988 PRESIDENTIAL ELECTION

PLURALITY VOTE

■ Republican (George Bush)
■ Democrat (Michael Dukakis)

Plurality vote for Bush recorded in 84 of 120 counties.

1992 PRESIDENTIAL ELECTION

PLURALITY VOTE

■ Republican (George Bush)
■ Democrat (Bill Clinton)

Plurality vote for Clinton recorded in 72 of 120 counties.

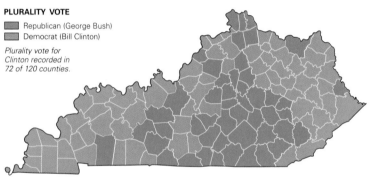

1996 PRESIDENTIAL ELECTION

PLURALITY VOTE

■ Republican (Bob Dole)
■ Democrat (Bill Clinton)

Plurality vote for Clinton recorded in 58 of 120 counties.

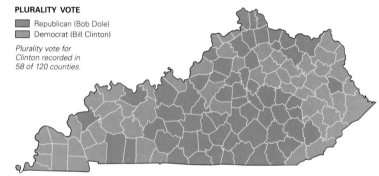

VOTE FOR INDEPENDENT CANDIDATE PEROT
1992 PRESIDENTIAL ELECTION

PERCENT OF TOTAL VOTE

■ Less than 10.00
■ 10.00 - 13.00
■ 13.01 - 15.00
■ 15.01 - 18.00
■ More than 18.00

County extremes:
Jackson = 7.52
Pendleton = 23.26

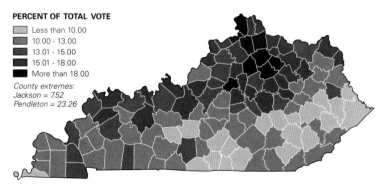

VOTE FOR INDEPENDENT CANDIDATE PEROT
1996 PRESIDENTIAL ELECTION

PERCENT OF TOTAL VOTE

■ Less than 5.0
■ 5.0 - 7.9
■ 8.0 - 9.9
■ 10.0 - 11.9
■ 12.0 or more

County extremes:
Hickman = 3.0
Elliott = 14.2

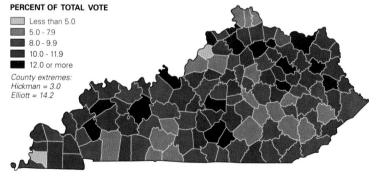

improved his performance there about 5 percent over 1980—but the incumbent president also carried suburban Bluegrass counties around both Louisville and Lexington, as well as suburban northern Kentucky. Reagan won many of the same counties he had won in 1980, but with higher percentages.

In 1988 Republican vice president George Bush faced Democrat Michael S. Dukakis for the presidency. Even though Bush defeated Dukakis in Kentucky by only about 12 percent, the voting pattern was strikingly similar to that for Reagan's 1984 win. Bush won all but thirty-six counties; those he lost were primarily in eastern Kentucky and in the western third of the state. Bush, like Reagan in 1984, won the five most populous counties. Most of the south central counties that had voted so strongly for Nixon and Reagan also voted for Bush.

The 1992 election involved a three-way race between Democrat Bill Clinton, incumbent Republican George Bush, and independent candidate Ross Perot. Kentucky voters did not support Bush's reelection: he lost to Clinton by 3 percent. Clinton won seventy-two of the state's counties compared with Bush's forty-eight. Clinton captured counties in eastern and western Kentucky, but the Arkansas governor also carried Jefferson County and some adjacent suburbs. Bush won the Inner Bluegrass, suburban northern Kentucky, and the traditionally Republican counties of south central Kentucky. The 1992 presidential election marked a departure from previous contests, as Democrats made inroads in the five most populous counties: Bush won Fayette, Kenton, and Campbell Counties, but Clinton carried Jefferson and Daviess Counties. Perot won 14 percent of the vote across the Commonwealth, but the Perot vote exhibited distinctive patterns. Perot's highest percentages appeared in urban and suburban counties, especially in the Bluegrass and western Kentucky. The Perot vote peaked in a corridor of counties from Covington to Lexington, at 19-23 percent.

In 1996 Democrat Bill Clinton ran for reelection against Republican challenger, Bob Dole. Clinton narrowly carried the state (by only 1 percent), as he did four years earlier. Third-party candidate Ross Perot fared less well nationally and in the state than in 1992. Clinton won fifty-eight counties in 1996 compared to seventy-two in 1992. Overall, the 1996 voting patterns mirrored closely those of 1992. Clinton won Jefferson County in both elections; he lost Fayette in 1992 but carried it in 1996. He lost Kenton and Campbell Counties in

President Bill Clinton's 1996 reelection campaign brought him to Louisville, where he spoke to Democratic supporters in front of the Hillerich & Bradsby Louisville Slugger headquarters building on west Main Street.

both elections. The Jackson Purchase region voted for Clinton in both years, as did most of the Western Kentucky Coal Field counties and most counties east of Interstate 75. The Pennyroyal, as it had in most previous presidential elections, remained solidly Republican. Perot's strongest support came in Elliott County, where he received 14.2 percent of the vote.

The state's record of voting for the national presidential winner since 1964 makes Kentucky a mirror of national sentiment and an important predictor of success in presidential election campaigning. Internally, some counties have exhibited considerable voter inertia in the eight most recent presidential elections. The south central Appalachian and Pennyroyal counties, as well as the five largest urban counties, are consistently conservative. Jefferson County voted Republican in six of eight elections; Democrats won the county only in 1992 and 1996. Fayette County voted Democrat in 1996, Republican in other years. Kenton and Campbell Counties consistently voted Republican. Daviess County has a mixed record, voting Republican in 1968, 1972, 1984, 1988, and 1996 (all years in which a Republican won the presidency). These five urban counties account for almost 35 percent of the state's total vote in presidential elections. Jefferson County alone is usually responsible for about one-fifth of the state's vote.

Senatorial Elections. Kentucky elects U.S. senators every two to four years, depending upon the expiration of particular terms. This section reviews voting patterns in the senatorial elections of 1962, 1966, 1990, 1992, and 1996.

In 1962 the Republican incumbent, Thruston B. Morton, received 51 percent of the vote, narrowly defeating Democratic nominee Wilson W. Wyatt. Morton had general support across the state, but especially in the central, west central, and northern counties. Morton won 64 of the state's 120 counties, and his strongest support came from the south central region, the traditional Republican stronghold. Here his winning margin across a dozen counties was between 65 and 85 percent. Morton also won the urban counties of Jefferson, Fayette, Kenton, Campbell, and Daviess. Wyatt's support centered in the Jackson Purchase, the western Pennyroyal, eastern Kentucky, and the rural Bluegrass between Louisville, Lexington, and Covington.

In 1966 Republican incumbent John Sherman Cooper faced Democrat John Y. Brown Sr. Cooper received 64 percent of the statewide vote, defeating Brown by a margin of almost two to one. Cooper, like Morton in 1962, won sixty-four counties. In fact, the 1966 voting pattern was much like that of the 1962 vote. The major difference was the higher percentage of Republican votes in counties that Cooper won. Cooper's strength was in south central Kentucky, the

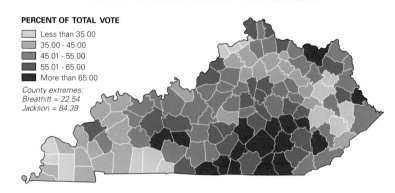

VOTE FOR REPUBLICAN CANDIDATE MORTON 1962 SENATORIAL ELECTION

PERCENT OF TOTAL VOTE
Less than 35.00
35.00 - 45.00
45.01 - 55.00
55.01 - 65.00
More than 65.00
County extremes:
Breathitt = 22.54
Jackson = 84.39

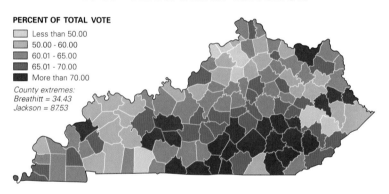

VOTE FOR REPUBLICAN CANDIDATE COOPER 1966 SENATORIAL ELECTION

PERCENT OF TOTAL VOTE
Less than 50.00
50.00 - 60.00
60.01 - 65.00
65.01 - 70.00
More than 70.00
County extremes:
Breathitt = 34.43
Jackson = 87.53

eastern Pennyroyal, and suburban counties near Cincinnati. He also carried the five most populous counties. Brown did best in the Jackson Purchase, in selected counties of eastern Kentucky, and in rural Bluegrass counties.

In 1990 Addison Mitchell ("Mitch") McConnell, the Republican incumbent, faced Democratic challenger Harvey Sloane. McConnell received 52 percent of the statewide vote. He won seventy-two counties, including Fayette, Kenton, and Campbell, but he lost Daviess County and his and Sloane's home county, Jefferson. McConnell won counties in each political region of the state, but the largest group of contiguous counties was located in south central Kentucky and the Pennyroyal. Republican support in the Bluegrass, the Jackson Purchase, and eastern Kentucky was scattered. McConnell's power base was similar to that of Republicans Morton

1990 SENATORIAL ELECTION

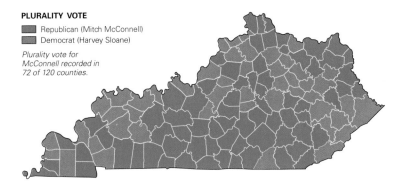

PLURALITY VOTE

■ Republican (Mitch McConnell)
■ Democrat (Harvey Sloane)

*Plurality vote for
McConnell recorded in
72 of 120 counties.*

1992 SENATORIAL ELECTION

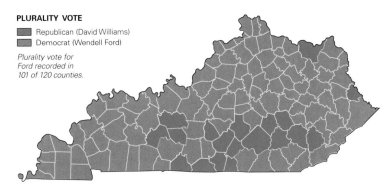

PLURALITY VOTE

■ Republican (David Williams)
■ Democrat (Wendell Ford)

*Plurality vote for
Ford recorded in
101 of 120 counties.*

1996 SENATORIAL ELECTION

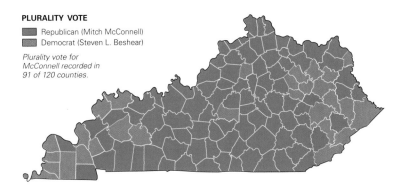

PLURALITY VOTE

■ Republican (Mitch McConnell)
■ Democrat (Steven L. Beshear)

*Plurality vote for
McConnell recorded in
91 of 120 counties.*

and Cooper. Democrat Sloane carried many counties in the Western Kentucky Coal Field Region, southeastern Kentucky, and the Bluegrass.

In 1992 longtime Senate Democrat Wendell H. Ford won 63 percent of the vote and 101 counties, handily defeating his Republican challenger, David Williams. The win reflected Ford's statewide appeal. He even won regions and cities previously won by Republicans McConnell, Cooper, and Morton. Seventeen of the nineteen counties that Williams won were in the traditional Republican stronghold of south central Kentucky. Ford enjoyed strong support from several eastern counties, and he won almost two-thirds of the vote in Jefferson County and in his home county, Daviess, as well as 57 percent in both Fayette and Campbell Counties.

In 1996, incumbent Republican Mitch McConnell faced Democrat Steven Beshear. The pattern of voting support evident in McConnell's 1990 election win was mirrored in the 1996 race, in which he defeated Beshear by winning 91 counties. The state's urban centers all supported the Republican senator's reelection bid, as did a broad swath of counties in the state's midsection. Only the Jackson Purchase in the west and coal counties in the Western and Eastern Coal Fields supported Beshear.

Gubernatorial Elections. Kentucky elects its governors every four years. Elections are held in odd-numbered years so as not to conflict with elections for president, members of Congress, or state senators and representatives. From 1800 on, when the second state constitution went into effect, governors could not succeed themselves, so many political unknowns established their gubernatorial credibility by first serving as lieutenant governor. During the 1992 state legislative session, however, the Kentucky Senate and House of Representatives approved amending the state constitution to allow governors to serve two consecutive terms, effective with the governor elected in 1995.

Gubernatorial elections are strongly influenced by party registration. Since the state commonly registers many more Democrats than Republicans, Democratic governors have predominated in the state's highest office. This section examines the gubernatorial voting patterns in six of the nine elections from 1963 to 1995.

In 1963 Democrat Edward T. ("Ned") Breathitt Jr. faced Republican challenger Louie B. Nunn. Breathitt won this close

election, polling 51 percent of the total vote and carrying sixty-eight counties. Breathitt's strength was in the Jackson Purchase, a number of counties in the Western Kentucky Coal Field, eastern Kentucky, and the rural Bluegrass. This pattern foreshadowed that for the 1976 Carter-Ford presidential contest. Breathitt received more than two-thirds of the vote in a half dozen Jackson Purchase counties and several in east central Kentucky. Nunn won almost two-thirds of the vote in the south central counties as well as the urban counties of Jefferson, Fayette, Kenton, and Campbell.

In 1967, in his second attempt at the governorship, Republican Louie Nunn defeated his Democratic opponent, Henry Ward. Nunn's total support (52 percent) was almost the same as Breathitt's had been four years earlier. In addition, the pattern of Nunn's 1967 win strongly resembles that for his 1963 loss. The deciding difference was a higher percentage of Republican votes in those areas he had lost earlier. Nunn won four of the five large urban counties; the exception was Daviess. His regional strength was again the conservative reservoir in south central Kentucky, where voters in eight counties contributed from 70 to 85 percent of their votes to the Republican column. Ward did best in rural and small-town counties in western Kentucky and in the northern Bluegrass.

In 1983 Kentucky elected its first woman governor. Democratic lieutenant governor Martha Layne Collins defeated Republican Jim Bunning with 55 percent of the vote. Winning handily throughout most of the state, Collins lost only thirty counties to Bunning. She won in several counties traditionally carried by Republicans in statewide races; she won Jefferson and Fayette Counties, for example, but not Kenton or Campbell. Her highest winning margins came in east central Kentucky, the Bluegrass between Louisville and Lexington, the western Pennyroyal, and the Jackson Purchase. Bunning ran well in his home area of northern Kentucky and the south central counties.

In 1987 Democrat Wallace G. Wilkinson faced Republican state representative John Harper for the governorship. This election illustrated the strength of the Democratic Party across the state. Wilkinson defeated Harper with 65 percent of the vote; the Democrat carried all but five counties. Wilkinson won several counties that traditionally vote Republican in U.S. Senate or gubernatorial elections. Harper's support exhibited no regional concentration; he was successful in Fayette County, Wilkinson's place of residence, and two suburban counties

1963 GUBERNATORIAL ELECTION

PLURALITY VOTE

- ▦ Republican (Louie Nunn)
- ▦ Democrat (Edward Breathitt)

*Plurality vote for
Breathitt recorded in
68 of 120 counties.*

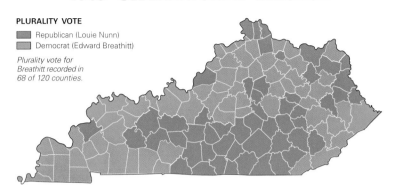

1967 GUBERNATORIAL ELECTION

PLURALITY VOTE

- ▦ Republican (Louie Nunn)
- ▦ Democrat (Henry Ward)

*Plurality vote for
Nunn recorded in
62 of 120 counties.*

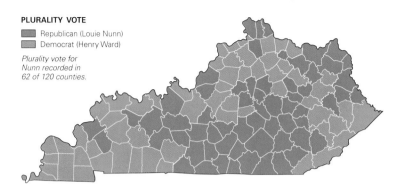

*Martha Layne Collins exits a voting booth after voting at
the Fairgrounds precinct in Woodford County. In the
November 1986 general election, Kentucky voters elected
her the Commonwealth's first woman governor.*

1983 GUBERNATORIAL ELECTION

PLURALITY VOTE

- ▦ Republican (Jim Bunning)
- ▦ Democrat (Martha Layne Collins)

*Plurality vote for
Collins recorded in
90 of 120 counties.*

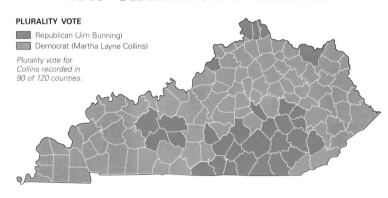

1987 GUBERNATORIAL ELECTION

PLURALITY VOTE

- ▦ Republican (John Harper)
- ▦ Democrat (Wallace Wilkinson)

*Plurality vote for
Wilkinson recorded in
115 of 120 counties.*

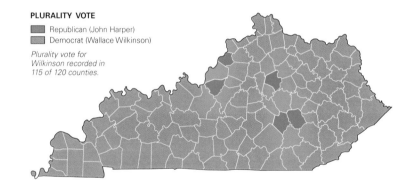

near Louisville. While Wilkinson had statewide support,
his highest vote percentages were recorded in the Jackson Purchase and in east central Kentucky. Here, he won
79-90 percent of the votes cast. He also accomplished a
rare Democratic sweep across the traditional south central counties, the Bluegrass region, and the northern Kentucky suburbs.

In 1991 Brereton C. Jones, lieutenant governor in
the Wilkinson administration, was the Democratic nominee running against Larry Hopkins, a Republican congressman elected from the Lexington area. Jones won
this election handily. His 64 percent did not quite match
Wilkinson's 65 percent, but the voting pattern was very
similar. Jones won 107 counties and carried counties
across all state regions, although the conservative south
central region largely voted for Hopkins. Jones won all
five of the most populous counties, a feat that had eluded
Wilkinson four years earlier. Jones even won in Hopkins'
congressional district and in Lexington, albeit by less than
10 percent of the vote. His highest vote percentages
came from eastern and western Kentucky.

In 1995 Paul E. Patton of Pike County, Jones's lieutenant governor, faced Republican nominee Larry Forgy

1991 GUBERNATORIAL ELECTION

PLURALITY VOTE

- ▦ Republican (Larry Hopkins)
- ▦ Democrat (Brereton Jones)

*Plurality vote for
Jones recorded in
107 of 120 counties.*

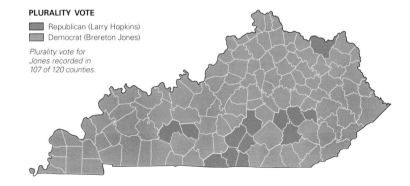

1995 GUBERNATORIAL ELECTION

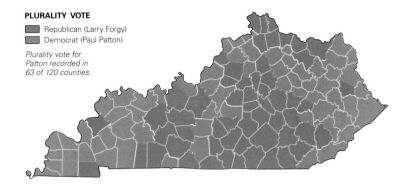

PLURALITY VOTE

- Republican (Larry Forgy)
- Democrat (Paul Patton)

Plurality vote for Patton recorded in 63 of 120 counties.

BIRTHPLACE OF KENTUCKY'S GOVERNORS

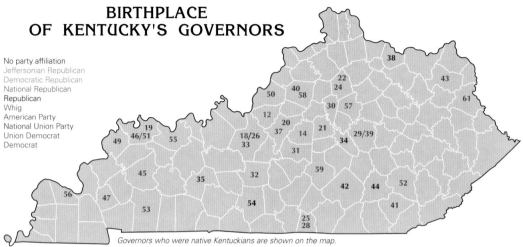

No party affiliation
Jeffersonian Republican
Democratic Republican
National Republican
Republican
Whig
American Party
National Union Party
Union Democrat
Democrat

Governors who were native Kentuckians are shown on the map.

of Lexington. This election, like those in 1963 and 1967, resulted in a very close vote: Patton won with only 51 percent. In another marked parallel to earlier closely contested elections, Patton won sixty-three counties. The pattern of counties that voted for Forgy was similar to the pattern of those Nunn carried in 1963 and 1967. Forgy was successful in the traditional Republican area of south central Kentucky, the Western Kentucky Coal Field, northern Kentucky, and selected rural Bluegrass counties. Forgy won Fayette, Kenton, and Campbell Counties but lost Jefferson and Daviess Counties to Patton. Patton's major strength was in eastern Kentucky, especially those half dozen counties near Pikeville, where he won between 65 and 85 percent of the votes cast. The striking similarities—and differences—between the 1995 gubernatorial voting pattern and those of 1963 and 1967 illustrate traditional areas of strength for both parties and also identify those counties that tend to shift party allegiance from one election to another.

GOVERNOR	TERM(S)	LIFE	BIRTH PLACE	PARTY AFFILIATION
1. Isaac Shelby	1792-96	1750-1826	Hagerstown, MD	None
2. James Garrard	1796-1804	1749-1822	Stafford Co. VA	Jeffersonian Republican
3. Christopher Greenup	1804-08	1750-1818	Loudoun Co. VA	Jeffersonian Republican
4. Charles Scott	1808-12	1739-1813	Powhatan Co. VA	Jeffersonian Republican
5. Isaac Shelby (2nd term)	1812-16	1750-1826	Hagerstown, MD	None
6. George Madison	1816	1763-1816	Rockingham Co. VA	Jeffersonian Republican
7. Gabriel Slaughter	1816-20	1767-1830	Culpepper Co. VA	Jeffersonian Republican
8. John Adair	1820-24	1757-1840	Chester District SC	Democratic Republican
9. Joseph Desha	1824-28	1768-1842	Monroe Co. PA	Democratic Republican
10. Thomas Metcalfe	1828-32	1780-1855	Fauquier Co. VA	National Republican
11. John Breathitt	1832-34	1786-1834	New London, Henry Co. VA	Democrat
12. James Turner Morehead	1834-36	1797-1854	Shepherdsville, Bullitt Co. KY	National Republican
13. James Clark	1836-39	1779-1839	Bedford Co. VA	Whig
14. Charles Anderson Wickliffe	1839-40	1788-1869	Springfield, Washington Co. KY	Whig
15. Robert Perkins Letcher	1840-44	1788-1861	Goochland Co. VA	Whig
16. William Owsley	1844-48	1782-1862	VA	Whig
17. John Jordan Crittenden	1848-50	1787-1863	Versailles, Woodford Co. KY	Whig
18. John Larue Helm	1850-51	1802-1867	Elizabethtown, Hardin Co. KY	Whig
19. Lazarus Whitehead Powell	1851-55	1812-1867	Henderson Co. KY	Democrat
20. Charles Slaughter Morehead	1855-59	1802-1868	Bardstown, Nelson Co. KY	American Party
21. Beriah Magoffin	1859-62	1815-1885	Harrodsburg, Mercer Co. KY	Democrat
22. George W. Johnson (Confederate)	1861-62	1811-1862	Georgetown, Scott Co. KY	Democrat
23. Richard Hawes (Confederate)	1862-65	1797-1877	Caroline Co. VA	Democrat
24. James F. Robinson	1862-63	1800-1882	Scott Co. KY	Democrat
25. Thomas Elliott Bramlette	1863-67	1817-1875	Cumberland Co. (Now Clinton Co.) KY	Union Democrat
26. John Larue Helm (2nd term)	1867	1802-1867	Elizabethtown, Hardin Co. KY	Democrat
27. John White Stevenson	1867-71	1812-1886	Richmond, VA	National Union Party
28. Preston Hopkins Leslie	1871-75	1819-1907	Wayne Co. (Now Clinton Co.) KY	Democrat
29. James B. McCreary	1911-15	1838-1918	Madison Co. KY	Democrat
30. Luke Pryor Blackburn	1879-83	1816-1887	Woodford Co. KY	Democrat
31. J. Proctor Knott	1883-87	1830-1911	Marion Co. KY	Democrat
32. Simon Bolivar Buckner	1887-91	1823-1914	Munfordville, Hart Co. KY	Democrat
33. John Young Brown	1891-95	1835-1904	Elizabethtown, Hardin Co. KY	Democrat
34. William O'Connell Bradley	1895-99	1847-1914	Lancaster, Garrard Co. KY	Republican
35. William Sylvester Taylor	1899-1900	1853-1928	Morgantown, Butler Co. KY	Republican
36. William Goebel	1900	1856-1900	Sullivan Co. PA	Democrat
37. J.C.W. Beckham	1900-07	1869-1940	Bardstown, Nelson Co. KY	Democrat
38. Augustus Everett Willson	1907-11	1846-1931	Maysville, Mason Co. KY	Republican
39. James B. McCreary (2nd term)	1911-15	1838-1918	Madison Co. KY	Democrat
40. Augustus Owsley Stanley	1915-19	1867-1958	Shelbyville, Shelby Co. KY	Democrat
41. James Dixon Black	1919	1849-1938	Barbourville, Knox Co. KY	Democrat
42. Edwin Porch Morrow	1919-23	1877-1935	Somerset, Pulaski Co. KY	Republican
43. William Jason Fields	1923-27	1874-1954	Willard, Carter Co. KY	Democrat
44. Flem D. Sampson	1927-31	1875-1967	London, Laurel Co. KY	Republican
45. Ruby Laffoon	1931-35	1869-1941	Madisonville, Hopkins Co. KY	Democrat
46. Albert Benjamin Chandler	1935-39	1898-1991	Corydon, Henderson Co. KY	Democrat
47. Keen Johnson	1939-43	1896-1970	Branson's Chapel, Lyon Co. KY	Democrat
48. Simeon Willis	1943-47	1879-1965	Lawrence Co. OH	Republican
49. Earle Chester Clements	1947-50	1896-1985	Morganfield, Union Co. KY	Democrat
50. Lawrence W. Wetherby	1950-55	1908-	Middleton, Jefferson Co. KY	Democrat
51. Albert Benjamin Chandler (2nd term)	1955-59	1898-1991	Corydon, Henderson Co. KY	Democrat
52. Bert T. Combs	1959-63	1911-	Manchester, Clay Co. KY	Democrat
53. Edward Thompson Breathitt, Jr.	1963-67	1924-	Hopkinsville, Christian Co. KY	Democrat
54. Louie B. Nunn	1967-71	1924-	Park, Barren Co. KY	Republican
55. Wendell Hampton Ford	1971-74	1924-	Daviess Co. KY	Democrat
56. Julian Morton Carroll	1974-79	1931-	Paducah, McCracken Co. KY	Democrat
57. John Y. Brown, Jr.	1979-83	1933-	Lexington, Fayette Co. KY	Democrat
58. Martha Layne Collins	1983-87	1936-	Bagdad, Shelby Co. KY	Democrat
59. Wallace Wilkinson	1987-91	1941-	Casey Co. KY	Democrat
60. Brereton Jones	1991-95	1939-	Point Pleasant, Mason Co. WV	Democrat
61. Paul Patton	1995-	1937-	Fallsburg, Lawrence Co. KY	Democrat

Constitutional Amendments. Kentuckians have voted on a number of state constitutional amendments since the mid-1960s. Proposed amendments appear on the ballot during presidential election years and in those years when gubernatorial candidates stand for election.

In 1966 Kentucky voters elected not to call a state constitutional convention. Since Kentucky became a state, its residents have lived under four constitutions, adopted in 1792, 1799, 1850, and 1891. Changing the constitution is extremely difficult, in part because of obstacles set by the 1890-91 constitutional convention. Historian Thomas D. Clark, commenting on the convention's wish to protect the state's agrarian society from emerging industrial economies, observed: "One gets the impression . . . that many of the delegates were, in fact, little Red Riding Hoods trudging alone and frightened through the perplexing forest of constitutional law, hoping that the big bad wolves of industrial and progressive changes were mere fragments [or figments?] of their badly agitated imagination, and that a rigid constitution with static provisions would serve to dispel these threatening wraiths."

Calls for constitutional conventions in 1931, 1947, 1960, and 1977 were all defeated. In 1966 only 666,000 voters cast ballots on a measure to revise the state constitution. The referendum was one of only two from the mid-1960s to the mid-1990s in which all counties voted no. Statewide, 78 percent of the voters chose to reject a new convention. In only fourteen counties was opposition somewhat more modest. These included several counties with large populations—Jefferson, Fayette, Kenton, and Daviess Counties—as well as several counties of the western Pennyroyal. Opposition was strongest in rural counties in south central and eastern Kentucky. In fifteen counties voters cast more than 90 percent of their ballots against this referendum, and the highest negative votes occurred in Elliott, Jackson, Monroe, Owsley, and Metcalfe Counties. This vote, more than that on any other recent referendum, demonstrated the state's rural-urban dichotomy.

In 1975 voters were asked to consider judicial reform. The proposed amendment was to provide a modern and more efficient method of conducting the judicial duties of the state court system, including the election and selection of judges. Although only 395,000 votes were cast for this amendment, it passed with 54 percent of the vote, carrying only twenty-five counties. The strongest support came from large urban counties; 49 percent of the yes vote was recorded in the five largest counties. Opposition, again, came from rural areas and small towns.

In 1979 the independence of the state legislature was the issue before voters. This lengthy amendment contained several provisions intended to give the state House and Senate more responsibilities and influence. Proposed provisions included holding elections in even-numbered years, stipulating that the General Assembly would meet in January in odd-numbered years to elect leaders and committees, and ensuring that the General Assembly would meet no more than sixty legislative days each session. The measure also contained provisions affecting compensation and members' terms of office. Only 307,000 votes were cast on the amendment, but it passed, in large measure because of support from urban counties. Large sections of the state voted against the measure; of these, south central Kentucky recorded the strongest opposition.

Voters were asked to approve two amendments in 1988. Both passed, with more than one million votes cast on each measure. The lottery amendment authorized the General Assembly to establish a state lottery. More residents voted on this amendment than on any other since 1966. It passed with 61 percent of the vote, but the approval pattern was very different from that for other amendments. Eastern counties strongly supported the lottery, as did a series of urban Catholic counties along the Ohio River. The strongest opposition came from parts of the Bluegrass, the Pennyroyal, the Western Kentucky Coal Field, and the Jackson Purchase. To understand local and regional patterns of support or opposition on this issue, one would need to examine closely the maps of religious preference in Chapter 3, because religious groups took varying positions on the lottery.

The other amendment on the 1988 ballot concerned the broad form deed. This amendment was the only one in the 1966-92 period for which all Kentucky counties voted yes. The proposed amendment prohibited mineral extraction according to broad form deeds. Popularized in the late nineteenth century by entrepreneur John C. C. Mayo, this type of deed was used extensively in eastern Kentucky to transfer title of underground minerals, especially coal, from people who owned the land to speculators and mining companies. The landowner could continue living on the property and was required to pay the taxes on it, but the deeds granted the

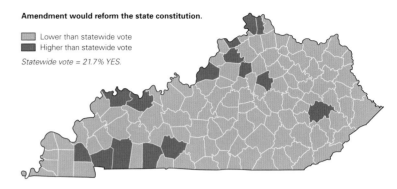

CONSTITUTIONAL REFORM AMENDMENT, 1966

Amendment would reform the state constitution.

Lower than statewide vote
Higher than statewide vote

Statewide vote = 21.7% YES.

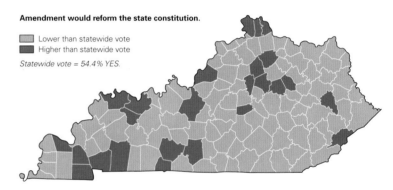

JUDICIAL SYSTEM REFORM AMENDMENT, 1975

Amendment would reform the state constitution.

Lower than statewide vote
Higher than statewide vote

Statewide vote = 54.4% YES.

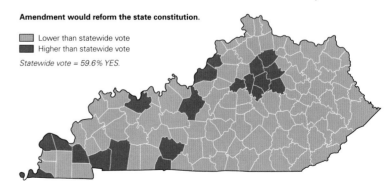

LEGISLATIVE REFORM AMENDMENT, 1979

Amendment would reform the state constitution.

Lower than statewide vote
Higher than statewide vote

Statewide vote = 59.6% YES.

STATE LOTTERY AMENDMENT, 1988

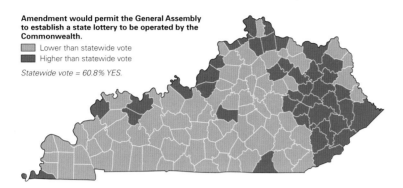

Amendment would permit the General Assembly to establish a state lottery to be operated by the Commonwealth.

▨ Lower than statewide vote
■ Higher than statewide vote

Statewide vote = 60.8% YES.

BROAD FORM DEED AMENDMENT, 1988

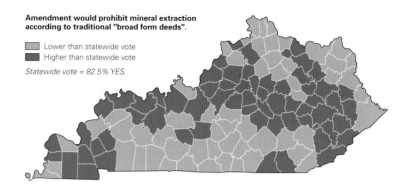

Amendment would prohibit mineral extraction according to traditional "broad form deeds".

▨ Lower than statewide vote
■ Higher than statewide vote

Statewide vote = 82.5% YES.

ELECTIVE OFFICE AMENDMENT, 1992

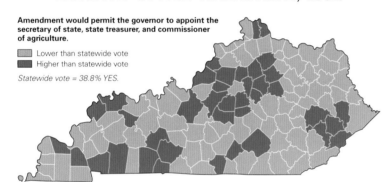

Amendment would permit the governor to appoint the secretary of state, state treasurer, and commissioner of agriculture.

▨ Lower than statewide vote
■ Higher than statewide vote

Statewide vote = 38.8% YES.

CHARITABLE LOTTERY AMENDMENT, 1992

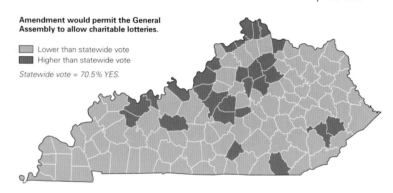

Amendment would permit the General Assembly to allow charitable lotteries.

▨ Lower than statewide vote
■ Higher than statewide vote

Statewide vote = 70.5% YES.

deed holder the minerals and the power to extract them using any means deemed necessary, including extensive surface mining. Mining could proceed at any time without the landowner's permission. The amendment abolishing this practice was approved statewide by 83 percent of the vote. Although the five largest counties were responsible for 37 percent of the total yes vote, support was equally strong in rural areas throughout the state. The highest yes votes tended to occur in coal mining counties, such as Leslie, Magoffin, Morgan, Letcher, Breathitt, and Floyd. Nevertheless, a few coal-producing counties, such as Pike and Martin in the Eastern Kentucky Coal Field and Hopkins and Muhlenberg in the west, recorded yes votes below the state average.

In 1992 more than one million voters registered their opinions on an amendment permitting charitable lotteries. Passed with 71 percent of the vote, this amendment acquired support across the state, much like the state lottery amendment of 1988. The amendment passed in the five most populous counties,

which accounted for 47 percent of the total yes vote. The strongest support came from the largest urban areas and those counties with large Catholic populations, especially along the Ohio River, in the Inner Bluegrass, and in the Outer Bluegrass around Bardstown. Only nine counties voted no. Opposition centered in southeastern and south central Kentucky, especially in the Pennyroyal. Counties with a small Catholic presence recorded the largest percentages of no votes.

Also in 1992 Kentucky voters considered the matter of gubernatorial appointments. Voters in all counties rejected this proposed amendment, which would have allowed the governor to appoint the secretary of state, the state treasurer, and the agricultural commissioner. The only other amendment that fared as poorly as this one during the previous thirty years was the 1966 constitutional reform initiative. Only 39 percent of voters supported the 1992 measure. The pattern of opposition and support was mixed, although large sections of the Bluegrass recorded

above-average affirmative votes. Many urban and suburban precincts supported the reform measure, whereas the five largest counties accounted for 37 percent of the no vote. South central and southeastern counties recorded the highest negative votes, but the amendment's strongest support often came from counties adjoining those where opposition was strong.

Future referenda will likely address the length of General Assembly sessions, funding for education, environmental quality, and financial support for economic growth. During the past three decades, patterns of support and opposition have varied from one referendum to the next. In order to gain statewide consent for a proposed amendment, promoters of the initiative must convince voters in rural areas and major cities to agree. Many issues lack local or regional appeal, however, and proposals must receive concerted publicity and public hearings if their sponsors realistically expect support from voters across the state's diverse regions.

The Urban Landscape

CHAPTER TWELVE

The Urban Landscape

Historic engraving of Louisville, late nineteenth century.

U NLIKE MANY STATES in the eastern part of the United States, Kentucky has retained a relatively rural population. Whereas more than three-quarters of the U.S. population live in or near cities, about half of all Kentuckians reside in urban areas. Nevertheless, Kentucky cities have been—and continue to be—important elements of the state's landscape. These cities act as focal points for commerce and trade, offer a diversity of entertainment and culture not found in outlying areas, and often serve as a gauge of advances and shortcomings for the entire Commonwealth. Consequently, gaining an appreciation for Kentucky's cities is vital in order to understand the state as a whole.

It is easy to imagine that living in a city is often different from living in the country. Cities provide public water, sewerage, and trash collection, while rural inhabitants frequently take care of these matters on their own. As centers of commerce, cities experience traffic problems and noise not found in rural locales. Access to shopping and workplaces is commonly much easier in cities, and dramatic variations in housing and people are often found within small areas. Such statements, however, are only generalizations of city life. Furthermore, it is not easy to make clear distinctions between "city" and "country." We must first clarify the terms *city, urban,* and *metropolitan.*

Defining Urban

Kentucky ranks thirty-eighth among the states in the proportion of its residents who live in an urban setting. In 1992 just under half (48.5 percent) of Kentucky's population lived in counties that are part of Metropolitan Areas (MAs) as defined by the U.S. Bureau of the Census. The Census Bureau describes an MA as "a large population nucleus, together with adjacent communities that have a high degree of economic and

METROPOLITAN AREA POPULATIONS

Metropolitan Statistical Area	Population			Percent in Kentucky	Percent Change	
	1980	1990	1994		1980-1990	1990-1994
Cincinnati, OH-KY-IN	1,467,643	1,526,090	1,581,216	20.7	4.0	3.6
Louisville, KY-IN	953,520	949,012	980,851	78.6	-0.5	3.4
Lexington, KY	370,900	405,936	430,842	100.0	9.4	6.1
Huntington-Ashland, WV-KY-OH	336,410	312,529	316,414	35.9	-7.1	1.2
Evansville-Henderson, IN-KY	276,252	278,990	286,624	15.4	1.0	2.7
Clarksville-Hopkinsville, TN-KY	150,220	169,439	186,016	40.7	12.8	9.8
Owensboro, KY	85,949	87,189	90,138	100.0	1.4	3.4

Data have been revised to conform to MSAs as defined on June 30, 1993.

A city's entertainment district was often anchored by an opera house or, from the early twentieth century through the 1950s, by one or more motion picture theaters. In Ashland, the majestic Paramount Theater still operates at the north end of the business district.

METROPOLITAN POPULATION

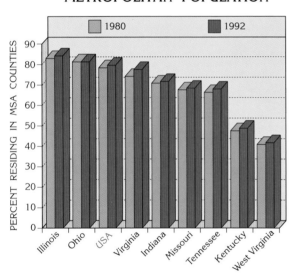

URBAN POPULATION

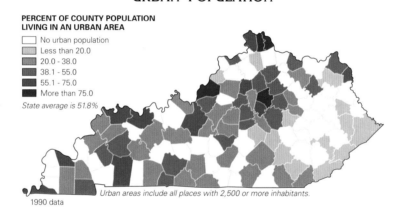

PERCENT OF COUNTY POPULATION
LIVING IN AN URBAN AREA

- No urban population
- Less than 20.0
- 20.0 - 38.0
- 38.1 - 55.0
- 55.1 - 75.0
- More than 75.0

State average is 51.8%

Urban areas include all places with 2,500 or more inhabitants.

1990 data

URBANIZATION, 1790-1990

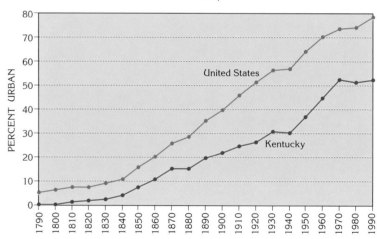

social integration with that nucleus." Various criteria are used to define MAs either as Metropolitan Statistical Areas (MSAs) or as Consolidated Metropolitan Statistical Areas (CMSAs). CMSAs are divided into one or more Primary Metropolitan Statistical Areas (PMSAs). A PMSA consists of a large urban county or cluster of counties that demonstrate strong, locally recognized internal economic and social links. Among those MAs wholly or partly in Kentucky, only Cincinnati-Hamilton is a CMSA. It is divided into two PMSAs, the Cincinnati, OH-KY-IN PMSA and the Hamilton-Middletown, OH PMSA.

Across the United States, almost 80 percent of the population lived in metropolitan counties in 1992. Among the seven states contiguous to Kentucky, only West Virginia had a smaller proportion of its population residing in metropolitan counties than did Kentucky.

Since 1950 the Census Bureau has also defined as *urban* those cities and towns with 2,500 or more inhabitants, excluding towns in rural portions of extended cities. In 1990, 51.8 percent of Kentucky's population resided in such cities and towns. Forty-five of Kentucky's 120 counties had no city or town that met this

definition, and eastern Kentucky was the least urbanized region in the state. At the other extreme, the urban populations of four counties—Jefferson, Fayette, Kenton, and Campbell—constituted more than 75 percent of the total county populations.

As of mid-1993, twenty-two Kentucky counties were part of seven MSAs located entirely or partly within the state. The largest metropolitan area is the Cincinnati PMSA, with a total population estimated to be 1.58 million in 1994. The next largest metropolitan areas by population size in 1994 were Louisville, Lexington, Huntington-Ashland, Evansville-Henderson, Clarksville-Hopkinsville, and Owensboro. The Cincinnati-Hamilton CMSA and the Louisville and Lexington MSAs in 1992 ranked twenty-third, forty-fourth, and ninetieth, respectively, among the nation's MAs by population. Only two MSAs, Lexington and Owensboro, were entirely in Kentucky. More than three-quarters of the population of the Louisville MSA lived in the state. The dominant central cities for the other four MSAs were in contiguous states.

During the 1980s two of the MSAs—Huntington-Ashland and Louisville—lost population. The populations of all the others increased, most rapidly in the Clarksville-Hopkinsville MSA (12.8 percent) and the Lexington-Fayette MSA (9.4 percent). (The Lexington-Fayette MSA was renamed Lexington in 1992.) In fact, these two MSAs continued to grow faster than any others in Kentucky during the first half of the 1990s. They grew at a rate far outpacing the overall growth rate of the state during the 1980s, which was just 0.7 percent. By contrast, the nation's overall growth rate was 9.8 percent during this period.

The Kentucky General Assembly also classifies certain places in the state as "cities." According to the state constitution, municipalities that are chartered or incorporated under Kentucky law are classified in one of six classes on the basis of their population and a certain minimum level of government services, which is set by state statute. First-class cities have populations of 100,000 or more; second-class, between 20,000 and 99,999; third-class, between 8,000 and 19,999; fourth-class, between 3,000 and 7,999; fifth-class, between 1,000 and 2,999; and sixth-class, fewer than 1,000.

The map "Classification of Incorporated Cities" depicts only the first five classes of cities in 1995; it is obvious that some cities are incorrectly classified. For example, Lexington, which should now be classified as a first-class city, was still listed by the state government as a second-class city in 1996. The town of Martin, in Floyd County, which had only 694 persons in 1990, was still classified as a fourth-class city. Such inaccurate classifications result from the fact that individual cities must be reclassified by legislative action, a process that requires time. In 1996 Kentucky had 431 classified cities: 1 first-class, 11 second-class, 19 third-class, 93 fourth-class, 118 fifth-class, and 189 sixth-class. Clearly, many of these were very small towns: 6 of the sixth-class cities each had a population of fewer than a hundred people.

As we saw in Chapter 3, the earliest settlers of European and black ancestry to arrive in Kentucky came either through the Cumberland Gap along the Wilderness Road or on boats down the Ohio River. Thus, the earliest towns emerged as river ports on the

The Cincinnati-Northern Kentucky metropolitan area focuses on the confluence of two Ohio River tributaries. Separated by the Licking River, Covington (foreground) and Newport (right) crowd a basin on the south bank of the Ohio River. Downtown Cincinnati stands along the Ohio on a low terrace, with access to the hinterland on both the Ohio and the Kentucky side of the river.

KENTUCKY'S METROPOLITAN AREAS

Metropolitan Statistical Area (MSA), 1990

County added to MSA between January 1, 1990 and June 30, 1993

County deleted from MSA between January 1, 1990 and June 30, 1993

Daviess County comprises a separate MSA (Owensboro). Carter County was deleted from the Huntington-Ashland MSA on December 28, 1992, and added back on June 30, 1993.

CLASSIFICATION OF INCORPORATED CITIES

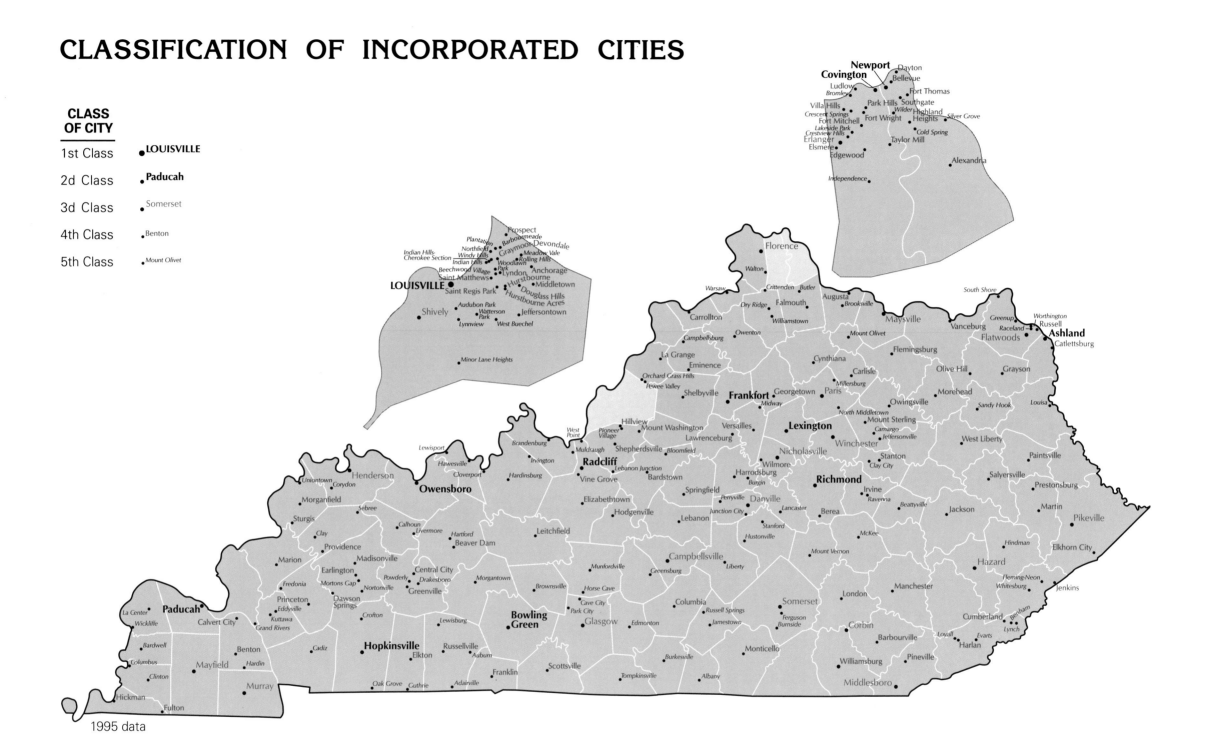

CLASS OF CITY

1st Class	● LOUISVILLE
2d Class	● **Paducah**
3d Class	● Somerset
4th Class	● Benton
5th Class	● Mount Olivet

1995 data

An enhanced Landsat Thematic Mapper (TM) false-color image (bands 1, 4, and 6) of the Owensboro area. Data were acquired on October 14, 1992. Residential sections of Owensboro are blue-green. High-density industrial/commercial activity is magenta. The Ohio River in the northeast and the Green River to the west are dark blue. Actively growing vegetation—forest, grass, and some late crops such as soybeans—are green. Mature crops ready for harvest, bare tilled cropland, and stubble comprise much of the rural landscape and are depicted in various shades of magenta.

Ohio or as settlements in the fertile farmlands of the Bluegrass Region, the destination for most of the earliest travelers who planned to make Kentucky their home. The first permanent settlement in Kentucky, founded in 1774, was Harrodsburg (originally called Harrod's Town). Soon thereafter, Lexington emerged at the center of an important crossroads, and it came to be the central city for the entire Bluegrass Region. Indeed, according to Richard Schein, "Lexington showed great promise at the turn of the nineteenth century and was perhaps second only to Lancaster, Pennsylvania, as the major 'inland' American urban place. It quickly emerged as the commercial center of not just the Bluegrass, but for much of the trans-Appalachian West." Permanent settlement of Lexington began in 1779, and by the time of the first national census in 1790, the town had a population of 834, making it by

far Kentucky's largest. As early as 1800 many people considered Lexington to be the most important place in the state, with its fine homes, landed estates, taverns and inns, and diverse businesses and manufacturing activities. By the early 1800s the Wilderness Road from the Cumberland Gap, the Limestone Road from Maysville (formerly Limestone) on the Ohio River, and the Midland Trail, which linked Ashland (then called Poage Settlement) and Louisville, all passed through or terminated in Lexington. With the presence of Transylvania University and other cultural and educational influences, Lexington by the 1820s could lay claim to the appellation "Athens of the West." By 1820 the city could still boast of being the largest place in Kentucky, twice as large as Louisville.

Even Frankfort was larger than Louisville in these early

years. Because it offered more land, material, and money for the construction of the statehouse than did other Kentucky towns, the legislature approved Frankfort as the state capital in 1792. (Actually, the debate over the location of the state capital persisted until 1904, when the legislature voted to allocate one million dollars to construct a new capitol building in Frankfort.) The census takers in 1800 counted 628 persons in Frankfort, making it the state's second-largest town after Lexington, which then had 1,795 people. Six other towns in or near the Bluegrass Region had populations in excess of 250 in 1800: Bardstown (579), Washington (570), Paris (377), Georgetown (348), Danville (270), and Shelbyville (262). The only river town with a population exceeding 250 in 1800 was Louisville, with 359 people. It was then the state's sixth largest place. Smaller Ohio River ports that

URBAN HIERARCHY ORDERING SCHEME

	1800-1840 Census Years	1880-1920 Census Years	1960-1990 Census Years
1st order	20,000 or more	50,000 or more	100,000 or more
2nd order	5,000-19,999	25,000-49,999	40,000-99,999
3rd order	2,000-4,999	10,000-24,999	15,000-39,999
4th order	1,000-1,999	2,500-9,999	5,000-14,999
5th order	250-999	1,000-2,499	2,500-4,999

emerged in these early years included Limestone, Newport, and Port William. By 1800 the only place of any size in western Kentucky was the town of Henderson, with a population of 205. It was laid out in 1797 and established as the seat of Henderson County in 1798.

The importance of Lexington and Frankfort would diminish through the 1800s with the rapid growth of Louisville. Its location at the Falls of the Ohio and the break in navigation caused by this physical feature, together with the invention of the steamboat, shifted the focus of urban life in Kentucky to Louisville. Muscle-powered keelboats with cargo required three or more months to move upstream from New Orleans to Louisville; the steamboat made the same trip in only one week. The first steamboat arrived at Louisville in 1811, and by 1830 the city's population numbered more than 10,000, making it the state's largest urban place. By 1840 Louisville's population numbered more than 21,000; the city was thus more than three times as large as Lexington. The population of Covington-Newport, also a growing river port, ranked that area third in population among the state's cities, and by 1840 Maysville had become another thriving river town, with 2,741 inhabitants. Frankfort, with fewer than 2,000 people, had fallen to fifth place in the state. Indeed, Frankfort has attained a size and importance today that belies its site and situation. It probably would not have emerged as the place it currently is except for its early selection as the state capital, its location on the Kentucky River,

and its situation between Louisville and Lexington.

By 1880 the towns and cities of western Kentucky had begun to grow, as people traveled farther west in search of new agricultural lands and as river ports began to expand in response to the need for the transport of agricultural and natural resources. Five of the state's ten largest places in 1880 were located in the western part of the state: Paducah, with 8,036 people, ranked fourth in the state; Owensboro, with 6,231, ranked sixth; Henderson, 5,365, was seventh; Bowling Green, 5,114, was ninth; and Hopkinsville, 4,229, ranked tenth. Only two cities, Louisville and Covington-Newport, had populations exceeding 50,000 in 1880. Lexington, Frankfort, and Maysville ranked third, fifth, and eighth in population size, respectively. Only one eastern Kentucky town, Ashland (incorporated in 1856), had a population in excess of 2,500.

The late 1800s and early 1900s, which saw the emergence of coal mining and associated rail transport, marked a time of growth among towns in eastern Kentucky. Formidable barriers to westward movement, including Cumberland and Pine Mountains, as well as the rugged topography of eastern Kentucky, meant that the eastern part of the state was settled from the Bluegrass Region. That is, eastern Kentucky was developed by settlers moving from west to east, contrary to the east-west pattern common in most American places during the settlement era. By 1920 at least nine places in eastern Kentucky had populations

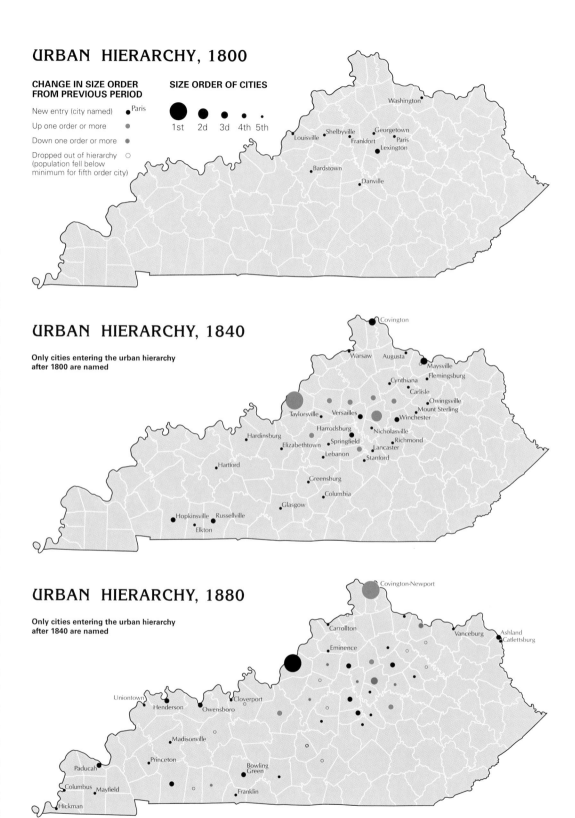

URBAN HIERARCHY, 1800

CHANGE IN SIZE ORDER FROM PREVIOUS PERIOD

New entry (city named) ● Paris

Up one order or more ●

Down one order or more ●

Dropped out of hierarchy ○
(population fell below
minimum for fifth order city)

SIZE ORDER OF CITIES

● ● ● · ·
1st 2d 3d 4th 5th

URBAN HIERARCHY, 1840

Only cities entering the urban hierarchy after 1800 are named

URBAN HIERARCHY, 1880

Only cities entering the urban hierarchy after 1840 are named

URBAN HIERARCHY, 1920

Only cities entering the urban hierarchy
after 1880 are named

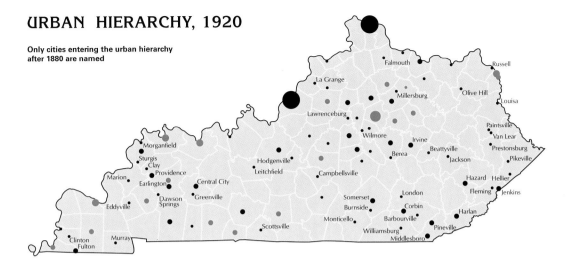

URBAN HIERARCHY, 1960

Only cities entering the urban hierarchy
after 1920 are named

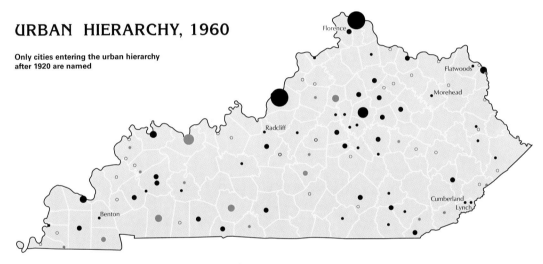

URBAN HIERARCHY, 1990

Only cities entering the urban hierarchy
after 1960 are named

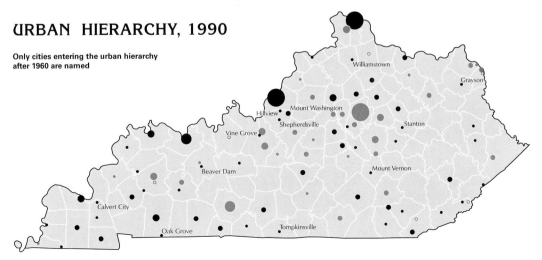

Level land for urban commercial or residential building is rare and expensive in the eastern mountains. The Levisa Fork carved the broad river terrace where Prestonsburg, the Floyd County seat, is sited.

of at least 2,500 people: Ashland-Catlettsburg (18,912), Middlesboro (8,041), Jenkins (4,707), Somerset (4,672), Hazard (4,348), Corbin (3,406), Pineville (2,908), Irvine (2,705), and Harlan (2,647). Additionally, company towns emerged and grew quite rapidly, in part as a means of accommodating the influx of mine laborers. These towns also allowed mining companies to hold partial control over the residents' economic activities. In 1920 some of the larger company towns included places like Hellier, Seco, McVeigh, and Van Lear. It was not unusual, however, for company towns to disappear almost as quickly as they had appeared, and today what towns are left tend to be much smaller than they were in the early 1900s.

By 1960 almost 45 percent of Kentucky residents lived in a place with a population that exceeded 2,500; the national figure was almost 70 percent. Excluding incorporated places that were part of larger urbanized areas—such as Anchorage and Shively near Louisville, Fort Mitchell and Highland Heights in the Covington-Newport area, and Catlettsburg and Flatwoods near Ashland—sixty-eight cities and towns in the state in 1960 had at least 2,500 inhabitants. By the time of the 1990 census, this number had increased to eighty-four. The modern urban hierarchy reflects that of earlier years: eastern Kentucky, with the exception of Ashland-Catlettsburg (now with Flatwoods and Russell) and Middlesboro, still has no cities with more than 10,000 people; the Bluegrass Region and the larger Golden Triangle, defined by the vertices of Louisville, Lexington, and Covington-Newport, contain the largest places and most of the state's urban population; and the Ohio River ports of earlier years still perform important transportation functions.

In the second half of the twentieth century some new cities and towns have emerged in Kentucky, and existing ones have grown rapidly, in response to three events. First, the construction of the interstate highways and state parkways has encouraged the growth of industry and other commercial activities in places located on or near such roads. Places that have experienced growth rates in excess of the state figure (4.5 percent) during the 1980-94 period include Williamstown (38 percent), Georgetown (18 percent), and Berea (24 percent)—all places on Interstate 75 that have attracted industry. Second, the construction of offices, industries, and retail malls away from central cities, coupled with improvements in transportation and communications, have meant that the journey to work is much easier and faster today than it was previously. Thus, smaller places, more distant from the large metropolitan areas, have experienced rapid growth: Florence, in northern Kentucky, saw a 40 percent increase in population between 1980 and 1994; Nicholasville, south of Lexington, grew by 55 percent; and Shepherdsville, south of Louisville, grew by 25 percent. Third, towns have emerged near the two major military installations in Kentucky, Fort Campbell (opened in 1942) and Fort Knox (established in 1918 and made the branch headquarters of the U.S. Armored Force in 1940). Such emerging towns include the second-class city of Radcliff near Fort Knox, which had a population of nearly 20,000 in 1990, and Oak Grove, near Fort Campbell. Between 1980 and 1994 Radcliff's population increased by 36 percent, and Oak Grove's grew by 33 percent. Elizabethtown, which grew by 32 percent during the same period, is an example of a city affected by all three growth factors: it is located not far from Louisville, a large metropolitan area; it is served by an interstate highway (I-65) and two state parkways (the Blue Grass Parkway and the Western Kentucky Parkway); and it is near the major military installation of Fort Knox.

Kentucky's Major Metropolitan Areas

Like most urban places in North America, Louisville, Lexington, and northern Kentucky all expanded in area between 1950 and 1990. The downtown, or Central Business District (CBD), is at the center of each metropolitan area. With the development and improvement of the road network, advances in automotive technology and in public

Fourth Street in downtown Louisville in the 1930s.

A court day at Cheapside on Lexington's courthouse square, 1897.

Lexington's Romanesque courthouse was completed in 1900.

transportation, and widespread automobile ownership, each metropolitan area has gradually expanded. Today, suburban developments that are part of the urban area may be fifteen or twenty miles or more from the CBD. Anchorage, for example, about ten miles east of downtown Louisville, became part of the urban area by the 1970 census, and incorporated places such as Hillview and Pioneer Village in Bullitt County joined the Louisville urban area in the 1990 census. The expansion will certainly continue, and Shepherdsville and Mount Washington will likely become part of Louisville's urban area by the census of 2000 or 2010.

The Ohio town of Cincinnati was laid out in 1788 and incorporated in 1802. The first settlement on the south bank of the Ohio River across from Cincinnati, Newport, was laid out in the early 1790s and incorporated as a village in 1795. Covington, today northern Kentucky's largest incorporated city, was created as a town by the Kentucky legislature in 1815. Cincinnati and Covington-Newport grew more or less independently of one another until 1867, when the first permanent bridge, the Roebling Suspension Bridge, was completed. At that time it was the long-

est suspension bridge in the world. Several other bridges have since been constructed across the Ohio River linking Cincinnati with northern Kentucky, including a railroad bridge built in 1888 by the Chesapeake and Ohio Railway. Thus, since the late nineteenth century, towns like Covington, Newport, Dayton, Fort Thomas, and more recently Florence, Independence, and Alexandria have become workplaces and residential suburbs integrated with the Cincinnati urban area.

Industry was also important to the early development of the northern Kentucky river ports. According to Will Frank Steely, writing about Newport in *The Kentucky Encyclopedia,* "The large German immigration in the 1880s and 1890s and the completion of bridges from northern Kentucky to Cincinnati promoted the development of Newport as a residential suburb of the Ohio metropolis. One of the largest steel mills in the South and the largest brewery in the United States south of the Ohio River, Wiedemann, were industrial bases for Newport's development in the early twentieth century." In recent years both the areal extent and the population of the northern Kentucky metropolis have grown rap-

idly as a result of the development of the Greater Cincinnati International Airport (located near Florence in Boone County), nearby business and industrial parks, service activities, and retail malls.

Lexington has expanded in a star-shaped pattern along the major highways that radiate out from the CBD. These roads connect Lexington with other county seats of the Bluegrass Region: Georgetown, Paris, Winchester, Richmond, Nicholasville, Harrodsburg, and Versailles. Lexington's present-day street system can thus be likened to the spokes of a wheel; each of the spokes is a road that connects a nearby county seat with downtown Lexington, and each is named for the town it links with the central city of the Bluegrass. In the 1960s an outer ring was added to the wheel with the construction of a four-lane highway around the city, New Circle Road (State Route 4). In the 1980s Man O' War Boulevard, a four-lane semicircle, was added farther out from the CBD to accommodate the expanding southern suburbs.

On a larger scale, circular or semicircular loops have also been constructed around both Cincinnati (I-275) and Louisville (I-264 and I-265) in order to manage traffic flows and more readily

John A. Roebling's great suspension bridge linking Covington and Cincinnati, opened in 1867, shown here as it looked in the early 1950s.

Main Street and the courthouse green, Paris, circa 1915.

connect the old downtown and other parts of the city with the rapidly growing suburbs. What is happening in Cincinnati and Louisville, as well as in the largest MAs across the nation, is that new "downtowns"—called "edge cities" by Joel Garreau—are emerging at critical junctions on these outer loops. Such places, in the middle of new residential subdivisions, are where the new malls, multistory office complexes, health care centers, and workplaces are being built. In short, such MAs now have multiple downtowns, and the original CBD often no longer performs the same vital function that it did before the 1950s, when travel took much more time and suburban expansion had not yet occurred. The central cities of both the Cincinnati and the Louisville MAs are still vibrant (at least during the daytime), and the tallest buildings are still being built there. Kentucky's tallest skyscraper, the postmodern thirty-five-story Capital Holding Center building, was completed in Louisville in 1992. Nonetheless, factors such as highway construction, widespread use of the automobile, and the supplanting of railroads with trucks have resulted in spreading of the metropolitan area and the consequent decline of the CBD. The addition of five counties to the Cincinnati PMSA between 1990 and 1993, three of which are in Kentucky (Gallatin, Grant, and Pendleton), indicates further the relationship between high-speed transportation routes and the expansion of an urban area.

URBAN EXPANSION

URBAN LIMIT

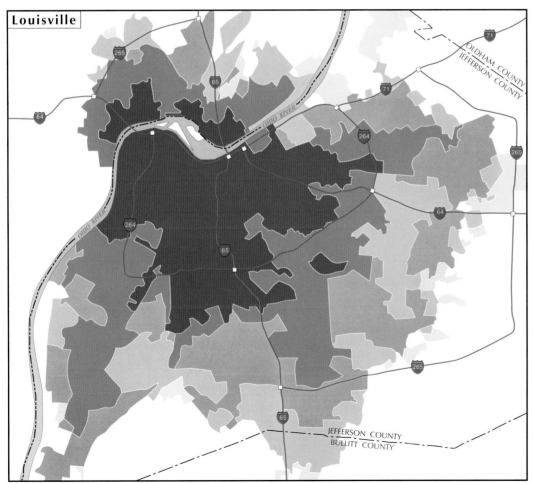

■ 1950 ▨ 1970 □ 1990

▦ 1960 ▨ 1980

0 ——————— 5 Miles

Scale

Interstate highways shown for reference

Louisville

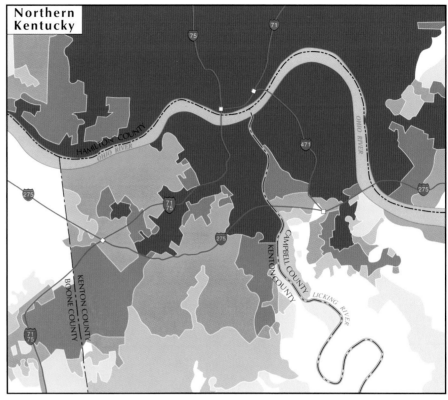

Northern Kentucky

Lexington

Spatial Patterns within Cities

In order to understand the more detailed spatial patterns of social and economic characteristics within an MA, one must examine it at the level of the census tract. Census tracts are the sections into which metropolitan areas have been divided for the purpose of reporting census data. Such tracts typically have populations between 2,500 and 8,000 and, when first delineated, are designed to be relatively uniform with regard to population characteristics, economic status, and living conditions. Census tracts that are within urban areas are further subdivided into census blocks, each of which corresponds to a city block. In 1990 there were about fifty thousand census tracts and seven million census blocks in the United States.

Because population densities are highest near the central city, census tracts there are generally smallest in areal size. As we move away from the central city and out of the urban core, the tracts become larger in area. The maps of the three major MAs in Kentucky—Louisville, Lexington, and northern Kentucky—showing the percentage of the population that resides in urban areas suggest that urban intensity is greatest in and near the central city and declines outward, as expected. Thus, for example, the southern portions of Boone, Kenton, and Campbell Counties were the least urbanized parts of those counties in 1990. As one would expect, in each of the MAs the oldest housing is located in and around the old CBD; there is clearly an inverse relationship between the age of housing and distance from the central city. The expanding residential suburbs continue to be built farther and farther from the old downtown.

With a few notable exceptions, the relationship between the value of housing and the distance from the central city is a direct one: as distance increases, so does the value of housing. In the Cincinnati PMSA, the eastern portion of Hamilton County, Ohio, had the highest housing values in 1990; in the northern Kentucky portion of the PMSA, newer residential suburbs such as Villa Hills and Crestview Hills had the highest housing values. In the Louisville MSA, Oldham County and the eastern portion of Jefferson County had some of the highest housing values; included in these areas are the fourth-class Kentucky cities of Prospect, Anchorage, and Middletown. In the Lexington-Fayette MSA, the southern and western portions of Fayette County had the highest housing values. Some of these census tracts are large and have a low population density; such tracts include some of the area's best-known thoroughbred horse farms.

Text continues on page 270

KENTUCKY'S MAJOR METROPOLITAN AREAS

– – – – –	State boundary
————	County boundary
—🅗—	Interstate highway
—⑥⓪—	U.S. highway
—④—	Other highway

BOONE County name

Census tract

0   5   10 miles
Scale

Cincinnati OH–KY–IN Primary Metropolitan Statistical Area (PMSA), 1990

Louisville KY–IN Metropolitan Statistical Area (MSA), 1990

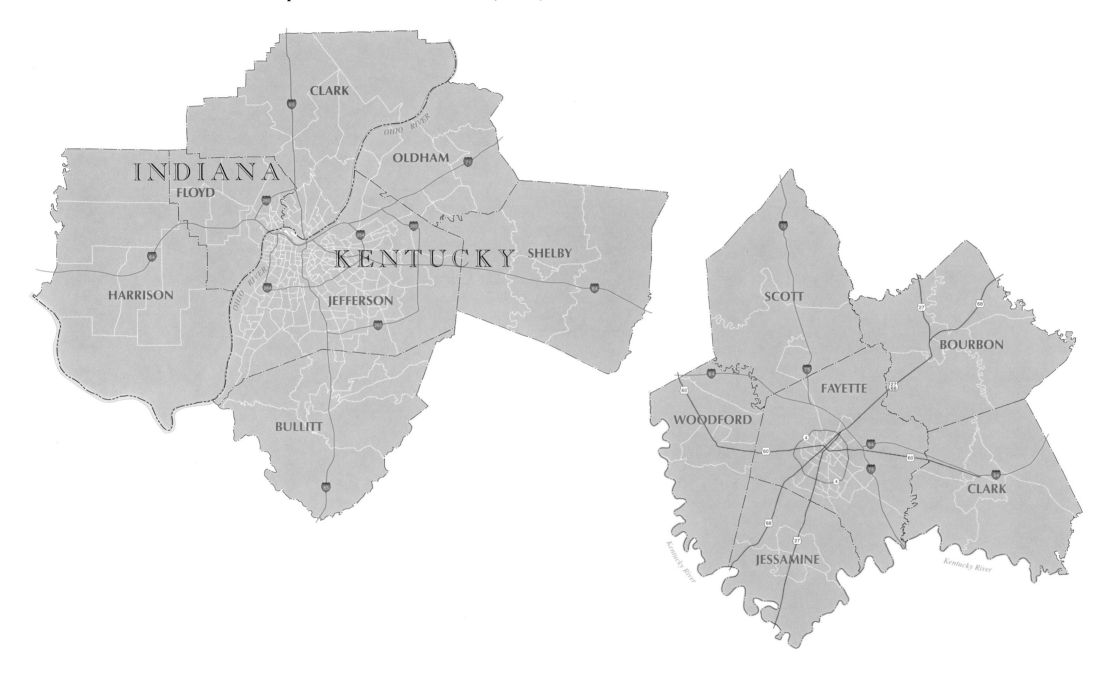

Lexington–Fayette KY Metropolitan Statistical Area (MSA), 1990

The Falls of the Ohio was a major navigational hazard for nineteenth-century river traffic. Here riverboats had to be unloaded and goods moved overland to a point above or below the Falls and reloaded for shipment beyond Louisville. With a location at this important trans-shipment point, Louisville grew to become the state's largest city.

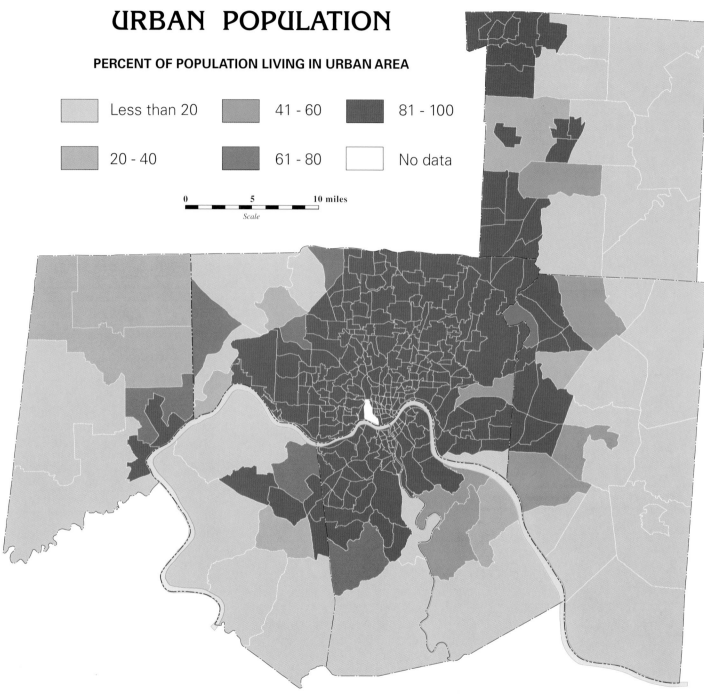

URBAN POPULATION

PERCENT OF POPULATION LIVING IN URBAN AREA

Less than 20	41 - 60	81 - 100
20 - 40	61 - 80	No data

0   5   10 miles
Scale

Cincinnati OH–KY–IN PMSA, 1990

Louisville KY–IN MSA, 1990

Lexington–Fayette KY MSA, 1990

Mid-nineteenth-century buildings in Covington's business district often had commercial businesses on the first floor and space for residential occupancy on the upper floors.

A contemporary apartment complex in Richmond.

AGE OF HOUSING

MEDIAN YEAR STRUCTURE WAS BUILT

■ Before 1940	■ 1950 - 1959	□ 1970 - 1979
■ 1940 - 1949	□ 1960 - 1969	□ 1980 - 1989

0   5   10 miles
Scale

□ No data

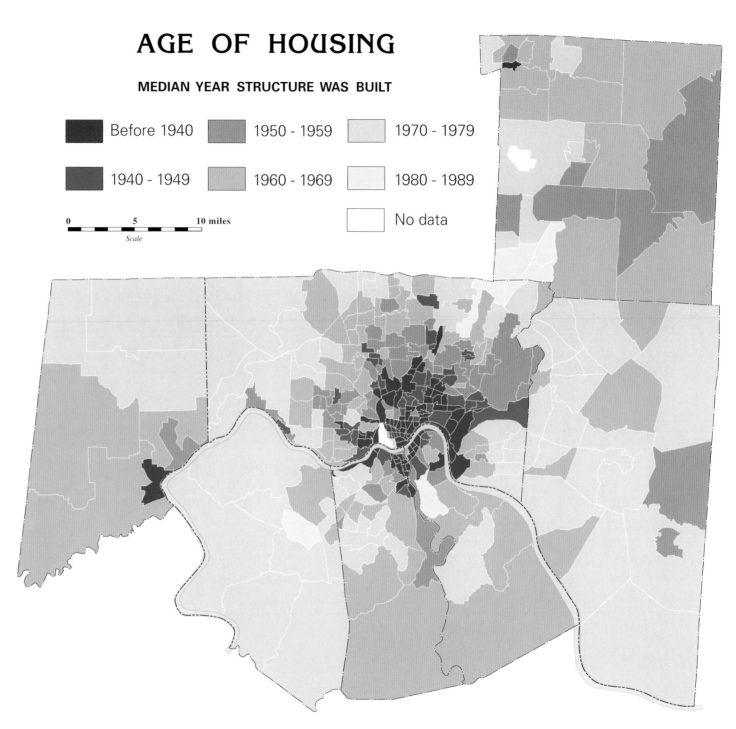

Cincinnati OH–KY–IN PMSA, 1990

Louisville KY–IN MSA, 1990

Lexington–Fayette KY MSA, 1990

Metropolitan median housing values often peak in new subdivisions such as this one near U.S. 60 in eastern Jefferson County.

Older residential buildings hold their value or appreciate if maintained and not used consumptively. This Andrew Jackson Downing-style Gothic cottage stands on Bath Avenue in Ashland.

MEDIAN HOUSE VALUE

VALUE OF HOUSE (in thousands of dollars)

Less than 50   76 - 100   More than 150

50 - 75   101 - 150   No data

0   5   10 miles
Scale

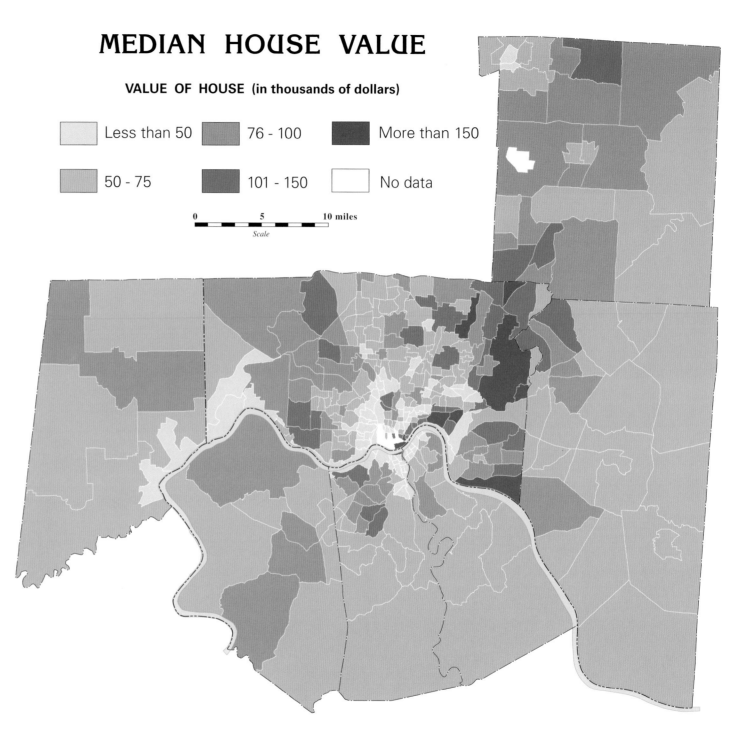

Cincinnati OH–KY–IN PMSA, 1990

Louisville KY–IN MSA, 1990

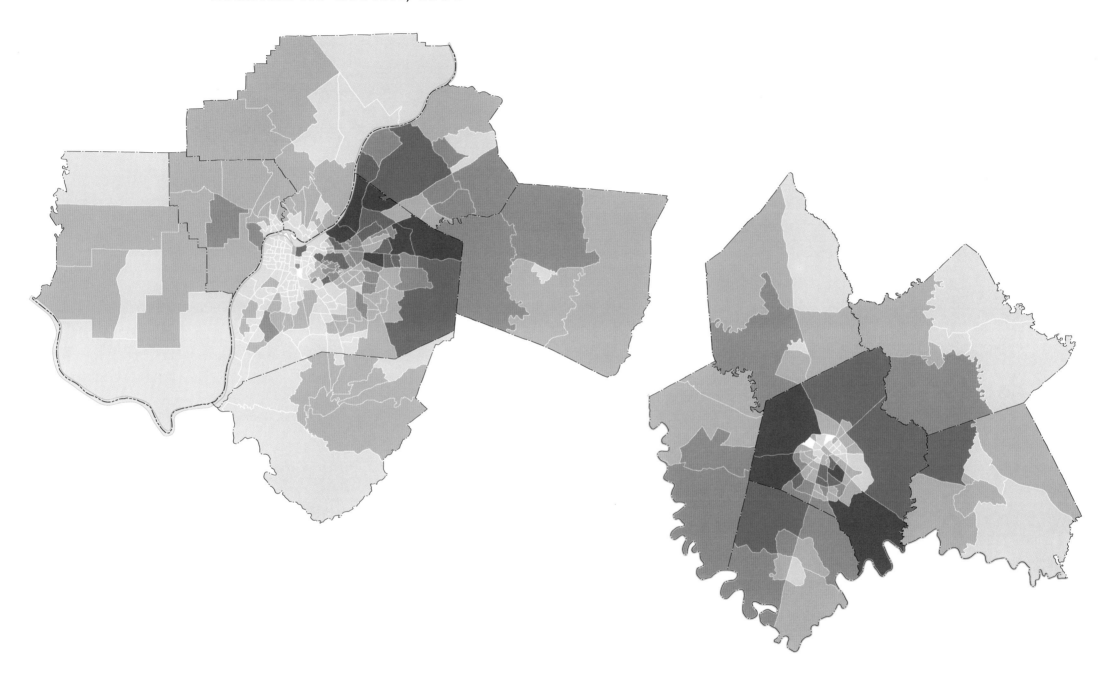

Lexington–Fayette KY MSA, 1990

SELECTED SOCIOECONOMIC CHARACTERISTICS

	% Black	% 65 or Over	Median Housing Value	% College Graduates
Lexington-Fayette MSA	10.7	10.3	$70,000	15.5
Fayette County	13.4	9.9	$73,900	30.6
Louisville MSA	13.1	12.6	$56,000	17.3
Kentucky portion	15.2	12.7	$57,100	18.6
Jefferson County	17.1	13.5	$57,000	19.3
Louisville city	29.7	16.5	$44,300	17.2
Cincinnati CMSA	11.7	11.8	$71,000	20.2
Kentucky portion	1.9	11.2	$66,100	16.1
Hamilton County	20.9	13.3	$72,200	23.7
Cincinnati city	37.9	13.9	$61,900	22.2
Kentucky	7.1	12.7	$50,500	13.6
United States	12.1	12.6	$79,100	20.3

Median housing value includes owner-occupied only (excludes mobile homes, homes with business or medical offices, and housing units in multi-unit buildings). Percent college graduates includes persons twenty-five years of age or older with at least a bachelor's degree.

1990 data

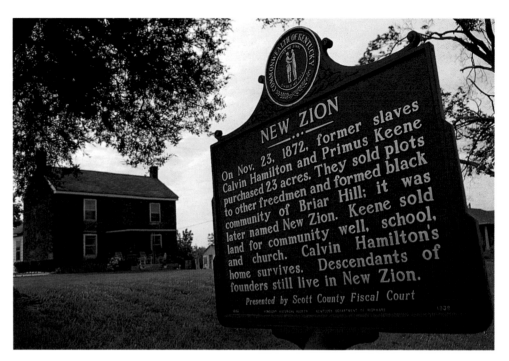

New Zion, in Scott County, is one of more than thirty black hamlets created in the Bluegrass Region after the Civil War to house former slaves.

Examining selected socioeconomic characteristics in the central areas of each of Kentucky's three MAs offers a clearer understanding of spatial patterns. The characteristics mapped are the distribution of the black population, median household income, the location of the elderly, and the residences of college graduates. The general patterns that surface here are similar to those in other large MAs in the United States: racial and social divisions are clear, and blacks and the elderly, who tend to have lower incomes, live in or near the older sections of the central city, where housing costs are lower than in the newer suburbs. For workers, access to employment is also less costly, since most jobs are still located in or near the CBD. College graduates are found disproportionately in the newer suburbs of the cities, where housing values are highest. There are, however, exceptions to these spatial patterns. For example, some parts of central cities—often historic areas once occupied by the upper class—are undergoing a process of gentrification whereby middle-income homeowners (and so-called yuppies) move into and restore dilapidated inner-city areas, often displacing lower-income, long-term residents.

In order to begin to understand the contemporary spatial distribution of a group such as black Americans, one must examine the group's history of population growth and settlement. Before the twentieth century, most blacks lived in the rural South. Not until the first few decades of the 1900s did a major black migration stream from the rural South to northern cities begin. Between 1900 and 1920 some 600,000 blacks left the rural South to pursue opportunities in northern cities. Depressed farming conditions in southern cotton areas and elsewhere, as well as the lure of factory jobs in the North, contributed to this migration. The only affordable housing for such newcomers to the city was located in and around the central core, where houses were being vacated by the upwardly mobile middle class.

Louisville and Lexington, usually considered southern cities, also experienced black population growth during this period. Like other cities of the South, Louisville and Lexington began to see the growth of the black population earlier. After the Civil War hundreds of freed slaves from rural areas of Kentucky migrated to the state's two largest cities. Richard Schein reports that in Lexington "the end of the Civil War and manumission in Ken-

tucky brought an exodus of African Americans from the surrounding countryside. . . . Many entrepreneurial land developers subdivided properties around the city's then-periphery in order to accommodate the increased housing demand. The result was a series of 'black enclaves,' creating a micro-geography of segregation scattered throughout the larger city. The placement of these black residential areas in part shaped the city's subsequent racial geography."

Text continues on page 275

BLACK POPULATION

PERCENT OF TOTAL POPULATION

10.0 or less	20.1 - 40.0	More than 80.0
10.1 - 20.0	40.1 - 80.0	No data

0   5 Miles

Scale

1990 data

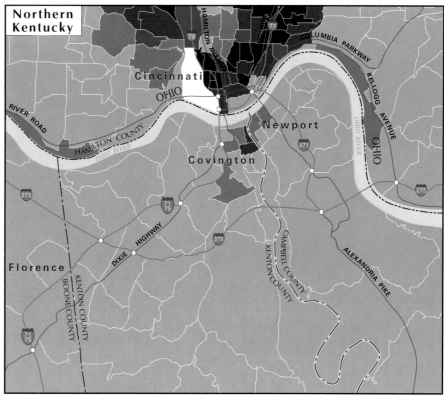

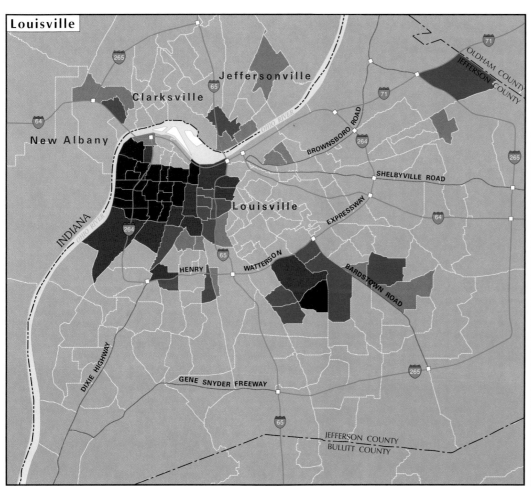

ELDERLY POPULATION

PERCENT OF POPULATION OVER AGE SIXTY-FIVE

- Less than 10
- 10 - 15
- 16 - 20
- 21 - 25
- More than 25
- No data

0 _____ 5 Miles
Scale

1990 data

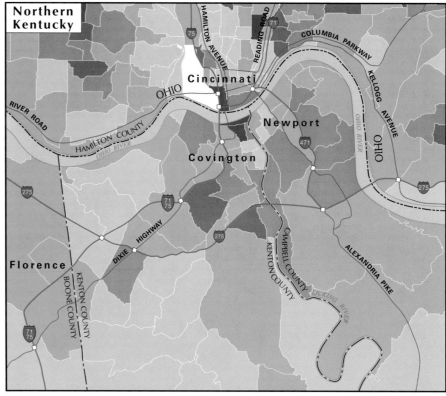

Northern Kentucky

Lexington

Louisville

MEDIAN HOUSEHOLD INCOME

INCOME IN DOLLARS

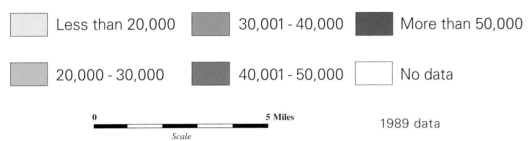

Less than 20,000   30,001 - 40,000   More than 50,000

20,000 - 30,000   40,001 - 50,000   No data

0   5 Miles

Scale

1989 data

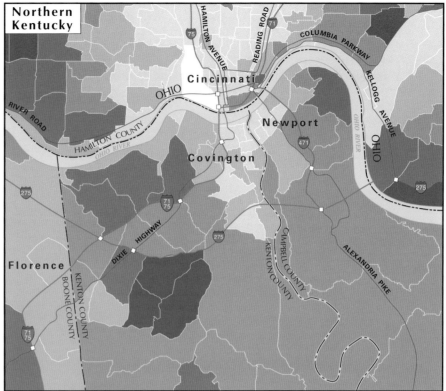

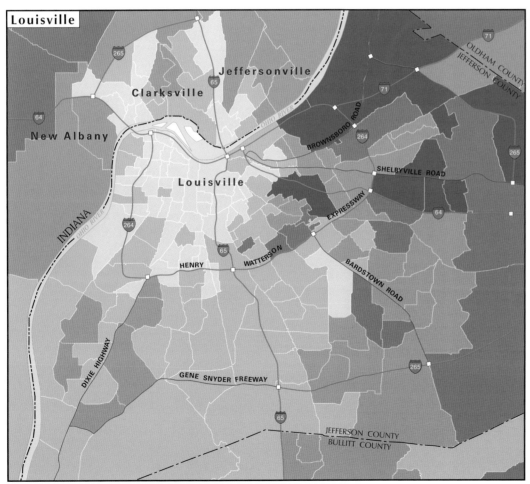

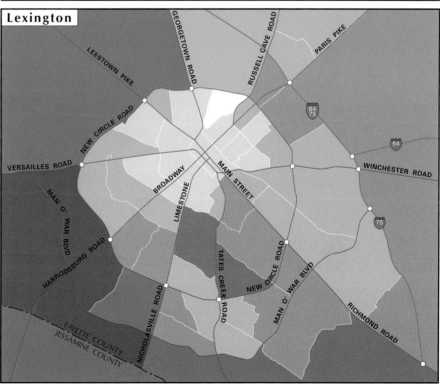

COLLEGE GRADUATES

PERCENT OF POPULATION TWENTY-FIVE YEARS OF AGE AND OLDER WITH A BACHELOR'S DEGREE OR HIGHER

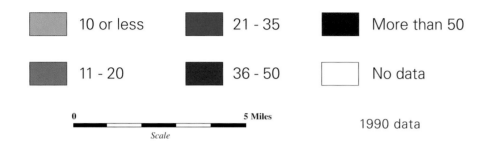

10 or less	21 - 35	More than 50
11 - 20	36 - 50	No data

0   5 Miles

Scale

1990 data

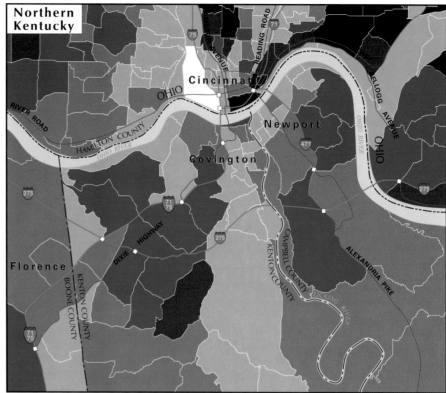

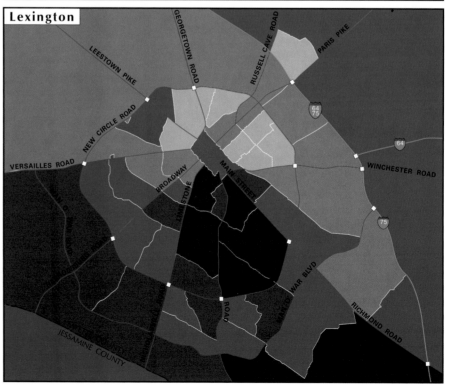

LOUISVILLE COMMUTER SHED, 1990

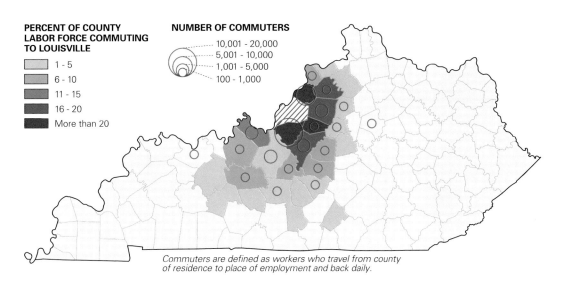

PERCENT OF COUNTY LABOR FORCE COMMUTING TO LOUISVILLE
- 1 - 5
- 6 - 10
- 11 - 15
- 16 - 20
- More than 20

NUMBER OF COMMUTERS
- 10,001 - 20,000
- 5,001 - 10,000
- 1,001 - 5,000
- 100 - 1,000

Commuters are defined as workers who travel from county of residence to place of employment and back daily.

CINCINNATI COMMUTER SHED, 1990

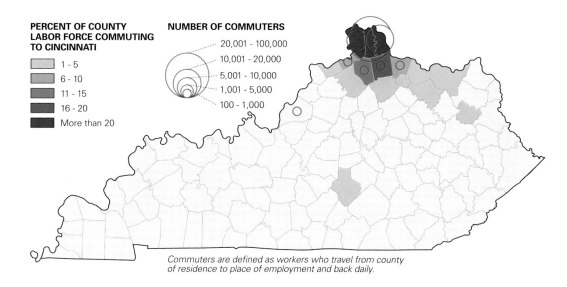

PERCENT OF COUNTY LABOR FORCE COMMUTING TO CINCINNATI
- 1 - 5
- 6 - 10
- 11 - 15
- 16 - 20
- More than 20

NUMBER OF COMMUTERS
- 20,001 - 100,000
- 10,001 - 20,000
- 5,001 - 10,000
- 1,001 - 5,000
- 100 - 1,000

Commuters are defined as workers who travel from county of residence to place of employment and back daily.

INCREASE IN THE NUMBER OF WORKERS COMMUTING TO LOUISVILLE, 1980-1990

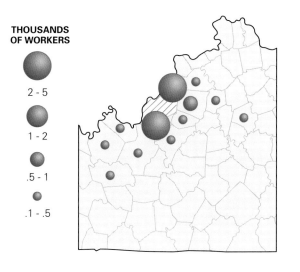

THOUSANDS OF WORKERS
- 2 - 5
- 1 - 2
- .5 - 1
- .1 - .5

CROSS-STATE COMMUTING

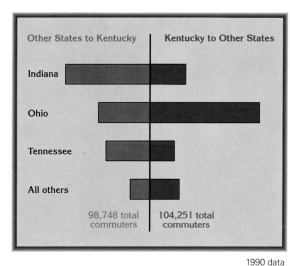

Other States to Kentucky	Kentucky to Other States
Indiana	
Ohio	
Tennessee	
All others	
98,748 total commuters	104,251 total commuters

1990 data

INCREASE IN THE NUMBER OF WORKERS COMMUTING TO CINCINNATI, 1980-1990

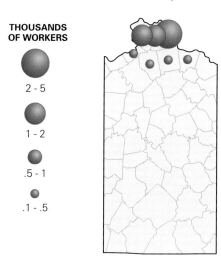

THOUSANDS OF WORKERS
- 2 - 5
- 1 - 2
- .5 - 1
- .1 - .5

The Journey to Work

For most adults, the choice of where to live is strongly influenced by the location of jobs, and the journey to work becomes a routine part of everyday life. Most workers desire to minimize the time spent each day traveling to a place of employment. Indeed, in Kentucky's three major metropolitan areas in 1990, a majority of all workers commuted daily to jobs within their county of residence (the figure for the Cincinnati PMSA was 69 percent; for the Louisville MSA, 79 percent; and for the Lexington MSA, 78 percent). Furthermore, nearly four-fifths, or more than one million, of the workers in these MAs drove alone in their automobiles to work each workday, rather than carpooling, using public transportation, walking, or bicycling. The automobile has clearly had a major impact on urban expansion and will certainly continue to do so in the decades to come.

Of course, many individuals travel across county lines, by choice or necessity, to their workplace. Examination of the commuter sheds for Kentucky's three major MAs demonstrates that distance remains an important factor in the journey to work, with

LEXINGTON COMMUTER SHED, 1990

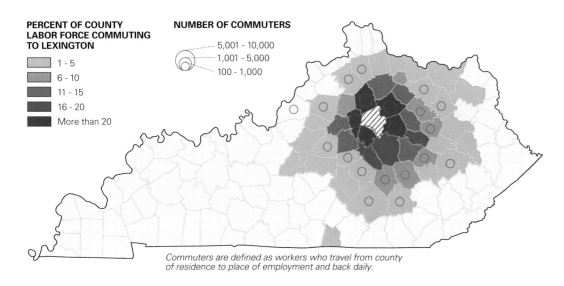

PERCENT OF COUNTY LABOR FORCE COMMUTING TO LEXINGTON

- 1 - 5
- 6 - 10
- 11 - 15
- 16 - 20
- More than 20

NUMBER OF COMMUTERS

- 5,001 - 10,000
- 1,001 - 5,000
- 100 - 1,000

Commuters are defined as workers who travel from county of residence to place of employment and back daily.

INCREASE IN THE NUMBER OF WORKERS COMMUTING TO LEXINGTON, 1980-1990

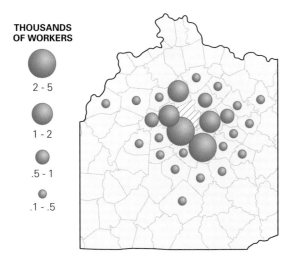

THOUSANDS OF WORKERS

- 2 - 5
- 1 - 2
- .5 - 1
- .1 - .5

more highly populated counties located near the MAs sending more commuters to the metropolitan centers than less populated and more distant counties. Since most people prefer to drive themselves to work, the major highways, especially interstates, influence the residential location of long-distance commuters. Workers have such long commutes for several reasons. Many have resided in smaller towns and rural areas all their lives but have taken jobs in urban places because of greater employment opportunities there and because local jobs in farming, mining, and even retailing have declined. Many residents in these small towns and rural areas remain in place and continue to spend anywhere from twenty minutes to several hours daily driving to their jobs. A residential exodus from large cities is yet another reason for long commutes. During the 1970s many city dwellers in the United States left urban areas to escape the crime, pollution, overcrowding, and other ills that were perceived to exist in the larger cities. This phenomenon has continued through the 1990s. It is most likely responsible, for example, for the slow growth in Louisville's population and the expansion of fringe cities and smaller towns in counties surrounding Louisville. Located on a state boundary, Louisville attracts more than

thirty thousand commuters from counties in Indiana. Cincinnati is similarly situated to receive commuters from Kentucky. Workers, especially those in high-paying jobs, are willing to sacrifice the expense of a long commute for the different lifestyle afforded by a residence away from a large metropolitan complex. Furthermore, advances in communication technologies now allow a growing number of people to work from their own homes while retaining ties to their place of work through telephone and fax lines and computer links.

Kentucky cities are, indeed, vital centers of economics, culture, and diversity. They are influenced not only by their myriad residents, commuters, and visitors but also by the histories that have shaped the landscapes we see. Cities often serve as focal connections to other states and other countries, and their periods of growth and decline often have impacts that reach far beyond their own boundaries. Compared with rural areas, Kentucky cities exhibit change that is visibly much more rapid. What will these cities be like in the future? The answer to this question is complex and will depend not only on the actions of city planners and politicians but also on the many people who live in, commute to, or visit cities each day.

Kentucky in the Future

CHAPTER THIRTEEN

Kentucky in the Future

Education is the key to Kentucky's future. The William T. Young Library, shown here under construction at the University of Kentucky in Lexington, will employ state-of-the-art information technology. The building also symbolizes the commitment of Commonwealth citizens to providing high-quality education for future generations.

THE *ATLAS OF KENTUCKY* IS DESIGNED to inform readers about the Commonwealth's diversity and complexity: physical landscapes that influenced early exploration and continued survival; early settlements that influenced the pattern and growth of our cities and the evolution of transportation networks; political elements that have shaped development; and the dynamics of cultural, demographic, and economic change that have produced the contemporary human environment.

But to what extent do the patterns and trends identified by the *Atlas* portend future conditions? It is impossible to say with certainty what Kentucky—or any place—will be like in the future. Demographers, for example, were taken by surprise by the post–World War II Baby Boom, and state officials and educators were ill-prepared to accommodate the consequent tidal wave of new students as they progressed from grade school through college. Economists did not foresee the repercussions of the 1970s oil embargo. Earth scientists cannot predict the exact nature of impending environmental change or the extent of damage and loss caused by hazards, either natural or human-induced. And few could have anticipated the suddenness and far-reaching consequences of the information age, in which fiber optics, satellites, computers, and the internet have transformed communication, learning, business, and even leisure and entertainment.

What does it mean to "look to the future"? We make no pretense here that we can accurately predict what Kentucky and its people will be like in times to come. Based on some recent national and state trends, however, we offer some insights on several pressing issues and make suggestions of scenarios for consideration during the next quarter of a century or so. In discussing these issues, we offer as one common "thread," or focus, the way in which changes in the size and composition of the state's population will affect the future and be affected by it. Our goal, then, is to assess what has gone before and what now exists, and to use this knowledge in order to identify some of the important issues that will confront Commonwealth residents as we move into the next century.

The Foundations of Change

The Kentucky of tomorrow will naturally be based on what now exists in the state, and to some degree on established trends that have evolved over time. Fundamental characteristics that will have powerful and wide-reaching implications for the state's future include population size and composition, the social and economic characteristics of its citizens, the structure of its families, and the persistence of traditions. A large number of people living today will still be alive by 2020. These people will require goods and services; their labor, spending, and investments will drive the economy; and their ideals, values, and decisions will shape the nature of change. And the state—because its population relies more heavily on government programs than do citizens of many other states—must plan carefully how best to use its limited resources to provide for its citizens, especially in a time when the cost of public programs continues to expand.

Demographic trends that have gathered momentum over the past two decades suggest that communities of the twenty-first century will look quite different from those of today. They will have adapted in many ways to meet the needs of more older citizens, fewer young people, changing social and economic characteristics, and changes in family structures.

Changing Population Size and Composition.   While there was an unexpected increase in Kentucky's population during the first half of the 1990s, demographers do not expect population growth to persist at such levels over the next quarter century. Instead, underlying trends suggest that Kentucky will experience only moderate population growth in the years to come, rising about 23 percent from its 1990 level of almost 3.7 million to approximately 4.5 million by the year 2020. The small size of the projected increase is due in part to the anticipated decline in natural increase—the contribution of new births as offset by the loss from deaths. This change in the rate of natural increase will be caused by ongoing fertility declines (individual reproductive behavior) and the evolving age structure of the Commonwealth's population. Contrary to stereotypes,

Kentucky registered the forty-ninth-lowest birthrate in the nation during the 1980s, resulting in an accelerated transition toward an older age structure. More important, the state's population growth will also be affected by national migration trends; the Commonwealth as a whole should gain slightly more population from the arrival of newcomers than it loses by residents leaving the state over the next several decades. Migration will play an increasing role in some counties as the influence of natural increase progressively declines.

Fertility decline will slow the growth of the state's youngest age groups. Although this trend is commensurate with a slowing of overall population growth, it also suggests changing conditions for education and family economies. Few adults can forget the overcrowded schools and teacher shortages brought about in the 1950s by the Baby Boom generation, or in the 1970s by the Baby Boomers' children. Kentucky's near future should provide a clear departure from such problems. More than half of the state's counties (67 of 120) will experience a decline in the size of the zero-to-seventeen age group by 2020, a pattern that will be seen especially in Appalachian Kentucky and the state's rural western and south central sections. Indeed, only two states lost a greater percentage of young people than Kentucky during the 1980s, and this characteristic has not altered during the 1990s.

The relative homogeneity of Kentucky's population has also persisted in recent years, a characteristic that will change if the national movement toward globalization of the economy occurs in the state. Such globalization demands increasingly higher levels of comfort with diverse peoples and cultures. It is in the state's interests to foster greater cultural heterogeneity, defined here as racial and ethnic diversity, because states that are perceived as economically or socially unwelcoming of diverse groups may risk a continuing and accelerating decline in their minority populations that will, in turn, undermine their economic competitiveness.

Toward Lifelong Learning.   In spite of its ranking at or near the bottom of all states on many measures of educational attainment, Kentucky has ascended to national prominence in educational reform. In just a few short years the Kentucky Education Reform Act (KERA) has yielded measurable improvements in student performance and has engaged thousands of parents, teachers, and administrators in new ways of improving the quality of education.

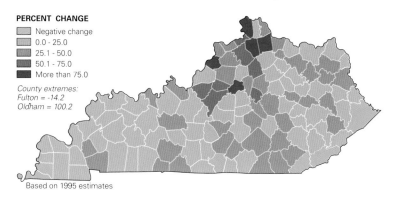

CHANGE IN TOTAL POPULATION, 1990-2020

PERCENT CHANGE
Negative change
0.0 - 25.0
25.1 - 50.0
50.1 - 75.0
More than 75.0

County extremes:
Fulton = -14.2
Oldham = 100.2

Based on 1995 estimates

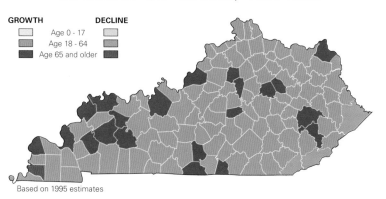

DOMINANT AGE GROUP
INFLUENCE ON CHANGE, 1990-2020

GROWTH   DECLINE
Age 0 - 17
Age 18 - 64
Age 65 and older

Based on 1995 estimates

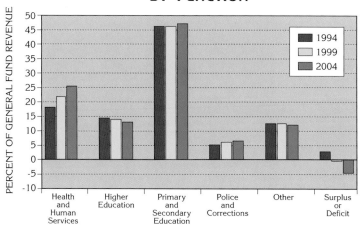

BASELINE EXPENDITURE PROJECTION
BY FUNCTION

1994
1999
2004

CHANGE IN POPULATION, AGE 0 TO 17
1990-2020

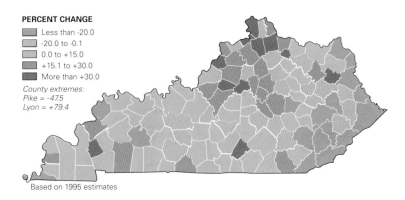

PERCENT CHANGE
- Less than -20.0
- -20.0 to -0.1
- 0.0 to +15.0
- +15.1 to +30.0
- More than +30.0

County extremes:
Pike = -47.5
Lyon = +79.4

Based on 1995 estimates

CHANGE IN POPULATION, AGE 18 TO 64
1990-2020

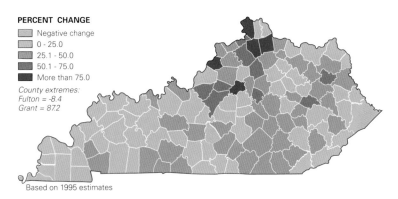

PERCENT CHANGE
- Negative change
- 0 - 25.0
- 25.1 - 50.0
- 50.1 - 75.0
- More than 75.0

County extremes:
Fulton = -8.4
Grant = 87.2

Based on 1995 estimates

COMPONENTS OF POPULATION CHANGE

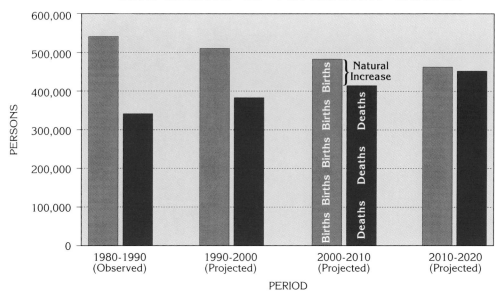

PERSONS

1980-1990 (Observed) · 1990-2000 (Projected) · 2000-2010 (Projected) · 2010-2020 (Projected)

PERIOD

As a new century unfolds, however, educational needs will increase, and the majority of American workers will require more than a high school diploma. The demand for highly trained workers who have a solid intellectual foundation will continue to expand, and the importance of updating and honing the skills of new and current workers will become critical to Kentucky's economic future. Although innovative models for integrated education exist in other states, the Commonwealth's current system of education and training remains relatively fragmented, frequently inaccessible or not affordable, poorly linked to business and industry, underused, and subject to much political manipulation. Voices of support for training our present and future workforce through vocational/technical schools, apprenticeships, school-to-work programs, literacy and adult education, on-the-job training,

and universities and community colleges, as well as other strategies, need to be united.

Knowledge and the analytical and creative skills higher education provides not only enhance employability but greatly enrich life and living. Such knowledge and skills will not only provide public and private institutions with direction, they will shape the work we do, the products we make, our quality of life, and the future we build. While higher education continues to provide the intellectual capital needed to meet future challenges, cost poses an increasingly formidable obstacle to its benefits. Although education remains both a public and a private responsibility, competing and compelling needs as well as local and regional politics have eroded public investment in higher education, challenging educators to accomplish more with less state support.

Poverty and the Evolving Family. A noteworthy result of improved education, one that will be facilitated by the slowing growth expected in the youngest population cohorts, may be a reduction in the number of children living in poverty, a problem that has been especially acute in the state's eastern counties. But such an outcome will be contingent on expected changes in family dynamics. Kentuckians now marry—and divorce—more often than Americans as a whole, which means greater economic uncertainty in the lives of children and parents. The increased incidence of divorce and a growing number of births to unmar-

ried women mean that more children live in households headed by single parents who are usually women and who are much more likely to be poor than in the past. As more Kentuckians postpone marriage or choose to live alone, and as families continue to fragment, the number of households has risen, and the number of people living in them has declined. While household formation continues to outpace population growth, the trend actually peaked in the 1970s. Nevertheless, the rapid expansion in the number of households has triggered widespread, if belated, attention to the implications of this trend for families, the environment, and a tenuous community fabric.

The Pursuit of Prosperity

Kentucky's economy will continue to expand; estimates indicate that the Gross State Product (GSP) will grow nearly 50 percent between 1995 and 2025. Geographically, the most rapid economic growth will still be centered in the Golden Triangle—comparable to the Bluegrass Region as defined by the Kentucky Long-Term Policy Research Center (LTPRC). Kentucky's other three regions, West, Central, and East (see accompanying map), may fare less well, and the economy of the state's eastern portion may experience the slowest expansion, mirroring past trends.

KENTUCKY'S POLICY REGIONS

BLUEGRASS

EAST

WEST   CENTRAL

1996 data   Kentucky's Policy Regions shown here were defined by the
Kentucky Long-Term Policy Research Center.

Horticultural crops such as cabbage offer an attractive alternative to tobacco as demonstrated on this family farm in Lee County.

Economic advancement will continue to depend upon the availability and education of the labor force. Population growth of those between eighteen and sixty-five years of age is expected in all but six Kentucky counties; the state's northern metropolitan counties stand out with the largest projected gains, a consequence of expected continued in-migration associated with rapid development in the Golden Triangle. Although the Commonwealth should not experience a shortfall of employable persons in coming decades, the question remains as to whether this labor force will be adequately trained for the tasks required by the rapidly evolving manufacturing and service economies.

Whither the Primary Sector. Historically, Kentucky's economy has depended heavily on the primary industries of agriculture and coal mining, both of which have recently been undergoing noticeable transformation. Whereas agriculture and coal mining remain vital to the state's economic well-being, both industries face uncertain long-term prospects. As was demonstrated in Chapter 7, small family farms are on the decline, with larger—and intensely farmed—landholdings becoming the cornerstone of agricultural production. Farm employment is falling, and fewer farm children are assuming the responsibilities of their retired parents in maintaining agricultural viability. Kentucky's farm economy nevertheless enjoys record output, as farming methods and equipment advance productivity. The vulnerability of its key crop—tobacco—however, creates significant uncertainty. The

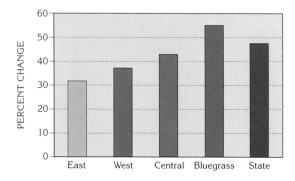

ESTIMATED GROSS STATE PRODUCT GAINS 1995-2025

(bar chart, PERCENT CHANGE on y-axis from 0 to 60, categories: East, West, Central, Bluegrass, State)

COUNTIES AT RISK FROM A SIGNIFICANT TOBACCO QUOTA DECLINE

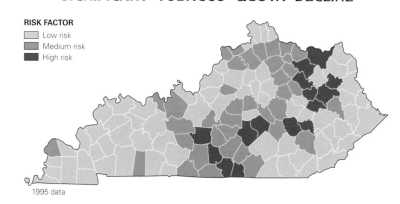

RISK FACTOR
- Low risk
- Medium risk
- High risk

1995 data

future of burley tobacco production, which is affected by declining use and increased leaf importation, presents the most significant potential for change for the state's farming and farmers. Crop and livestock alternatives to farmland now used for tobacco need to be considered without delay.

Although Kentucky's proven coal reserves could endure through at least the middle of the twenty-first century, the reluctance of young adults to seek employment in the industry, coupled with newly developed employment-reducing technologies and the vagaries of market demand, especially for coal with

high sulfur content, make continued extraction at current levels questionable.

Kentucky's hardwood forests provide the basis for another major primary industry, but the virtual absence of an important secondary tier of value-added wood products industries prevents state residents from capturing the full benefit of this abundant and renewable resource. Approximately 70 to 75 percent of the 700 million board feet of high-grade lumber cut annually is shipped out of state in an unprocessed condition. If the enormous potential of the secondary wood industry for long-term

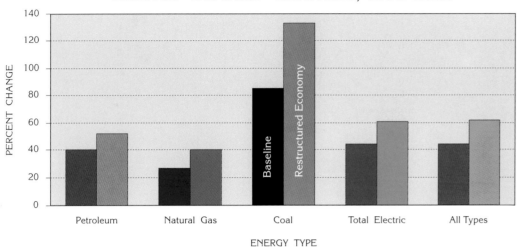

ENERGY DEMAND CHANGES, 1995-2025

Chart showing PERCENT CHANGE (y-axis, 0 to 140) versus ENERGY TYPE (x-axis). Categories: Petroleum, Natural Gas, Coal (with Baseline and Restructured Economy bars), Total Electric, All Types.

Interstate 75 has become a primary north–south artery linking Michigan to Florida by way of central Kentucky. Increased traffic flows near Lexington in Fayette County have necessitated widening the highway to six lanes, including the bridge over the Kentucky River palisades shown here.

economic prospects is to be developed, Kentuckians must invest in husbandry expertise, initiate entrepreneurial innovation, and cultivate a new ethic of land and forest stewardship that will help sustain Kentucky's forests for future generations.

Kentucky's Economy in the Post-Industrial Era. Employment in the secondary economic sector (manufacturing) is projected to decline nationally in coming years (indeed, its relative importance has already declined in many states), but manufacturing nevertheless is making dramatic gains in Kentucky's economy, and the state is outpacing the nation in terms of both jobs and share of the Gross Domestic Product (GDP). The automobile industry's focus on Kentucky bodes especially well for the future, and the state's manufacturing establishments produce a diverse range of products. There remains, however, a predominance of small firms, often insufficiently capitalized, and the state suffers from an inadequate base of intellectual capital, skilled workers, and technological know-how that may inhibit its ability to compete in the global marketplace.

New kinds of small enterprises are expected to be the engine of the state's future economy, fueled by technological and managerial expertise, an infusion of U.S. and international capital, and a broad-based commitment to the development of such enterprises. As the economy's small business sector creates a disproportionate share of new opportunities, the importance of new and enabling entrepreneurship rises. The state's future success in cultivating small enterprises and the opportunities they may provide hinges on the availability of competent and dedicated personnel, sufficient start-up capital, and suitable market exposure.

Technology and Globalization. The extraordinary era in which we now live is characterized by rapid change driven by technological innovation and the ongoing challenge to adapt to the consequences of the broad-based application of new technology. While human resources will be critical to Kentucky's ability to use and apply technology to create opportunities, existent and expanding infrastructure, particularly Kentucky's planned Information Superhighway, may offer points of leverage. They include the possibility of exploiting new advances in communications, pharmaceuticals, biotechnology, and environmental redemption. New technologies have brought about the virtual convergence of places around the world—the "shrinking world" notion, expanded global commerce, and forced changes in the organization, conduct, and location of the workplace. Kentucky's industries have fared quite well thus far in the global marketplace as its products have found markets across the world. The state has facilitated the "selling" of Kentucky; its boosterism has both attracted foreign industry and promoted overseas investment opportunities for Kentuckians. Global competition has introduced extraordinary demands on business and industry and, in turn, on workers, who have placed greater emphasis on making higher-quality products in less time and at a lower cost. Simultaneously, competition has elevated requisite employment skills to new heights that make flexibility and longterm education imperative.

The Information Age has arrived and it will exert an increasingly powerful influence on the way we work and live. The rapidly expanding technologies that store, organize, analyze, and communicate information promise to liberate and empower workers, yet the technical complexity involved threatens to diminish the ability of many to participate in the economic and social mainstream of our state and nation. Coping with rapid change is difficult at best, but those who can accept change and prepare themselves for it will find that accelerating technological innovation is creating an explosion of opportunities for highly skilled workers and innovative firms.

Growth in the Golden Triangle. Although Chapter 12 suggests that urbanization has not been as rapid in the Commonwealth as in much of the rest of the nation, Kentucky's population continues to urbanize. Foremost among the state's areas of growth has been the Golden Triangle, the area defined by the vertex cities of Louisville, Lexington, and Covington and roughly delimited by Interstate highways 64, 75, and 71. The rapid expansion of this region in population size, personal economic advancement, and industrial development was demonstrated in Chapters 4, 8, and 12. Urban growth is, in part, a consequence of the movement of rural Kentuckians toward the opportunities offered in cities, and in part the consequence of in-migration from outside the state. Continued expansion of commuter sheds, Interstate-oriented businesses, and the rapid growth of new communities and businesses at Interstate interchanges in the Golden Triangle all

NATIONAL LABOR FORCE STRUCTURE AND ECONOMIC DEVELOPMENT

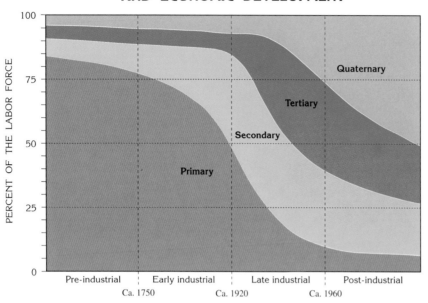

ADULTS WHO HAVE ACCESSED THE INTERNET

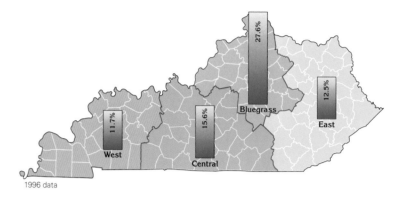

1996 data

CHANGE IN POPULATION, AGE 65 AND OLDER 1990-2020

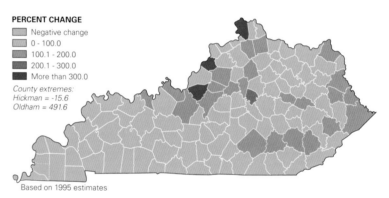

PERCENT CHANGE

- Negative change
- 0 - 100.0
- 100.1 - 200.0
- 200.1 - 300.0
- More than 300.0

County extremes:
Hickman = -15.6
Oldham = 491.6

Based on 1995 estimates

contribute to the expansion of urban land uses within the Triangle. In short, this area exemplifies the impact of adequate highway, rail, water, and air transport, and of concerted political and social efforts to promote economic expansion, combined with an ideal location and accessibility within a day's drive of two-thirds of the U.S. population.

Can the Golden Triangle alone carry the Commonwealth through the twenty-first century? While the Triangle acquires and spreads some benefits statewide, many distant counties remain poor, with high unemployment and welfare rates and many social needs that are underserved. These same counties also remain beyond the attendant urban problems of congestion, crime, and pollution. Kentucky is expected to retain its essentially rural character for some time and thereby delay or avoid the arrival of some of the problems that American urbanization brings. Natural beauty and the enhanced quality of life offered by the state's rural traditions may hold enormous future appeal for those in flight from urban congestion, consumer materialism, crime, and pollution elsewhere in America. These qualities become particularly attractive as economic forces shift away from manufacturing dominance toward professional and service-oriented jobs in the tertiary and quaternary sectors and as communication technologies provide increasingly strong links to places farther away from urban centers—links that encourage, for example, telecommuting.

The Community of Tomorrow

The aging of Kentucky's—and the nation's—population presents an odd potential for difficulty and success. Projections indicate a widespread pattern of population aging, with most Kentucky counties expecting an increase of at least 30 percent in the number of elderly residents by 2020. Common perspectives view this trend with pessimism; it represents a larger number of individuals beyond their so-called productive labor-force years and in need of social and health services, thus straining both family and state revenues. Furthermore, the oldest segment of our population, those aged eighty-five and above, will certainly increase in number over the next few decades; these people are alive now, and we can accurately track their progress over time. They are, by and large, Kentucky natives who have lived and worked in the state throughout their lifetimes. They will require assistance from the state, especially since many will be relatively poor. And they will require, when available, assistance from families who will be hard pressed, if current conditions prevail, to provide the necessary care.

The reality of aging is, however, quite different from common stereotypes. Tomorrow's older population, in actuality, will be dominated by recent retirees having larger savings and investments and higher educational levels than are found in former and even current elderly age groups. They will generally be in good health and highly independent. The central challenge will be not how to accommodate more disabled elders but how best to incorporate into development strategies the resources and needs of an active and vibrant population group.

As our nation's population ages, the large store of natural and historical sites in Kentucky (discussed in Chapter 10) and the state's transportation accessibility are expected to increase its appeal to tourists, magnifying the economic benefit of tourism to Kentucky. Tourism, when viewed over time, is often a vital precursor to permanent relocation, especially among retirees. While it is viewed as a viable, sustainable element for the development of rural communities, tourism has not proven to be an antidote to poverty. The seasonal and marginal jobs it creates often do not produce suitable returns in the lives of those who work in tourism. Successful development strategies should promote economic balance, maximizing local labor involvement and increasing opportunities for Kentuckians to own and operate businesses and create important complementary industries to counter tourism's cyclical nature. The Lake Cumberland area, especially Pulaski County, provides a successful case study of how more balanced economic development, which has accommodated a rapidly growing elderly population, can work to benefit areas situated some distance from major urban centers.

New Prescriptions for Well-Being. An aging population is commonly associated with increasing demands on health care services. Most studies indicate, however, that if such demands are met, the benefits extend naturally across all age categories. Remarkable changes, driven primarily by advancing technology and rapidly rising health care costs, are under way in our thinking about health and health care. Increasingly, emphasis will shift to health promotion and disease prevention as means for making health care more accessible and affordable while maintaining quality. Second, a higher level of cooperation among individuals, private service providers, and government is expected to emerge. Advanced applications of technology and more active involvement by communities will overarch these trends as we attempt to create a more manageable, useful, and equitable health care system.

Declining Representation and New Voting Behaviors. As we have noted, although the state's population will increase slowly in an absolute sense, Kentucky's share of the national population will decrease because the state's population will grow more slowly than that of the nation. Such a growth differential has national political implications because the state's representation in the U.S. Congress will decline. Today, Kentucky has six U.S. representatives, down from the eleven who represented the state in an earlier era. Depending on exactly how the population changes, the number of Kentucky's congressional representatives could decline to five, or even four, during the first quarter of the twenty-first century. This means the state will be less influential in introducing congressional legislation.

An aging population often implies new voting behaviors and priorities that reflect the needs of older voters. Thus, concerns about the accessibility and affordability of health care, access to leisure and recreational opportunities, and crime will gain importance in the platforms of individuals seeking election to national,

CONSOLIDATED COUNTIES

Present-day county boundary
County boundary after consolidation

state, and local offices. As we have seen, the number and proportion of elderly persons will continue to rise nearly everywhere in the state. Furthermore, the significance of the elderly vote is even greater than its proportion suggests because nationally the elderly are more likely to vote than are any other age group.

County Consolidation. As noted in Chapter 1, there are those who argue that Kentucky has too many counties, many of which are poor and have small populations, leading to low tax bases that often cannot support even minimal services. Except for Rhode Island, Kentucky has the smallest counties (the average size is 335 square miles) of any state having independently organized county governments. Proponents argue that one strategy the state could adopt to become more "efficient" economically is to reduce the number of counties from the current 120 to, say, as few as 12 or 15 through consolidation. This is not a new or revolutionary idea. Limiting the number of counties was first proposed in the state legislature in the mid-nineteenth century, and on a number of occasions since the 1930s the topic has reappeared. Opponents claim the cost of such consolidation would be too high, county seats would be too distant and thus would increase the time and inconvenience of travel, and local economies would be ruined as county seats lost their political function. One proposal has been to reduce the number of counties to thirteen based on the following criteria: the counties should have homogeneous social, economic, and political structures; they should be large enough to acquire economies of scale but small enough to be accessible; they should have an adequate tax base;

compact units should be created out of existing contiguous counties; the integrity of existing county boundaries should be respected; and multi-county urban areas should be maintained (Webster, 1984). Given the state's history and a political culture in which power often accrues to those who control county jobs, this change may be difficult, perhaps impossible, to realize.

Environmental Preservation and Degradation. If the Commonwealth's population is to enjoy some measure of well-being, environmental integrity must be promoted and maintained at a variety of scales. Living conditions within the home, for example, must be improved for the many state residents who now exist without the benefit of complete plumbing or adequate heating. Residents who now heat with coal or wood could opt for cleaner alternatives, such as electricity or natural gas, where available, to heat their homes. While Kentuckians have prospered greatly from the largesse of the state's natural resources and have made significant progress in restoring health to the environment, they have often done so in the interest of short-term monetary gain—bequeathing long-term environmental consequences to future generations at immeasurable cost.

The quality of Kentucky's water continues to improve, but substantial investment in infrastructure will be required in coming years to treat wastewater safely and to extend public drinking water to more residents. Continuing the strong, consistent, and targeted enforcement of environmental regulations, which has enabled Kentucky to make substantial gains in the restoration of its water resources, will be critical to maintaining the progress

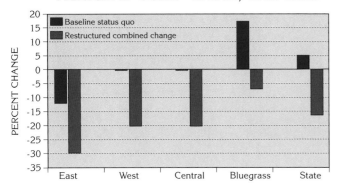

CHANGE IN ANNUAL GENERATION OF MUNICIPAL SOLID WASTE, 1995-2025

Legend:
- Baseline status quo
- Restructured combined change

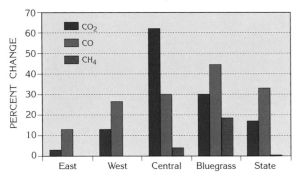

CHANGE IN GREENHOUSE GAS EMISSIONS, 1995-2025

Legend:
- CO_2
- CO
- CH_4

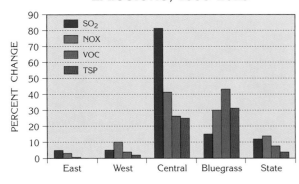

CHANGE IN GROUND-LEVEL EMISSIONS, 1995-2025

Legend:
- SO_2
- NOX
- VOC
- TSP

that has been achieved. New approaches to managing water pollution are likely to focus on the source, rather than the outcome. Yet it is important to recognize that the leading source of Kentucky's water pollution is resource extraction. If the potential of Kentucky's water resources is to be fully realized, then special attention must be paid to reducing the amount of pollutants from ongoing extraction of coal and timber that are allowed to enter water sources and to minimizing any adverse impact of land reclamation efforts. It is also worth noting that 500,000 Kentucky residents rely on private wells and springs for their water, and bacterial pollution remains a major problem in water quality for these people.

Kentucky's air quality has also improved dramatically since the enactment of the federal Clean Air Act in 1970. The act's provisions have led to significant reductions in levels of several harmful pollutants, such as lead and carbon monoxide. When fully enforced, the federal Clean Air Act Amendments of 1990 are expected to reduce air pollution dramatically but will have an as yet undetermined impact on the state's coal industry and manufacturing growth.

Timely legislation has checked the influx of out-of-state garbage, but management of Kentucky's solid waste continues to challenge policymakers at every level. During the early 1990s Kentuckians made significant headway in their recycling efforts, and progress was made in identifying and cleaning up illegal dumpsites, especially those sites that contained hazardous waste materials. Programs to promote further waste reduction and recycling will likely increase as landfill disposal costs rise and more markets for recyclables become available. Much more remains to be done to reverse the negative impact of decades of pollution and to prevent costly and destructive future problems. Significant evidence suggests that, by doing so, state residents will strengthen the foundation for progress and possibly expand business opportunity and income.

The Human Element and Change

The path the Commonwealth will take into the future is not clearly marked, nor is there an accurate map that will lead to prosperity and improved quality of life. One's vision of the future is necessarily blurred by the understanding that many different elements of change will be working at once, with alterations in any single element potentially causing unexpected results. Although

Kentucky's future will always be somewhat uncertain, there are definitive ways in which residents can shape progress and skillfully navigate the state into the next century.

Appropriate governmental actions will be central to a successful journey. The effectiveness of government in the future will depend upon the ability of leaders to overcome cynicism, parochialism, alienation, and even despair among the citizens they serve, and to build the social capital and citizen participation needed to meet future challenges. Communities and regions that enjoy high levels of civic engagement are far more prosperous. While hopeful signs of rising engagement can be detected in Kentucky's civic life, many community activists observe that, for a variety of reasons, citizen participation is inadequate and therefore self-limiting. Despite this, government must, as a matter of routine, turn to citizens and nongovernmental organizations for guidance, direction, and support if it hopes to build the social capital on which successful development relies.

Of critical importance is the element of personal initiative. Throughout this atlas, we have illustrated change: the transformation of Native American trade routes to modern-day highways; the growth of small riverside communities to vibrant urban centers; the rejuvenation of barren mine landscapes and clear-cut forests into healthy wildlife habitats and recreational areas; and the rise and fall of rural economic prosperity. Much of this change, although shown at the aggregate county, regional, or state level, is the product of individual actions. In short, *we,* the citizens of Kentucky, are essentially responsible for the situation in which we now live, and *we* will ultimately be responsible for altering the situation in ways that will prove either beneficial or detrimental for the Commonwealth and its citizenry.

Glossary

acid rain: Acidified precipitation (having a pH of less than 5.6) that may severely damage flora, fauna, and certain calcium-based building materials. It is formed from the oxides of sulfur and nitrogen released into the atmosphere when fossil fuels are burned, especially in urban regions that have manufacturing industries, or downwind from areas that have coal-fired power plants.

arch: A geologic formation that develops when a resistant rock layer (such as sandstone) overlies a less resistant rock layer (such as limestone) and erosional forces wear away a portion of the lower layer. If the worn-away portion becomes exposed from the sides (as in a narrow ridge bounded by cliffs), an arch, or "natural bridge," results.

Baby Boom: The dramatic increase in birthrates following World War II. The Baby Boom cohort refers generally to those individuals born between 1946 and 1964.

baseline projection: A projection of future occurrences calculated from current data, such as a projection of Kentucky's economic development that assumes no major shifts in economic policy or in the national or global economic systems. Cf. RESTRUCTURED PROJECTION.

basin: An area drained by a river system and its tributaries, or a broad depression in the surface of the land.

bedding plane: The boundary between two distinct sedimentary rock layers.

bituminous coal: Also called soft coal (as opposed to anthracite, or hard coal). Coals are ranked according to the amount of carbon they contain, which determines the amount of heat, or energy, they produce when burned. The energy value of coal is expressed as the number of British Thermal Units (BTUs) per pound. Nearly all of Kentucky's coal is bituminous, of medium to high volatility, ranging in energy value from about 12,000 to 15,000 BTUs per pound.

board foot: A unit of lumber measurement one foot long, one foot wide, and one inch thick, or the equivalent.

census tract: An administrative subdivision into which a metropolitan area has been divided for the purpose of reporting census data. A tract typically has a population between 2,500 and 8,000 and, when first delineated, is designed to be relatively uniform with regard to population characteristics, economic status, and living conditions.

company town: A town or village created by a resource extraction company (such as a coal company) to house workers in an area of dispersed rural settlement. Generally the company owned all property, including houses, roads, processing and storage facilities, and a store. Many companies paid workers with their own currency, or scrip, which could be spent only at the company store.

conglomerate: A coarse-grained sedimentary rock formed by the cementation of rounded gravel.

containerized freight: Goods moved inside metal boxes of standard lengths, usually either 20 or 40 feet. See also INTERMODAL TRANSPORTATION.

continuous miner or continuous mining machine: An electrically powered excavating machine for underground coal mining that combines a large toothed cylinder or drum and a conveyor belt. The operator steers the rotating drum into the coal seam to remove coal at high speed; the coal is then carried from the mine on a system of conveyor belts.

crude birthrate: A rough measure of a population's fertility level, calculated as the total number of births divided by the total population and usually expressed as a number per 1,000 population.

crude deathrate: A rough measure of a population's mortality level, calculated as the total number of deaths divided by the total population and usually expressed as a number per 1,000 population.

diffusion: The process whereby ideas or things (for example, people, tools, seeds) move or are carried from one place or region to another.

dimensional lumber: Lumber cut to pieces of a specified size or dimension for use in the construction industry or in secondary wood manufacturing.

dolomite: A compact limestone or marble rich in magnesium carbonate.

economy of scale: A reduction in unit costs through increased size of the production facilities.

ecosystem: The complex totality of interactions among all organisms and the environment in any region or area.

emigrant: A person who leaves his or her native country to establish residence in another country. Cf. IMMIGRANT; MIGRANT.

erosion: The movement of rock materials and soils through the action of wind, water, or ice. This process is distinct from mass wasting, the down-slope movement of materials in response to gravity, and from weathering, the initial breaking down of rock.

escarpment: A steep slope or cliff, sometimes marking the edge of a plateau.

ethnicity: The ancestral origins of a particular group, such as African American, English, Scots-Irish, or Mexican.

eutrophication: The process whereby a body of fresh water (for example, a lake) becomes enriched with dissolved nutrients, thereby reducing the oxygen content and possibly causing the extermination of some life forms.

family: A unit composed of a householder and one or more other persons living in the same household who are related to the householder by birth, marriage, or adoption. Cf. HOUSEHOLD.

fault: A fracture in bedrock along which movement of the rock has taken place.

fecundity: The physical ability to give birth.

fertility: Reproductive performance as measured by the number of live births.

galena: An ore mineral of commercial value composed of lead sulfide and mined and refined to yield lead.

gentrification: A process whereby middle-income homeowners move into and restore dilapidated inner-city areas, often displacing lower-income, long-term residents. Such areas are usually located in or near central cities and are often historic areas that were once occupied by the upper class.

gerrymandering: The geographic division of an area into election districts so as to give one group (such as a political party) a majority while minimizing the voting strength of the opposition group or groups.

Golden Triangle: In Kentucky, the triangle formed by the cities of Louisville, Lexington, and Covington-Newport that contains Kentucky's three largest metropolitan areas and most of the state's urban population.

greenhouse gases: A group of gases that includes water vapor, carbon dioxide, methane, nitrous oxide, ozone, and chlorofluorocarbons. Increase in the concentration of these gases in the atmosphere has caused temperatures to increase globally.

Gross Domestic Product (GDP): The total value of goods and services produced by labor and property in a defined area. For a particular state it is referred to as **Gross State Product (GSP)**.

groundwater: Water found in the pore spaces of rock and soil in the zone of saturation within the ground and potentially a major source of water for human use.

hemp: The durable fiber extracted from the inner bark of *Cannabis sativa*, an annual plant, and used to manufacture rope, cloth, sailcloth, sheeting, and floor covering. It was raised in Kentucky first in the 1770s.

hogshead: A round wooden packing container of 100-140 gallons resembling a large barrel.

household: A unit made up of all persons who occupy a housing unit regardless of their relationship to each other. Cf. FAMILY.

housing unit or **dwelling unit**: A house, an apartment, a mobile home, a group of rooms, or a single room occupied or intended for occupancy as separate living quarters.

iconography: A collection of representations or symbols that illustrate a subject or a theme.

immigrant: A person who enters a country from another country and establishes residence. Cf. EMIGRANT; MIGRANT.

income: Money acquired as wages or salary or from self-employment, interest, dividends, rents or royalties, Social Security or railroad retirement, public assistance or welfare, or disability. Income levels are reported at the personal, family, or household level.

industry: The primary activity engaged in by an establishment. The major industrial groups reported in the U.S. Census are based on the STANDARD INDUSTRIAL CLASSIFICATION (SIC).

information superhighway: The advanced infrastructural elements needed to exchange data, audio, and visual messages in digital format at local, national, and international scales. Such infrastructure includes fiber optic trunk and feeder lines along with computer hubs and digital switching technology.

in-lot: A small, centrally located lot in a town plan, generally one-half acre or less, intended for development as business or commercial property. Cf. OUT-LOT.

intermodal transportation: The use of several different modes of transportation in door-to-door shipment of goods. The transferability of goods from one mode to another is facilitated by the use of standard-sized containers, usually 20 or 40 feet long, that can be easily off-loaded and onloaded from ships with specially designed cranes. Their use encourages greater globalization of economic activity. Ship-to-rail-to-truck transfers are a common sequence.

karst topography: A landform that develops as a result of subsurface chemical weathering and erosion of calcium carbonate bedrock, characterized by caves, underground streams, and sinkholes.

Kentucky Education Reform Act (KERA): An act passed by the Kentucky General Assembly in 1989 and signed into law in April 1990 that mandates changes in three major areas of education: finance, governance, and curriculum. It resulted from a 1989 Kentucky Su-

preme Court decision that declared Kentucky's entire system of public elementary and secondary education unconstitutional because it was characterized by inequity and inefficiency.

Kentucky Educational Television (KET): An innovative public broadcast system initiated in 1969 that provides public and instructional programming statewide and supports all levels of Kentucky's educational system.

liquefaction: A quicksand-like condition in wet soil, caused by seismic (shock) waves, in which soils lose strength as particles lose frictional contact with each other.

loam: A category of soil texture in which none of the three component soil particle sizes—sand, silt, or clay—predominates. When used in combination with one of the component sizes (such as sandy loam), the term describes a texture in which the component size stated has a slight dominance.

location quotient: An index used to measure the importance of a variable (such as tourism receipts or manufacturing employment) in a given area (such as a county) to some established norm. For example, the location quotient for manufacturing employment in a Kentucky county can be determined by dividing the percentage of such employment by the state or national percentage. Thus, if a county has a location quotient greater than 1.0, it has a greater share of manufacturing employment than does the state (or the nation).

loess: Fine grains of silt that are carried long distances by wind and laid down elsewhere and eventually forms a thick layer of sediment. With irrigation, it forms a fertile soil.

longwall mining: A method of underground coalmining that employs a cutting machine mounted on rails that run parallel to an extended coal seam. The machine moves along the rails, removing a slice of coal from the seam, creating a "long wall" of exposed coal. Powerful hydraulic jacks support the roof over the machine and working face. As the long wall is extended, the jacks are moved, allowing the roof to collapse behind the mining operation.

macadamized road: A road surfaced using a method developed by Scottish inventor J.L. McAdam in the early 1800s in which successive layers of crushed stone are laid and rolled. The method predates the asphalt paving commonly used today.

manumission: Formal emancipation from slavery.

mass wasting. *See* erosion.

mean: The calculated **average** of a series of numbers obtained by dividing the sum of the series by the number of observations in the series.

median: The value that divides a series of numbers arranged in order of magnitude from highest to lowest whereby half the numbers lie below it and half above it.

metes and bounds: A manner of surveying land, used during the settlement process, that utilized natural features such as trees, streams, or rock outcrops, together with road alignments and survey lines, as corner markers and boundary lines. Each land parcel had to be laboriously surveyed, and each parcel had a unique size and shape that required elaborate records.

Metropolitan Area (MA): As defined by the U.S. Bureau of the Census,

a large population nucleus together with adjacent communities that have a high degree of economic and social integration with that nucleus. Various criteria are used to define an MA as either a **Metropolitan Statistical Area (MSA)** or a **Consolidated Metropolitan Statistical Area (CMSA)**. A CMSA is divided into one or more **Primary Metropolitan Statistical Areas (PMSAs)**, each of which comprises a large urban county or cluster of counties that demonstrate strong, locally recognized internal economic and social links. Cf. URBANIZED AREA.

migrant: A person who makes a permanent change of residence across a political or administrative boundary, such as between counties. This term most commonly refers to people who move within a country (**internal migrants**). **In-migrants** are counted as those who enter a specific area; **out-migrants** are counted as those who leave a specific area. *See also* EMIGRANT; IMMIGRANT.

morbidity: The incidence or rate of disease in a population.

mortality: The number of deaths in a population. Measures of mortality may be specified by the cause of death.

nativity: An individual's place of birth.

occupation: The primary activity engaged in by an individual. The major occupational groups reported in the U.S. Census are based on the *Standard Occupational Classification (SOC) Manual*.

orogeny: The process of mountain building through folding or faulting of the earth's crust or by volcanic activity.

out-lot: An urban property lot on the edge of an eighteenth- or nineteenth-century town, generally four acres or larger, intended to be used for gardening or pasture. Cf. IN-LOT.

ozone: A gas composed of molecules having three bonded oxygen atoms. As a pollutant, ozone forms from industrial activities that generate high levels of energy. Naturally occurring ozone develops in an atmospheric layer ranging from about 12 to 20 miles above the earth and serves to filter out harmful ultraviolet energy from the sun.

palisade: A line of bold cliffs. In Kentucky, the popular term for the nearly vertical walled gorge that the Kentucky River has eroded 300 feet or more into the Inner Bluegrass Region limestones.

particulates: Very small solid byproducts of combustion (such as ash). **Suspended particulates** are sufficiently small that any movement of air will keep them from settling to the ground.

physiographic region: A region of distinct and relatively homogeneous landform features that also has some uniformity in climate, natural vegetation, and soils.

plateau: A generally flat and horizontal upland surface often bounded by steep cliffs or escarpments. The Eastern Kentucky Coal Field Region is part of the Appalachian Plateau; over geological time this plateau has been deeply incised by stream erosion and has the appearance of a rugged mountainous area.

point of presence (POP): In advanced telecommunications, a high-capacity voice and data interchange between local telephone systems and long distance carriers that serves as an on/off ramp to the nation's INFORMATION SUPERHIGHWAY system.

post roads: Roads constructed in the early nineteenth century to promote the reliable delivery of mail.

precipitation: Mist, drizzle, rain, hail, sleet, or snow. When measured, ten inches of snow is generally equivalent to about one inch of liquid precipitation.

primary economic sector: A category encompassing those jobs and economic activities engaged in the direct extraction of resources from the environment, including mining, fishing, lumbering, and farming.

primary wood production: A manufacturing activity that involves basic processing of logs into low-value-added products such as rough-sawn lumber, dimensional lumber, posts, or pulp.

pulpwood: Timber converted into 4- or 5-foot lengths and chipped plant by-products that are prepared for manufacture into pulp, which is then used in the production of various types of paper.

quaternary economic sector: A category encompassing those jobs and economic activities concerned with the collection, processing, and manipulation of information.

race: A group of people who may be characterized by a generally distinctive combination of inheritable physical traits, such as Asian, black, or white.

restructured projection: A projection of future occurrences constructed to deviate from the baseline (current data) in particular sectors. Very generally, as used here, it reflects possible future changes in the five sectors identified as priorities by the Kentucky Long-Term Policy Research Center: tobacco and production agriculture; coal mining; secondary wood processing; manufacturing; and tourism. Cf. BASELINE PROJECTION.

sawn lumber: The initial manufacture of rough lumber from logs, typically done in small-scale lumber mills located near commercial forestlands. This process is followed by the manufacture of DIMENSIONAL LUMBER used in the construction industry or in secondary wood manufacturing, such as furniture making.

Scots-Irish: Descendants of Scots people who settled in Northern Ireland, especially during the reign of James I (1406-37).

secondary economic sector: A category encompassing those jobs and economic activities concerned with the processing and transformation of raw materials into finished products. The manufacturing sector.

secondary wood producer: A manufacturing activity that involves sophisticated processing of primary wood products into higher-value-added products such as wood furniture and moldings.

sediment: Rock particles or soil originally deposited by water, wind, or ice.

serious crime: A category of crime, also known as Crime Index Offenses, that includes murder and non-negligent manslaughter, forcible rape, robbery, aggravated assault, burglary, larceny-theft, and motor vehicle theft. Arson was added as the eighth index offense in 1979. Cf. VIOLENT CRIME.

site: The internal locational attributes of a place (such as a city or town) that include its physical attributes.

situation: The external attributes of a place or area, especially its location relative to other places or areas.

Standard Industrial Classification (SIC): An international classification system for all types of industrial activities. About 85 percent of the SICs for manufacturing industries are found in Kentucky.

station: A strongly built log house with heavy doors and other defensive provisions, such as slots through which rifles could be fired. Most early Kentucky settlements began as stations.

Superfund: A common name for the Comprehensive Environmental Response, Compensation, and Liability Act, a federal law enacted in 1980 and reauthorized in 1986. This act, a coordinated effort between federal and state agencies, enables the Environmental Protection Agency to respond to hazardous waste sites that threaten public health and the environment.

tectonic force: A force that originates within the earth's crust, including warping, folding, faulting, earthquakes, and volcanic activity.

telecommuting: Working at home through an electronic linkup with a central office.

telemedicine: The application of audio, visual, and digital communication technologies in the provision of health care services in remote areas. Such technology, for example, allows health care experts at a major urban hospital to participate actively in medical examinations and procedures taking place simultaneously in a distant rural county.

tertiary economic sector: Those jobs and economic activities concerned with services, such as banking, transportation, retail and wholesale sales, and routine office-based jobs.

timberland: Forestland producing or capable of producing crops of industrial wood and not withdrawn from timber utilization.

tipple: A metal superstructure containing bins, sieves, and machinery for breaking, sizing, and cleaning coal. It is usually built beside or astride railroad tracks to facilitate the loading of coal directly into specialized coal cars by gravity flow.

transpiration: The transfer of water in vapor form from vegetation to the atmosphere.

Urbanized Area (UA): A U.S. Census term defined as an area that comprises one or more central places and the surrounding densely settled territory (the "urban fringe") that together have a minimum of 50,000 population. Cf. METROPOLITAN AREA.

value-added: Relating to a product or resource that has been changed from its original form through a manufacturing process whereby its market value has been increased, such as wood furniture manufactured from timber resources.

violent crime: A category of crimes that includes murder and non-negligent manslaughter, forcible rape, robbery, and aggravated assault. Cf. SERIOUS CRIME.

warrant: A written authorization or certificate, as of land ownership, often given to Revolutionary War veterans in payment for military service.

weathering: Any of several processes that break rock materials down into smaller particles through physical disintegration or chemical decomposition.

wetland: A land area that has poor surface drainage, including marshes and swamps.

Selected Bibliography

The following list includes primary works that have been used throughout the *Atlas*. They are cited in abbreviated form in the Sources of Maps, Tables, and Graphs .

Karan, P.P., and Cotton Mather, eds. *Atlas of Kentucky*. Lexington: Univ. Press of Kentucky, 1977.

Kentucky, Commonwealth of
    Cabinet for Economic Development.
        Kentucky Deskbook of Economic Statistics. Frankfort,1995, 1996.
    Cabinet for Health Services. Kentucky State Center for Health Statistics. *Kentucky Annual Vital Statistics Reports, 1991-1993*. Frankfort, 1991-1993.

Kleber, John E., ed. *The Kentucky Encyclopedia*. Lexington: Univ. Press of Kentucky, 1992.

U.S. Bureau of the Census
    Eighth Census of the United States, 1860. Washington, D.C.: GPO, 1864.
    Tenth Census of the United States, 1880. Washington, D.C.: GPO, 1883.
    Twelfth Census of the United States, 1900. Washington, D.C.: GPO, 1902.
    Thirteenth Census of the United States, 1910. Washington, D.C.: GPO, 1913.
    Fourteenth Census of the United States, 1920. Washington, D.C.: GPO, 1921-22.
    Fifteenth Census of the United States: 1930. Washington, D.C.: GPO, 1932-33.
    Sixteenth Census of the United States: 1940. Washington, D.C.: GPO, 1943.
    Census of Population: 1950. Washington, D.C.: GPO, 1952.
    Census of Housing: 1950. Washington, D.C.: GPO, 1952-53.
    Census of Population: 1960. Washington, D.C.: GPO, 1960-64.
    Census of Housing: 1960. Washington D.C.: GPO, 1963.
    Census of Population: 1970. Washington, D.C.: GPO, 1972-73.
    Census of Housing: 1970. Washington, D.C.: GPO, 1972.
    Census of Population: 1980. Washington, D.C.: GPO, 1982-84.
    Census of Housing: 1980. Washington, D.C.: GPO, 1982.
    Gross Migration for Counties: 1975-1980. Supplementary report PC80-21-17. Washington, D.C.: GPO, 1984.
    1990 Census of Population and Housing.
        County-to-County Migration Flow Files. Special Project 312, CD90-MIG-02. On CD-ROM. 1995.
        Population and Housing Unit Counts. Washington, D.C.: GPO, 1993.
        General Population Characteristics. CP-1-19. Washington, D.C. 1992.
        Selected Place of Birth and Migration Statistics: 1990. CPH-L-121. Access on Internet. 1990.
        Summary Social, Economic, and Housing Characteristics: United States. CPH-5-1. Washington, D.C.: GPO, 1992.
    County and City Data Book, 1994. Washington, D.C.: GPO, 1994.
    Statistical Abstract of the United States. Washington, D.C.: GPO, various issues, 1920-1994.

U.S. Department of Health and Human Services. *Vital Statistics of the United States: 1990*. Hyattsville, Md., 1994.

Sources of Maps, Tables, and Graphs

ABBREVIATIONS

KCED	Kentucky Cabinet for Economic Development, Frankfort
KCHR	Kentucky Cabinet for Human Resources, Frankfort
KDF	Kentucky Division of Forestry, Frankfort
KDFWR	Kentucky Department of Fish and Wildlife Resources, Frankfort
KDMM	Kentucky Department of Mines and Minerals, Lexington
KDT	Kentucky Department of Travel, Frankfort
KNREPC	Kentucky Natural Resources and Environmental Protection Cabinet, Frankfort
KGS	Kentucky Geological Survey, Lexington
KLTPRC	Kentucky Long-Term Policy Research Center, Frankfort
KSCHS	Kentucky State Center for Health Statistics, Lexington
KSNPC	Kentucky State Nature Preserves Commission, Frankfort
KTC	Kentucky Transportation Cabinet
STF	Summary Tape File
USDA	United States Department of Agriculture, Washington, D.C.
USFS	United States Forest Service, Washington, D.C.
USGS	United States Geological Survey, Washington, D.C.

CHAPTER 1. KENTUCKY: ITS SETTING

PAGE 4

A Global Perspective. Compiled from an azimuthal equidistant projection provided by Will Fontanez, Geography Dept., Cartographic Services Lab, Univ. of Tennessee, Knoxville.

PAGE 5

Kentucky's Place in the Western Hemisphere. Globe compiled from digital file found in MapArt, Cartesia Software.

Travel Time from Lexington, Kentucky. Adapted from "Mileages and Driving-Times," map in *Rand McNally Road Atlas* (Chicago: Rand McNally, 1995), A18.

PAGES 6-7

Physical Framework. Gyula Pauer, Center for Cartography and Geographic

Information, Univ. of Kentucky; produced using the traditional photo-mechanical cartographic process.

PAGES 8-9

Commonwealth of Kentucky. Ibid.

PAGE 10

Mean Center of Population, 1790-1990. "Mean Center of Population for the United States: 1790 to 1990," map in *U.S. Census, 1990, Population and Housing Unit Counts*, vol. 4, chap. 10.

PAGE 11

State Populations, 1990. U.S. Bureau of the Census, *County and City Data Book, 1994*, p. 2.

Population Change, 1980-1990. Ibid.

Population Density, 1990. Ibid.

Population Living in Metropolitan Areas, 1992. U.S. Bureau of the Census, *Statistical Abstract: 1994*, p. 38.

PAGE 12

County Seats. *Kentucky County Maps* (Lyndon Station, Wis.: Thomas Publications, n.d.).

PAGE 13

Area Development Districts. Map in *Kentucky's Area Development Districts* (Frankfort: Kentucky Council of Area Development Districts, Kentucky Association of District Directors, 1996).

Population below Poverty Level, 1989 [U.S.]. U.S. Bureau of the Census, *Statistical Abstract: 1994*, p. 477.

Population below Poverty Level (Kentucky and Adjacent States). Ibid.

Median Household Income (Kentucky and Adjacent States). Ibid., p. 468.

PAGE 14

Comparative Cost of Living, 1992. U.S. Bureau of the Census, *Statistical Abstract: 1993*, pp. 487-88.

Serious Crimes (Kentucky and Adjacent States). U.S. Bureau of the Census, *Statistical Abstract: 1994*, p. 199.

State and Local Taxes Paid (Family of Four in Selected Cities). Ibid., p. 310.

Gross State Product for Major Industries, 1963-1990. Bureau of Economic Analysis, Economics and Statistics Division, U.S. Dept. of Commerce, *Survey of Current Business* 68, no. 5 (May 1988): 43; 73, no. 12 (Dec. 1993): 46.

OTHER SOURCES

"Area Development Districts," Bluegrass Area Development District Home Page <http://andromeda.mis.net/bgadd/index1.html>.

Cole, Leslie A., and Peggy Pauley, *State of Kentucky's Environment: 1994 Status Report* (Frankfort: Environmental Quality Commission, 1995).

Garriott, William C., "Area Development Districts," in Kleber, *Kentucky Encyclopedia*.

Ireland, Robert M., "Counties," in ibid.

Savageau, David, and Richard Boyer, *Places Rated Almanac: Your Guide to Finding the Best Places to Live in North America* (New York: Prentice-Hall Travel, 1993).

CHAPTER 2. THE NATURAL ENVIRONMENT

PAGE 18

Generalized Geology. Adapted from "Generalized Geologic Map of Kentucky," KGS, and "Geologic Map of Kentucky" sesquicentennial ed., USGS in cooperation with KGS, 1988.

PAGE 19

Geologic Timetable. Adapted from "Geologic Time Scale," table in Alan Strahler and Arthur Strahler, *Physical Geography: Science and Systems of the Human Environment* (New York: Wiley, 1997), p. 277.

PAGE 20

Major Tectonic Features. "Major Tectonic Features of Kentucky," map provided by Terry Hounshell, KGS.

PAGE 21

Physiographic Regions. Adapted from "Physiographic Diagram of Kentucky," map, KGS.

PAGE 22

General Soils. Adapted from "General Soils, Kentucky," map, 1:750,000, USDA, May 1990. Soil type descriptions provided by David Howarth, Dept. of Geography and Geosciences, Univ. of Louisville.

PAGE 23

Areas with Potential for Karstic Development. Adapted from "Generalized Geologic Map of Kentucky," KGS.

Karst Areas in the United States. Adapted from W.E. Davies and H.E. Legrand, "Karst of the United States," map in M. Herak and V.T. Stringfield, eds., *Karst: Important Karst Regions of the Northern Hemisphere* (New York: Elsevier, 1972), p. 470. Arthur N. Palmer, Dept. of Earth Sciences, State Univ. College, Oneonta, N.Y., was instrumental in researching this topic.

PAGE 24

Cave Distribution. Adapted from "Distribution of Caves in Kentucky," map by Angelo I. George (Louisville, 1985).

Generalized Karst Landscape. Created by Richard A. Gilbreath and John Watkins, Univ. of Kentucky, 1997.

PAGE 25

Mammoth Cave System. Mammoth Cave Quadrangle, Kentucky, 7.5 minute series topographic map, KGS, 1965. Cave passages compiled from "Mammoth Cave System," 1:5,000 map, Cave Research Foundation, Louisville.

PAGE 26

Drainage Basins. Compiled from digital information provided by Terry Hounshell, KGS, 1996.

PAGE 27

Groundwater Sensitivity Regions. Adapted from "Groundwater Sensitivity Regions of Kentucky," 1:500,000 map, Division of Water, Groundwater Branch, KNREPC, Frankfort, 1994.

PAGE 28

Wetlands. Adapted from "Wetland Resources of the U.S., 1991," map, National Wetlands Inventory, U.S. Fish and Wildlife Service, St. Petersburg, Fla.

PAGE 29

Domestic Water Well Locations. Compiled from digital information provided by Bart Davidson, Kentucky Ground-Water Data Repository, KGS, 1996.

Domestic Water Well Flow Rate. Ibid.

Depth to Groundwater. Ibid.

PAGE 30

United States Climatic System. Adapted from "The Köppen Climate System," map in Alan Strahler and Arthur Strahler, *Physical Geography: Science and Systems of the Human Environment* (New York: Wiley, 1997), pp. 188-89.

PAGE 31

Climates Similar to Kentucky's. Ibid.

Average Annual Temperature. Data provided by D. Glen Conner, Kentucky Climate Center, Western Kentucky Univ., 1996.

Average Growing Season. Ibid.

PAGE 32

Average Maximum January Temperature. Ibid.

Average Minimum January Temperature. Ibid.

Mean Monthly January Sunshine. Adapted from "Monthly Sunshine: January," map in *The National Atlas of the United States of America* (Washington, D.C.: USGS, 1970), p. 94.

Average First Fall 32°F Days after September 30. Data provided by D. Glen Conner, Kentucky Climate Center, Western Kentucky Univ., 1996.

Average Number of Days with Temperature of 32°F or Lower. Ibid.

Average Annual Snowfall. Ibid.

PAGE 33

Average Maximum July Temperature. Ibid.

Average Minimum July Temperature. Ibid.

Mean Monthly July Sunshine. Adapted from "Monthly Sunshine: July," map in *The National Atlas of the United States of America*, p. 95.

Average Last Spring 32°F Days after February 28. Data provided by D. Glen Conner, Kentucky Climate Center, Western Kentucky Univ., 1996.

Average Number of Days with Temperature of 90°F or Higher. Ibid.

Average Annual Heating Degree Days. Ibid.

Average Annual Cooling Degree Days. Ibid.

PAGE 34

Average Annual Precipitation. Ibid.

Average Number of Precipitation Days. Ibid.

Mean Annual Precipitation. Adapted from "Mean Annual Precipitation," map in *The National Atlas of the United States of America*, p. 97.

Average March Precipitation. Data provided by D. Glen Conner, Kentucky Climate Center, Western Kentucky Univ., 1996.

Average October Precipitation. Ibid.

Average Annual Precipitation and Temperature for Kentucky, 1895-1995. Compiled from digital data provided by David Howarth, Dept. of Geography and Geosciences, Univ. of Louisville, 1996.

PAGE 35

Climographs for Selected Cities. *Climatological Data: Kentucky, 1994,* National Oceanic and Atmospheric Administration, Environmental Data Service, National Climatic Center, Asheville, N.C.

PAGE 36

Acid Rain. Adapted from "pH Values of Acid Rain in the Midwestern and Eastern United States in 1987," map in Ralph C. Scott, *Introduction to Physical Geography* (St. Paul, Minn.: West Publishing, 1996), p. 264.

Air Quality Control Regions. Compiled from information provided by L. Michael Trapasso, Kentucky Climate Center, Western Kentucky Univ., 1996.

Toxic Air Emissions. Ibid.

Greenhouse Gas Emissions. Hugh T. Spencer, *Kentucky Greenhouse Gas Inventory: Estimated Sources and Emissions for the Year 1990* (Frankfort: KNREPC, Division of Energy, 1996), p. 137.

Air Concentration of Ozone. Ibid.

Ozone Air Pollution Problems. Ibid.

Ozone Standard Violations. Ibid.

PAGE 37

Air Concentrations of Sulfur Dioxide. Compiled from information provided by L. Michael Trapasso, Kentucky Climate Center, Western Kentucky Univ., 1996.

Sulfur Dioxide Emissions. Adapted from *The State of Kentucky's Environment: 1994 Status Report* (Frankfort: Kentucky Environmental Quality Commission, Feb. 1995), p. 42, and from *Electric Power Annual, 1993* (Washington, D.C.: Energy Information Administration, Dec. 1994), table 52, pp. 81-85.

Air Concentrations of Particulates. Compiled from information provided by L. Michael Trapasso, Kentucky Climate Center, Western Kentucky Univ., 1996.

Air Concentrations of Total Suspended Particulates. Ibid.

Air Concentrations of Carbon Monoxide. Ibid.

Air Concentrations of Nitrogen Dioxide. Ibid.

PAGE 38

Total Releases/Transfers of Toxic Chemicals. Ibid.

New Madrid Seismic Zone. Adapted from "Approximate Boundaries of New Madrid Seismic Zone with Recent Epicenters, 1974-1987," map in David M. Stewart and Ray Knox, *The Earthquake That Never Went Away: The Shaking Stopped in 1812—But the Impact Goes On* (Marble Hill, Mo.: Gutenberg-Richter Publications, 1996), p. 31.

PAGE 39

Maximum Seismic Intensities. Adapted from "Strong Earthquake Motion Map" in *Damages and Losses from Future New Madrid Earthquakes* (Cape Girardeau, Mo.: Center for Earthquake Studies, Southeast Missouri State Univ., 1994), and from "Earthquake Hazards Map," Central United States Earthquake Consortium, 1995.

Seismic Risk. "United States Seismic Risk," map, National Oceanic and Atmospheric Administration, Asheville, N.C., 1989.

PAGE 40

Radon Level Regions. Compiled from information provided by L. Michael Trapasso, Kentucky Climate Center, Western Kentucky Univ., 1996.

Highest Recorded Radon Levels in Homes. Ibid.

Tornado Tracks, 1950-1986. Adapted from "Tornado Tracks: 1950-1986, Kentucky," map supplied by Kentucky Climate Center, Western Kentucky Univ.; Fujita scale descriptions taken from "Tornado-Intensity Rating System," table in *The Weather Almanac*, 6th ed. (Detroit: Gale Research, 1992), p. 90.

Confirmed Tornados, 1950-1993. Adapted from "Tornadoes, Kentucky, 1950-1993," graph supplied by David Howarth, Dept. of Geography and Geosciences, Univ. of Louisville, 1995.

PAGE 41

Underground Storage Tanks. Adapted from "Underground Storage Tanks by County," table in *Kentucky Outlook 2000: A Strategy for Kentucky's Third Century* (Frankfort: NREPC and KLTPRC, 1995), vol. 3, Appendix 6, pp. 7-11.

Hazardous Waste Facilities. Adapted from "Permitted and Illegal Hazardous Waste (RCRA) Facilities by County," table in ibid., 2:688.

Superfund Sites. Adapted from "Active Superfund Sites (State and Federal) by County," table in ibid., pp. 691-93.

PAGE 42

Natural Regions. Compiled in cooperation with Julian Campbell, Kentucky Nature Conservancy, Lexington, 1997.

PAGE 43

Diversity of Selected Animals. World Resources Institute, *The Information Please Environmental Almanac* (Boston: Houghton Mifflin, 1994), pp. 229-82 (data from Nature Conservancy, Natural Heritage Data Center Network, 1993).

PAGE 44

Diffusion of Coyotes. Compiled from information provided by Lauren Schaaf, KDFWR, 1994.
Distribution of Ruffed Grouse. Adapted from "Current Ruffed Grouse Range, Restoration and Trapping Sites in Kentucky," map provided by Pamela Renner, Kentucky Fish and Wildlife Information System, KDFWR, 1994.
Poisonous Snakes. Information from Roger Conant, *A Field Guide to Reptiles and Amphibians of the United States and Canada East of the 100th Meridian* (Boston: Houghton Mifflin, 1958), pp. 336-37.

PAGE 45

Selected Endangered or Threatened Species, 1996. Compiled from information provided by Pamela Renner, Kentucky Fish and Wildlife Information System, KDFWR, 1994.

PAGE 46

Kentucky's Plant and Animal Species. Adapted from "Plants and Animals Presumed Extinct or Extirpated from Kentucky," unpublished report, KSNPC, Oct. 1992; "Number of Known Species in the World, the United States, and Kentucky," table, *Biodiversity* (Lexington: Cooperative Extension Service, Univ. of Kentucky, 1993), p. 3.; Brooks M. Burr and Melvin L. Warren Jr., "A Distributional Atlas of Kentucky Fishes," Scientific and Technical Series, no. 4 (Frankfort: KSNPC, 1986).

OTHER SOURCES

Aldrich, James R. "Kentucky's Treasured Wildlife," *Kentucky News: The Nature Conservancy* (Summer 1996), pp. 10-11.
Barbour, Roger W. *Amphibians and Reptiles of Kentucky.* Lexington: Univ. Press of Kentucky, 1971.
Barbour, Roger W., and Wayne H. Davis. *Mammals of Kentucky.* Lexington: Univ. Press of Kentucky, 1974.
Brown, William S. "Hidden Life of the Timber Rattler." *National Geographic* 172, no. 1 (July 1987): 128-38.
Campbell, Julian. "The Big Picture: Ecological Regions of Kentucky." *Kentucky News: The Nature Conservancy* (Summer 1996), pp. 14-15.
Carey, Daniel I., et al. *Quality of Private Groundwater Supplies in Kentucky.* Information Circular 44. Lexington: KGS, 1993.
Dougherty, Percy H., ed. *Caves and Karst of Kentucky.* Special Publication 12, ser. 11. Lexington: KGS, 1985.
Hopper, Margaret G., ed. *Estimation of Earthquake Effects Associated with Large Earthquakes in the New Madrid Seismic Zone.* Open-File Report 85-457 (Washington, D.C.: USGS, 1985).

KGS. *Ground-Water Supplies in Kentucky.* Ser. 11, Information Circular 44. Lexington: KGS, 1993.
KNREPC. 1994 Kentucky Report to Congress on Water Quality. Frankfort: Division of Water, KNREPC, 1994.
Mengel, Robert M. *The Birds of Kentucky.* Ornithological Monograph no. 3. Lawrence, Kans.: Allen Press for American Ornithologists' Union, 1965.
Sheldon, Andrew L. "Conservation of Stream Fishes: Patterns of Diversity, Rarity, and Risk." *Conservation Biology* 2, no. 2 (June 1988): 149-56.
Spencer, Hugh T. *Kentucky Greenhouse Gas Inventory: Estimated Sources and Emissions for the Year 1990.* Frankfort: Division of Energy, KNREPC, 1996.
"Wetlands: This Ecosystem of Water-Logged Soil Serves Many Functions." *Kentucky: Land, Air and Water* 8, no. 2 (Spring 1996): 8-11.
Winberry, John J., and David M. Jones. "Rise and Decline of the 'Miracle Vine': Kudzu in the Southern Landscape." *Southeastern Geographer* 13, no. 2 (Nov. 1973): 61-70.

INDIVIDUALS

Garrison, Lynn. Director, Division of Information and Education, KDFWR.
McCance, Robert. Director, Kentucky State Nature Preserves Commission.
Renner, Pamela. Project Leader, Kentucky Fish and Wildlife Information System, KDFWR.
Schaaf, Lauren. Wildlife Division Director, KDFWR.

INTERNET REFERENCES

Anderson, Raymond R., and Paul E. VanDorpe. "Iowa Perspective on Midwestern Earthquakes." <http://samuel.igsb.uiowa.edu/htmls/browse/quakes/quakes.htm>.
"Geologic Time." <http://www.uky.edu/KGS/coal/webfossl/pages/time.htm>.
"Geology of Mammoth Cave." <http://www.nps.gov/maca/geology.htm>.
"Karst Topography." <http://www.nps.gov/maca/karst.htm>.
"Water in Kentucky." <http://www.uky.edu/KGS/water/water1.htm#WRS_1_1>.

CHAPTER 3. HISTORICAL AND CULTURAL LANDSCAPES

PAGE 48

Kentucky. John Melish, *Travels through the United States of America, in the Years 1805 & 1807, and 1809, 1810, & 1811.* Philadelphia, 1812.

PAGE 50

Known Archaeological Sites, 1994. Data compiled by Berle Clay, Dept. of Anthropology, Univ. of Kentucky, based on data collected by the Kentucky State Archaeologist's Office, 1995.
Woodland Period Culture Areas. Michael Coe, Dean Snow, and Elizabeth Benson, *Atlas of Ancient America* (New York: Facts on File, 1986), p. 51; National Geographic Society, *Ohio Valley,* in map series The Making of America, 1985.

PAGE 52

Late Prehistoric Period Culture Areas. Coe, Snow, and Benson, *Atlas of Ancient America,* p. 54.
Tribal Territories at Time of European Contact. Ibid., p. 45; Bruce G. Trigger, ed., *Handbook of North American Indians,* vol. 15, *Northeast* (Washington, D.C.: Smithsonian Institution, 1978), p. ix; "Native Tribes of North America," map (in sleeve) in A.L. Kroeber, *Cultural and Natural Areas of*

Native North America (Berkeley: Univ. of California Press, 1963); National Geographic Society, *Ohio Valley,* in map series The Making of America, 1985.

PAGE 53

Pioneer and Indian Trails. Thomas P. Field, *Kentucky and the Southwest Territory, 1794,* Kentucky Study Series, no. 7 (Lexington: Dept. of Geography, Univ. of Kentucky, 1966); shaded relief provided courtesy of Julsun Pacheco, Univ. of Hawaii-Honolulu, 1995.

PAGE 54

Tribal Land Claims, circa 1787. "Land Claims by Tribe," map supplement by Sam B. Hilliard in *Annals of the Association of American Geographers* 62, no. 2 (1972).
Travel Routes of the Cherokee Removal (Trail of Tears, 1838). *Trail of Tears National Historic Trail,* Comprehensive Management and Use Plan, National Park Service (Denver Service Center, 1992); Glen Fleishmann, *The Cherokee Removal, 1838* (New York: Franklin Watts, 1971), p. 8.

PAGE 55

"A New Map of North America agreeable to the most approved Maps and Charts." Thomas Conder (London, 1794). Copy in Special Collections, M.I. King Library, Univ. of Kentucky.

PAGE 56

Proposed States West of the Allegheny Mountains, 1775-1785. "Colonies or States Proposed or Organized by Settlers West of the Allegheny Mountains, 1775-1785," map in Charles O. Paullin and John K. Wright, *Atlas of the Historical Geography of the United States* (Baltimore: Carnegie Institution of Washington and American Geographical Society, 1932), plates 41-45.

PAGE 57

County Evolution. "Evolution of Counties," maps in Karan and Mather, *Atlas of Kentucky;* "Map of Kentucky, 1787-1792 to Statehood," map by Wendell H. Rone Sr. in *An Historical Atlas of Kentucky and Her Counties* (Owensboro, Ky.: Progress Printing Co., 1965).

PAGE 58

Settlement Patterns, 1790-1910. "Density of Population," maps in Paullin and Wright, *Atlas of the Historical Geography of the United States,* plates 76-79.

PAGE 59

"Road from Limestone to Frankfort in the State of Kentucky." Georges Henri Victor Callot, *A Journey in North America Containing a Survey of the Counties Watered by the Mississippi, Ohio, Missouri, and Other Affluing Rivers . . .* (Paris, France, 1826), copy in Kentucky Library, Western Kentucky Univ.

PAGE 60

Pioneer Stations of the Inner Bluegrass. Nancy O'Malley, *Stockading Up: A Study of Pioneer Stations in the Inner Bluegrass Region of Kentucky,* Archaeological Report no. 127 (Frankfort: Kentucky Heritage Council, 1987).

PAGE 61

Boonesborough, Kentucky, circa 1790. Nancy O'Malley, *Searching for Boonesborough*, Archaeological Report no. 193, rev. ed., App. B (Lexington: Program for Cultural Resource Assessment, Kentucky Anthropological Research Facility, Univ. of Kentucky, 1990).
Boonesborough Town Plan, 1809. Ibid.

PAGE 62

Early Town Plan of Harrodsburg, circa 1786-1787. Untitled map in Clay Lancaster, "Planning the First Two Towns in Central Kentucky: Harrodsburg and Lexington," *Kentucky Review* 9, no. 3 (1989): 20.
Harrodsburg, 1952. USGS 1:24,000 7.5 minute topographic map, 1952.
Paris, 1877. D.G. Beers, *Atlas of Bourbon, Clark, Fayette, Jessamine, and Woodford Counties, Kentucky* (Philadelphia: D.G. Beers, 1877).

PAGE 63

Lancaster, 1952. USGS 1:24,000 7.5 minute topographic map, 1952.
Plan of Lystra, 1795. Paullin and Wright, *Atlas of the Historical Geography of the United States*, plate 48.
Main Post Roads, 1804. Plate 138 J in Paullin and Wright, *Atlas of the Historical Geography of the United States*.
Kentucky with Adjoining Territories, 1800. "The State of Kentucky with the adjoining Territories from the best Authorities 1800." In John Payne, *A New and Complete System of Universal Geography; . . . Describing Asia, Africa, Europe and America . . .* (New York, 1798-1800), copy in The Filson Club, Louisville.

PAGE 64

Natural Vegetation prior to Statehood. Compiled by Kent Anness, Bluegrass Area Development District, Lexington, Ky., from digital data provided by Dept. of Natural Resources, KNREPC, Frankfort.

PAGE 65

Shingled Land Claims, Shelby County. "*Todd vs Fry*, Shelby Co Bld 80 #1," map by Neal Hammon in *Early Kentucky Land Records, 1773-1780* (Louisville: Filson Club Press, n.d.).
Slaves Held, 1860. *U.S. Census, 1860*, vol. 1, table 2, pp. 180-81.
Tobacco Inspection Points. Leland Smith, "A History of the Tobacco Industry in Kentucky from 1783 to 1860," Master's thesis, Univ. of Kentucky, 1950, p. 30.

PAGE 66

Warehouses on the Kentucky River and Its Tributaries. . . . Mary Verhoeff, *The Kentucky River Navigation* (Louisville: John P. Morton, 1917), App., pp. 231-32.
Tobacco Produced, 1860. *U.S. Census, 1860*, vol. 6, pp. 58-65.
Tobacco Produced, 1900. *U.S. Census, 1900*, vol. 6, pt. 2, Crops and Irrigation, sect. 6, table 10, pp. 553-55.

PAGE 67

Wheat Produced, 1860. *U.S. Census, 1860*, vol. 6, pp. 58-65.
Wheat Produced, 1900. *U.S. Census, 1900*, vol. 6, pt. 2, Crops and Irrigation, sect. 1, table 55, pp. 165-66.
Hemp Produced, 1860. *U.S. Census, 1860*, vol. 6, pp. 58-65.
Hemp Produced, 1900. *U.S. Census, 1900*, vol. 6, pt. 2, Crops and Irrigation, sect. 4, table 11, p. 436.

PAGE 68

The Civil War and Kentucky. "The Western Theater," map in James Lee McDonough, *War in Kentucky: From Shiloh to Perryville* (Knoxville: Univ. of Tennessee Press, 1994), p. 13; "Confederate Invasion in the West (Perryville)," map 19 in Craig L. Symonds, *A Battlefield Atlas of the Civil War* (Annapolis, Md.: Nautical and Aviation Publishing Co., 1983), p. 44; map in Thomas E. Griess, ed., *Atlas for the American Civil War* (Wayne, N.J.: Avery Publishing, 1986), pl. 16.

PAGE 69

Immigrants and Rock Fences: Drystone Building Tradition. "Probable Source Areas for Kentucky Rock Fence Types: Drystone Building Tradition (Diagrammatic)," map compiled by Marvin Rinck, in Carolyn Murray-Wooley and Karl Raitz, *Rock Fences of the Bluegrass* (Lexington: Univ. Press of Kentucky, 1992), p. 85.

PAGE 70

Foreign-Born Population, 1880. *U.S. Census, 1880*, vol. 1, table 14, pp. 509-11.
Foreign-Born Population, 1920. *U.S. Census, 1920*, vol. 3, table 9, p. 384.

PAGE 71

Railroad Construction in Eastern Kentucky. "Railroad Construction in Eastern Kentucky by Decades, 1885-1905," map in Darrell Haug Davis, *The Geography of the Mountains of Eastern Kentucky* (Frankfort: KGS, 1924), p. 119.
Charcoal Iron Furnaces. J. Winston Coleman, *Old Kentucky Iron Furnaces* (Lexington, Ky.: Winburn Press, 1957); Sallie C. Eubank, "The Iron Industry in Kentucky," Master's thesis, Univ. of Kentucky, 1927; Donald E. Rist, *Kentucky Iron Furnaces of the Hanging Rock Iron Region* (Ashland, Ky.: Hanging Rock Press, 1974).

PAGE 72

Wheelwright, 1954. From USGS 1:24,000 7.5 minute topographic map, 1954.

PAGE 73

Church Members. Martin B. Bradley et al., "Churches and Church Membership in the United States, 1990" (Atlanta: Glenmary Research Center, 1992), tables 3 and 4. Table 3 is a listing of the number of church members by denomination for the state; Table 4 is a listing of the number of members by county.

PAGES 73-74

Church Members [by Denomination]. Ibid.

PAGE 75

Synagogues. "Synagogue Buildings in Kentucky," maps in Lee Shai Weissbach, *The Synagogues of Kentucky: Architecture and History* (Lexington: Univ. Press of Kentucky, 1995), p. 6.

PAGE 76

Black Hamlets in the Inner Bluegrass Region. "Negro Hamlets in Kentucky's Inner Bluegrass Region," map in Peter Smith and Karl B. Raitz, "Negro Hamlets and Agricultural Estates in Kentucky's Inner Bluegrass," *Geographical Review* 64, no. 2 (1974): 219.

Postbellum Inner Bluegrass Farm and Black Hamlet. Map, "The Contemporary Bluegrass Estate," in ibid., p. 227.

PAGE 77

European Ancestry. *U.S. Census, 1990, Social and Economic Characteristics*, table 137.
Irish. Ibid.
German. Ibid.
Scots-Irish. Ibid.
French. Ibid.
English. Ibid.

OTHER SOURCES

Axton, W.F. *Tobacco and Kentucky*. Lexington: Univ. Press of Kentucky, 1975.
Channing, Steven A. *Kentucky: A Bicentennial History*. New York: Norton, 1977.
Friis, Herman R. *A Series of Population Maps of the Colonies and the United States*. American Geographical Society Offset Publication no. 3. New York, 1968.
Hilliard, Sam B. *Atlas of Antebellum Southern Agriculture*. Baton Rouge: Louisiana State Univ. Press, 1984.
Kellogg, John. "The Formation of Black Residential Areas in Lexington, Kentucky, 1865-1887." *Journal of Southern History* 48, no. 1 (1982): 21-52.
Klee, John. "Tobacco Cultivation." Kleber, *Kentucky Encyclopedia*, pp. 884-87.
Turner, William H., and Edward J. Cabbell, eds. *Blacks in Appalachia*. Lexington: Univ. Press of Kentucky, 1985.
Ulack, Richard, Karl B. Raitz, and Hilary Lambert Hopper, eds. *Lexington and Kentucky's Inner Bluegrass Region*. Pathways in Geography Series no. 10. Indiana, Pa.: National Council for Geographic Information, 1994.
Williams, Michael. *Americans and Their Forests*. New York: Cambridge Univ. Press, 1989.

CHAPTER 4. POPULATION

PAGE 82

Kentucky's Population Growth. Karan and Mather, *Atlas of Kentucky*, p. 17; *U.S. Census, 1980*, vol. 1, chap. D, pt. 19, sec. 1, p. 11; *U.S. Census, 1990*, STF—3A (1992).
Population, 1930/1960/1990. *U.S. Census, 1930*, vol. 3, pt. 1; *U.S. Census, 1960*, vol. 1, pt. 19; *U.S. Census, 1990*, STF—3A (1992).

PAGE 83

Population Density, 1930/1960/1990. Same sources as above.
Change in Population Density. *U.S. Census, 1990, Population and Housing Unit Counts*, pp. 2, 124-25.

PAGE 84

Population Change [Kentucky]. Karan and Mather, *Atlas of Kentucky*, p. 17; *U.S. Census, 1980*, vol. 1, chap. D, pt. 19, sec. 1, p. 7; *U.S. Census, 1990*, STF—3A (1992).
Decennial Growth Rates. *U.S. Census, 1990, Population and Housing Unit Counts*, pp. 29-30.
United States Population Change. Ibid., p. 26.

PAGE 107 CONT.

Origin of College Students. Ibid.
Tuition at State-Supported Institutions. Ibid.
Funding Sources for Public Higher Education in Kentucky. Ibid.

PAGE 108

Private Postsecondary Institutions. *Accredited Institutions of Postsecondary Education,* 1991-92 (Washington, D.C., 1991-92).

PAGE 109

Median Household Income in Kentucky. *U.S. Census, 1960,* vol. 1, pt. 19, table 86; *U.S. Census, 1970,* vol. 1, pt. 19, table 124; *U.S. Census, 1980,* vol. 1, chap. C, pt. 19, table 180; *U.S. Census, 1990, Summary Social, Economic, and Housing Characteristics,* CPH-5-1, table 10.
Personal Income. KCED, *1995 Deskbook,* pp. 63-65.
Households with Annual Income below $10,000. *U.S. Census, 1990,* STF—3A (1992).
Households with Annual Income over $100,000. Ibid.

PAGE 110

Persons below Poverty Level. *U.S. Census, 1990,* STF—3A (1992).
Population Five Years Old or Younger below Poverty Level. Ibid.
Population Sixty-five or Older below Poverty Level. Ibid.
United States Poverty Thresholds. KCED, *1995 Deskbook.*
Population below Poverty Level [Kentucky and U.S.]. *U.S. Census, 1970,* vol. 1, chap. C, pt. 1, sec. 1, table 182; *U.S. Census, 1980,* vol. 1, chap. C, pt. 1, summary, tables 96 and 245; *U.S. Census, 1990, Summary Social, Economic, and Housing Characteristics,* CPH-5-1, table 5.

PAGE 111

Medicaid Eligibility. KCHR, *Medicaid Services in Kentucky,* MS-264, report series for fiscal year 1995 (Frankfort, 1996).
Households Receiving Supplemental Security Income. *U.S. Census, 1990,* STF—3A (1992).
AFDC Recipients. KCHR, *Public Assistance in Kentucky,* PA-264, report series for calendar year 1995 (Frankfort, 1996).
AFDC Payments. Ibid.
Food Stamp Recipients. KCHR, unpublished internal computer file, 1996.
Food Stamp Payments. Ibid.
Households with Retirement Income. *U.S. Census, 1990,* STF—3A (1992).
Households with Public Assistance Income. Ibid.

PAGE 112

Unemployment. KCHR, unpublished internal document, 1996.
Female Unemployment. *U.S. Census, 1990,* STF—3A (1992).
Male Unemployment. Ibid.
Labor Surplus. KCHR, unpublished computer data, 1994.

PAGE 113

Employment in Agriculture. *U.S. Census, 1990,* STF—3A (1992).
Employment in Mining. Ibid.
Employment in Government. Ibid.
Employment in Services. Ibid.
Persons Who Work at Home. Ibid.
Nonagricultural Employment. KCED, *1995 Deskbook.*

PAGE 114

Age of Housing. *U.S. Census, 1990,* STF—3A (1992).
Number of Rooms in Residential Structures. Ibid.
Housing Units in Kentucky. Census Office, *U.S. Census, 1900,* vol. 2, pt. 2, p. cvii; *U.S. Census, 1910,* vol. 1, p. 1287; *U.S. Census,* 1920, vol. 2, p. 1267; *U.S. Census, 1930, Population,* vol. 6, p. 55; *U.S. Census of Housing, 1940,* vol. 1, pt. 1, p. 15; *U.S. Census of Housing, 1950,* vol. 1, pt. 3, pp. 1723; *U.S. Census of Housing, 1960,* vol. 1, pt. 4, pp. 19-24; *U.S. Census of Housing, 1970,* vol. 1, pt. 19, pp. 19-28; *U.S. Census of Housing, 1980,* vol. 1, chap. C, pt. 19, p. 55; *U.S. Census, 1990,* STF—3A (1992).
Change in Number of Households. U.S. Bureau of the Census, *County and City Data Book, 1994.*
Dwellings with Public Sewer. *U.S. Census, 1990,* STF—3A (1992).
Dwellings with Public Water. Ibid.

PAGE 115

Sewage Disposal. Ibid.
Water Source. Ibid.
Dwellings with Public Electric or Gas Heat. Ibid.
Dwellings without a Complete Kitchen. Ibid.
Dwellings without Complete Plumbing. Ibid.
Kentucky Housing Quality Indicators. Ibid.

PAGE 116

Fuel Source by Area Development District. Ibid.
Change in Number of Occupied Dwellings, 1980-1990. *U.S. Census of Housing, 1980,* vol. 1, chap. C, pt. 19, p. 55; *U.S. Census, 1990,* STF—3A (1992).
Residential Vacancy Rate. *U.S. Census, 1990,* STF—3A (1992).

PAGE 117

Owner-Occupied Dwellings. Ibid.
Change in Number of Owner-Occupied Dwellings, 1980-1990. *U.S. Census of Housing, 1980,* vol. 1, chap. B, pt. 19, pp. 143-52; *U.S. Census, 1990,* STF—3A (1992).
Median Owner-Occupied Home Value. *U.S. Census of Housing, 1950,* vol. 1, pt. 3, tables 16 and 21; *U.S. Census of Housing, 1970,* vol. 1, pt. 19, table 5; *U.S. Census, 1990,* STF—3A (1992).
Value of Owned Residential Structures. *U.S. Census, 1990,* STF—3A (1992).
Monthly Owner Costs. Ibid.
Owner-Occupied Housing in Kentucky. *U.S. Census of Housing, 1940,* vol. 2, pt. 1, p. 7; *U.S. Census of Housing, 1950,* vol. 1, pt. 3, pp. 17-23; *U.S. Census of Housing, 1960,* vol. 1, pt. 4, pp. 19-25; *U.S. Census of Housing, 1970,* vol. 1, pt. 19, pp. 19-28; *U.S. Census of Housing, 1980,* vol. 1, chap. C, pt. 19, p. 55; *U.S. Census, 1990,* STF—3A (1992).
Monthly Gross Rent. *U.S. Census, 1990,* STF—3A (1992).
Monthly Gross Rental Costs. Ibid.

PAGE 118

Approval Rate for Home Purchase Loan Applications. Federal Reserve Board, Home Mortgage Disclosure Act data (accessed via Internet).
Mobile Homes in Kentucky. *U.S. Census of Housing, 1950,* vol. 1, pt. 3, pp. 17-26; *U.S. Census of Housing, 1960,* vol. 1, pt. 4, pp. 19-27; *U.S. Census of Housing, 1970,* vol. 1, pt. 19, pp. 19-28; *U.S. Census of Housing, 1980,* vol. 1, chap. C, pt. 19, p. 55; *U.S. Census, 1990,* STF—3A (1992).

Mobile Homes. *U.S. Census, 1990,* STF—3A (1992).
Change in Number of Mobile Homes, 1980-1990. *U.S. Census of Housing, 1980,* vol. 1, chap. C, pt. 19, p. 55; *U.S. Census, 1990,* STF—3A (1992).
Mobile Homes in the United States. *U.S. Census, 1990,* STF—3A (1992).

PAGE 119

Persons Living in Group Quarters in Kentucky, 1990 (Institutional and Noninstitutional). *U.S. Census, 1990,* STF—3A (1992).
Persons Living in Group Quarters. Ibid.
Military Areas. Information provided by David Altom, State Public Information Officer, Kentucky Dept. of Military Affairs, Public Affairs Office, Frankfort.

PAGE 120

Banks. KCED, *1996 Deskbook,* pp. 128-29.
Bank Deposits. Ibid.
Bank Deposits per Capita. Ibid.; 1994 population estimates from KCHR, 1996.

PAGE 121

Wholesale Sales. KCED, *1995 Deskbook,* pp. 108-13; 1992 population estimates from KCHR, 1996.
Retail Sales. Same sources as above.
Retail Sales by Type. KCED, *1995 Deskbook,* pp. 108-13.
Food Sales. Ibid.; 1992 population estimates from KCHR, 1996.
Service Industry Receipts. KCED, *1995 Deskbook,* pp. 116-21; 1992 population estimates from KCHR, 1996.
Service Industry Receipts by Type. KCED, *1995 Deskbook,* pp. 116-21.

PAGE 122

Local Tax Payments per Capita. U.S. Bureau of the Census, *County and City Data Book, 1994,* pp. 226, 240.
Federal Funding/Grants. Ibid.
Property Taxes. Ibid.
Local Government General Finances. Ibid.

PAGE 123

Hospital Admissions. KSCHS, *Hospital Inpatient Origin Report, 1994.*
Health Service Receipts. KCED, *1995 Deskbook,* pp. 116-21; 1992 population estimates from KCHR, 1996.
Health Care Payments per Family. *Skyrocketing Health Inflation, 1980-1993-2000: The Burden on Family and Businesses* (Washington, D.C.: Families USA Foundation, 1993).

PAGE 124

Number of Physicians. Kentucky Board of Medical Licensure, *Kentucky Medical Directory, 1993* (Louisville, 1993).
Availability of Physicians. Ibid.
Gender of Kentucky Physicians. Ibid.
Status of Kentucky Physicians. Ibid.
General Acute Care Hospitals. Health Care Investment Analysts, *Profiles of U.S. Hospitals* (Baltimore, 1994), pp. 258-71.
General Acute Care Hospital Beds. Ibid.
Origin of Kentucky Physicians' Medical Degrees. American Medical Association, *Directory of Physicians in the United States, 1994* (Chicago, 1994).

Timberland [Kentucky]. USFS, *Forest Statistics for Kentucky, 1975 and 1988*, Resource Bulletin NE-117.
Timberland Acreage [Kentucky]. Ibid. Compilation provided by Conrad T. Moore, Western Kentucky Univ.
Privately Owned Land within Daniel Boone National Forest. Compilation by Richard A. Gilbreath, Center for Cartography and Geographic Information, Univ. of Kentucky, map section from USFS, *Daniel Boone National Forest (South Half)*, 1984.

PAGE 149

Private Timberland. USFS, *Forest Resources of the United States, 1992*.
Ownership of Timberland. Ibid.; KDF, *1990 Sawn Lumber Production for Kentucky* (Frankfort, 1994).
Kentucky Sawn Lumber Volumes. KDF, *1990 Sawn Lumber Production for Kentucky*.

PAGE 150

Kentucky Sawn Lumber Production. Ibid.
Sawn Lumber. Ibid.
Primary Lumber Industries. KDF, *Primary Wood Industries of Kentucky: A Directory* (Frankfort, 1994).
Lumber and Wood Product Value. U.S. Bureau of the Census, *U.S. Census of Manufactures* (Washington, D.C.: GPO, 1977, 1982, 1987).

PAGE 151

White Oak. *The Eastwide Forest Inventory Data Base*, USFS computer file (St. Paul, Minn., 1992).
Red Oak. Ibid.
Hard Maple. Ibid.
Soft Maple. Ibid.
Ash. Ibid.
Poplar. Ibid.
Red Cedar. Ibid.

PAGE 152

Forest Fires. Kentucky Division of Forestry, Office of Fire Management, Frankfort, Forest Fire Data Base (unpublished).

CHAPTER 7: THE AGRICULTURAL LANDSCAPE

PAGE 156

Land Use Suitability. "Land Use Suitability," map in Karan and Mather, *Atlas of Kentucky*, pp. 112-13.

PAGE 157

Prime Farmland. Soil Conservation Service in cooperation with the Kentucky Agricultural Experiment Station and the Division of Conservation, Kentucky Dept. for Natural Resources and Environmental Protection, map, "Prime Farmland," Dec. 1979.

PAGE 158

Land In Farms, 1950. U.S. Bureau of the Census with USDA, *U.S. Census of Agriculture, 1978* (Washington, D.C., 1980), vol. 1, pt. 17, pp. 118-20.
Average Farm Size, 1950. Ibid.

PAGE 158 (right column)

Land In Farms, 1992. *U.S. Census of Agriculture, 1992* (Washington, D.C.: U.S. Bureau of the Census with USDA, 1994), vol. 1, pt. 17, pp. 162-77.
Average Farm Size, 1992. Ibid.
Land in Farms and Average Farm Size. *U.S. Census of Agriculture, 1978*, vol. 1, pt. 17, p. 1; *U.S. Census of Agriculture, 1992*, vol. 1, pt. 17, p. 8.

PAGE 159

Acres of Harvested Cropland. *U.S. Census of Agriculture, 1992*, vol. 1, pt. 17, p. 8.
Change in Harvested Cropland, 1950-1992. *U.S. Census of Agriculture, 1978*, vol. 1, pt. 17, p. 1; *U.S. Census of Agriculture, 1992*, vol. 1, pt. 17, p. 8.
Harvested Cropland. *U.S. Census of Agriculture, 1992*, vol. 1, pt. 17, pp. 162-77, 264-95.

PAGE 160

Principal Crops. Ibid., pp. 9-10.
Kentucky Agricultural Land Use. Ibid., p. 6.

PAGE 161

Tobacco Production. Ibid., pp. 162-77.
Dark Tobacco. Kentucky Agricultural Statistics Service in cooperation with Kentucky Dept. of Agriculture, Louisville, and USDA, *1991-1992 Kentucky Agricultural Statistics*, (1992), p. 28.

PAGE 162

Corn. *U.S. Census of Agriculture, 1992*, vol. 1, pt. 17, pp. 162-77.
Corn Harvested for Grain. Ibid.
Corn Harvested for Silage. Ibid.
Soybeans. Ibid.
Wheat. Ibid.
Corn, Soybean, and Wheat Harvests. Ibid., p. 9.

PAGE 163

Pasture. Ibid., pp. 280-95.
Hay. Ibid., pp. 162-77.

PAGE 164

Cattle and Calves Sold. Ibid.
Beef Cattle and Dairy Cows. Ibid.
Beef Cattle and Dairy Cows in Inventory. Ibid., p. 8.

PAGE 165

Hogs and Pigs in Inventory. Ibid.
Hogs and Pigs. Ibid., pp. 162-77.
Hogs and Pigs Sold. Ibid.
Broilers in Inventory. Ibid., p. 8.
Chickens. Ibid., pp. 162-77.
Broilers and Other Meat-Type Chickens Sold. Ibid.

PAGE 166

Registered Thoroughbred Foals. Jockey Club, *1995 Fact Book: A Guide to the Thoroughbred Industry in North America* (Lexington, Ky., 1995), p. 11.
Horses. *U.S. Census of Agriculture, 1992*, vol. 1, pt. 17, pp. 562-63.
Horses to Harvested Cropland. Ibid., pp. 162-77, 562-63.

PAGE 167

Market Value of All Agricultural Products Sold. Ibid., pp. 194-209.
Crop Yield. Ibid., p. 9.
Crop Value. Ibid., p. 8.
Value of All Agricultural Products Sold. Ibid.
Market Value of All Kentucky Agricultural Products Sold. Ibid., pp. 194-209.

PAGE 168

Dairy Products Sold. Ibid.
Market Value of Dairy Products Sold. Ibid.
Hogs and Pigs Sold. Ibid.
Market Value of Hogs and Pigs Sold. Ibid.
Cattle and Calves Sold. Ibid.
Market Value of Cattle and Calves Sold. Ibid.

PAGE 169

Tobacco Sold. Ibid.
Market Value of Tobacco Sold. Ibid.
Vegetables Sold. Ibid.

PAGE 170

Farm Earnings [U.S.]. U.S. Bureau of the Census, *County and City Data Book, 1994*, p. 11.
Government Payments Received. *U.S. Census of Agriculture, 1992*, vol. 1, pt. 17, pp. 226-41.
Net Cash Returns. Ibid.
Farm Income, 1992. U.S. Bureau of the Census, *Statistical Abstract of the United States: 1994*, p. 673.

PAGE 171

Total Farm Expenses. Ibid., pp. 210-25.
Hired Labor Expenditures. Ibid.
Feed Expenditures. Ibid.
Chemical Expenditures. Ibid.
Fertilizer Expenditures. Ibid.
Hispanic Migrant Farm Workers. Gil Rosenberg, Dept. of Rural Sociology, Univ. of Kentucky, compiled from a survey of county agents.
Farm Production Expenses. *U.S. Census of Agriculture, 1992*, vol. 1, pt. 17, p. 7.

PAGE 172

Tractors. Ibid., pp. 328-43.
New Farm Tractors. Ibid.
Value of Farmland and Buildings. *U.S. Census of Agriculture, 1978*, vol. 1, pt. 17, p. 1; *U.S. Census of Agriculture, 1992*, vol. 1, pt. 17, p. 8.
Value of Kentucky Farmland and Buildings [1945-1992]. *U.S. Census of Agriculture, 1992*, vol. 1, pt. 17, pp. 264-79.

PAGE 173

Farm Population [U.S.]. U.S. Bureau of the Census, *County and City Data Book, 1994*, p. 11.
Farming as a Principal Occupation. Ibid., pp. 352-67.
Part-Time Farmers. Ibid.
Kentucky's Part-Time Farmers [1959-1992]. Ibid., p. 8.

PAGE 174

Tenant Farms. Ibid., pp. 352-67.

Farmers Aged 65 and Older. Ibid., pp. 368-83.
Kentucky Farm Tenure, 1992. Ibid., p. 72.
Marijuana. "Fields of Grass" (article), "Task Force Eradication" map in *Lexington Herald-Leader*, Dec. 13, 1992, p. A1.

CHAPTER 8. MANUFACTURING

PAGE 178

Manufacturing Establishments. *1993 Kentucky Industrial Directory* (Twinsburg, Ohio: Harris Publishing, 1993), table 2, pp. 26-27.

PAGE 179

Employment in Manufacturing. KCED, "Average Monthly Workers in Manufacturing Industries Covered by Kentucky Unemployment Insurance by Industrial Division and County," calendar year 1993, unpublished.
Location Quotient: Manufacturing. Ibid.
Civilian Employment in Manufacturing. U.S. Bureau of the Census, *County and City Data Book, 1994*.
Kentucky's Manufacturing Employment, 1972-1993. Office of Employment and Unemployment Statistics, Bureau of Labor Statistics, U.S. Department of Labor, *BLS Report of Employment, Hours and Earnings, 1990-1993*, CD-ROM 1 (Washington, D.C.: GPO, 1993); table "Kentucky Manufacturing Employment: 1972-1993, by Industry Division" provided by KCED.

PAGE 180

Manufacturing Jobs as Share of Nonagricultural Employment. KCED, "Average Monthly Workers in Manufacturing Industries Covered by Kentucky Unemployment Insurance by Industrial Division and County," calendar year 1993, unpublished.
Change in Manufacturing Employment, 1984-1993. KCED, "Average Monthly Workers in Manufacturing Industries Covered by Kentucky Unemployment Insurance by Industrial Division and County," calendar years 1984, 1993, unpublished.
Employment Generated by KIDA-Approved Projects. KCED, Dept. of Financial Incentives, "Approved Incentive Projects by County," unpublished data on KIDA, KJDA, and KREDA projects, March 7, 1995.
Employment Generated by KJDA-Approved Projects. Ibid.
Employment Generated by KREDA-Approved Projects. Ibid.

PAGE 181

Kentucky's "Golden Triangle." Richard Ulack, Dept. of Geography, Univ. of Kentucky, 1995.
Kentucky's Twenty Largest Plants, 1994. KCED, *1994 Kentucky Directory of Manufacturers*, pp. x-xv.

PAGE 182

Average Weekly Wages of Workers in Manufacturing. KCED, "Average Weekly Wages of Workers Covered by Kentucky Unemployment Insurance Law by Industrial Division and County," calendar year 1993, unpublished.
Average Earnings in Manufacturing in Selected Cities, 1993. U.S. Dept. of Labor, "Average Hours and Earnings of Production Workers on Manufacturing Payrolls in States and Selected Areas," table in *Employment and Earnings* 41, no. 5 (May 1994): 162-65.

Average Weekly Earnings of Manufacturing Employees. Ibid.

PAGE 183

Number of Kentucky Employees by SIC Code, 1993. *1993 Kentucky Industrial Directory*, tables 2 and 3, pp. 26-27, 28-29.
Kentucky and U.S. Manufacturing Wages, 1992. KCED, *1994 Kentucky Directory of Manufacturers*, p. viii.
Manufacturing Firms with Unions. KCED, *Why Kentucky?* (New York: BPI Communications, Location Strategies Marketing Group, 1993), pp. 81-120; data available for each county under listing "% Firms with Unions (1992)."

PAGES 184-86

Manufacturing Employment by Type of Industry. *1993 Kentucky Industrial Directory*, tables 2 and 3, pp. 26-27, 28-29.

PAGE 187

Change in Employment in Kentucky by Selected Industries, 1972-1993. U.S. Dept. of Labor, *BLS Report of Employment, Hours and Earnings, 1990-1993*; table, "Kentucky Manufacturing Employment: 1972-1993, by Industry Division" provided by KCED.
Distilled Spirits: Value and Employment in the United States. U.S. Bureau of the Census, *1987 Census of Manufactures*, table 1a, p. 6; U.S. Bureau of the Census, *1991 Annual Survey of Manufactures* (Dec. 1992), table 2, pp. 10-11.
Employment in the Distilled Spirits Industry. *1993 Kentucky Industrial Directory*; U.S. Bureau of the Census, *1992 Census of Manufactures* (April 1995), table 2, p. 10.

PAGE 188

Employment in Foreign-Owned Plants. Job Development Dept., KCED, "Reported Foreign Industrial Investment in Kentucky (at least 10% Foreign Ownership), November 1994," unpublished.
Foreign Investment in Kentucky Manufacturing. Ibid.
Employment in Foreign-Owned Industries in Kentucky. Ibid.
Manufacturing Labor Force Employed in Foreign Companies [U.S.]. U.S. Bureau of the Census, *Statistical Abstract: 1993*, p. 799.

PAGE 189

Foreign-Owned Producers of Automotive Parts. Toyota Motor Manufacturing, U.S.A., "Production Materials—November 1992," unpublished press packet.
Employment in Foreign-Owned Automotive Parts Plants Opened since 1985. KCED, Research Division, "Automotive Parts Plants Announced in Kentucky since Announcement of Toyota," unpublished, dated June 1993; Job Development Dept., KCED, "Reported Foreign Industrial Investment in Kentucky (at least 10% Foreign Ownership), November 1994," unpublished.
Employment in All Foreign-Owned Industries in Kentucky. Same sources as above.

PAGE 190

Kentucky Exports by Industry. KCED, Research Division, "Kentucky Exports," unpublished, dated Aug. 1994.
Kentucky Exports, 1989-1993. Ibid.

Kentucky Export Destinations. U.S. Bureau of the Census, "1993 Trade Inflo," unpublished; table, "Exports Originating in Kentucky by Country of Destination, January-December, 1993, Ranked by Value of Exports, Total for All SIC Numbers," provided by KCED.

OTHER SOURCES

Kleber, *Kentucky Encyclopedia*.
Rubenstein, James M., *The Changing U.S. Auto Industry: A Geographical Analysis* (New York: Routledge, 1992), pp. 171-72.

CHAPTER 9. TRANSPORTATION AND COMMUNICATIONS

PAGE 194

Navigable Waterways. KCED, "Kentucky's Commercially Navigable Waterways," *Kentucky's Transportation Resources* (Frankfort, 1994).

PAGE 195

Ohio River Transportation Corridor. U.S. Army Corps of Engineers, *1994 Ohio River Navigation System Report* (Cincinnati, 1995).
Ohio River Freight Traffic, 1920-1990. U.S. Army Corps of Engineers, *Commerce on the Ohio River and Its Tributaries, 1993 Statistical Supplement* (Cincinnati, 1994).
Ohio River Waterway Commodity Distribution. U.S. Army Corps of Engineers, *1994 Ohio River Navigation System Report*.

PAGE 196

The Antebellum Rail System in the East Central United States. Designed after "Growth of Railroads in the United States before the Civil War," map in John F. Stover, *American Railroads* (Chicago: Univ. of Chicago Press, 1961), pp. 48-49.

PAGE 197

Railroads Serving Kentucky in 1938. "Kentucky Transportation Map, 1938," Map Dept., M.I. King Library, Univ. of Kentucky.
Recent Change in Kentucky's Rail Network. Kentucky Dept. of Transportation, Railway-Highway Section, "Railroads Serving Kentucky," maps, revised 1980, 1993.

PAGE 198

Evolution of Paved Roads. Kentucky Highway Dept., *Map of Kentucky Showing the Conditions of State Roads, Jan. 1, 1927*; Kentucky Highway Commission, *Road Condition Map of Kentucky, Oct. 1, 1935*; Kentucky Dept. of Highways, *Official Kentucky Road Map, Jan. 1, 1949*; Rand McNally, *Kentucky Detailed Highway Map* (Chicago, 1963).
Change in Kentucky's Road Surface. U.S. Bureau of the Census, *Statistical Abstract*, various issues, 1920-94.

PAGE 199

Interstate Highway System. U.S. Bureau of the Census, TIGER/line census files, 1992, Kentucky (computer file).

PAGE 200

Traffic Flows. Dept. of Highways, KTC, *Traffic Flow Map of Kentucky*, 1992.

PAGE 200 CONT.

Road Density. Division of Planning—Highway Systems Section, KTC, "State Road System Report," unpublished, 1994; Kentucky Transportation Center, Univ. of Kentucky, *A Report on County Road Program Finance, 1992-93* (Lexington, 1993).

Highway Coal Traffic. KTC, *Kentucky's Coal Haul Highway System* (Frankfort, 1993).

Intercity Bus Routes, 1975 and 1994. Developed from Karan and Mather, *Atlas of Kentucky;* Greyhound Lines, tables 1-74 (Dallas, 1994), pp. 36-37.

PAGE 201

Change in Vehicle Registration, 1982-1991. KTC, *Kentucky Motor Vehicle Registration Statistical Report,* 1982 and 1991 issues (Frankfort, 1983, 1992).

Households with No Vehicle. *U.S. Census, 1990,* STF—3A (1992).

Traffic Accidents. Kentucky Transportation Center, Univ. of Kentucky, *Analysis of Traffic Accident Data in Kentucky* (Lexington, 1993).

Vehicles Registered. U.S. Bureau of the Census, *Statistical Abstract,* various issues, 1920-94.

Mean Travel Time to Work. *U.S. Census, 1990,* STF—3A (1992).

People Who Work Outside County of Residence. Ibid.

Carpooling to Work. Ibid.

PAGE 202

Public Use Airports. G.R. Brandy and Associates, *Kentucky Aviation System Plan: Executive Summary, 1990* (Frankfort, 1990).

Air Passenger Traffic. Federal Aviation Administration, Airport Activity Statistics, 1975, 1980, 1985, 1989, 1992.

Air Freight Traffic. Same sources as above.

Direct Connection Commuter Flights from Kentucky Airports. Flight schedules from Blue Grass Airport, 1994; Cincinnati/Northern Kentucky International Airport, 1994; Louisville International Airport at Standiford Field, 1994.

PAGE 203

Scheduled Flights from Louisville International Airport at Standiford Field, 1978 (Before Deregulation). Donnelley Corp., *Official Airline Guide,* North American ed., Sept. 1978.

Scheduled Flights from Louisville International Airport at Standiford Field, 1994 (After Deregulation). Flight schedule from Standiford Field, 1994.

Scheduled Flights from Cincinnati/Northern Kentucky International Airport, 1978 (Before Deregulation). Donnelley Corp., *Official Airline Guide,* North American ed., Sept. 1978.

Scheduled Flights from Cincinnati/Northern Kentucky International Airport, 1994 (After Deregulation). Flight schedule from Cincinnati/Northern Kentucky International Airport, 1994.

PAGE 204

International Direct Flights from Cincinnati/Northern Kentucky International Airport. Ibid.

PAGE 205

Kentucky Educational Television (KET) Service Areas. "KET: The Kentucky Network," *Kentucky Journal of Politics and Issues* 1 (April 1993): 9.

Dominant Television Market Areas. "Arbitron ADI Market Atlas," in *Broadcasting and Cable Yearbook, 1994,* vol. 1 (New Providence, N.J.: Bowker, 1994), sec. C, pp. 132-71.

Television Stations. "Directory of Television Stations in the U.S." in ibid., pp. C29-30.

PAGE 206

FM Radio Stations. "Directory of Radio Stations in the U.S." in ibid., sec. B, pp. 145-56.

AM Radio Stations. Ibid.

PAGE 207

Weekly Circulation Comparison of Kentucky's Two Leading Newspapers. Standard Rate and Data Service, *Circulation 94* (Wilmette, Ill., 1994).

Newspaper with the Largest Market Share in Each County. Ibid.

Weekly Circulation Comparison of Kentucky's Other Regional Newspapers. Ibid.

United States Post Offices. U.S. Postal Service, *State List of Post Offices, Kentucky* (Washington, D.C.: GPO, 1994).

PAGE 208

Subscribers to TV Guide. Standard Rate and Data Service, *Circulation 94* (Wilmette, Ill., 1994).

Subscribers to Better Homes and Gardens. Ibid.

Subscribers to Time Magazine. Ibid.

Subscribers to Ebony Magazine. Ibid.

Public Library Books. Kentucky Dept. of Libraries and Archives, *Statistical Report of Kentucky Public Libraries, 1993* (Frankfort, 1994).

PAGE 209

Information Highways. Appalachian Regional Commission, *Telecommunications in Appalachia* (Washington, D.C., 1991).

Status of 911 Service. Kentucky State Police, Emergency Medical Services Branch, unpublished data, 1994.

PAGE 210

Service Areas of Telephone Companies. Kentucky Telephone Association, *Kentucky* (Frankfort, 1992).

Cellular Telecommunication Markets. Information provided by Kentucky Public Service Commission (Frankfort, 1996).

Households with No Telephone. *U.S. Census, 1990,* STF—3A (1992).

OTHER SOURCE

KTC, *Kentucky Transportation: A 200-Year Calendar of History, 1792 to 1992* (Frankfort: TRANSPO, 1992).

CHAPTER 10: TOURISM AND RECREATION

PAGE 214

Travel and Tourism Regions. KDT, *Economic Impact of Kentucky's Tourism and Travel Industry, 1994 and 1995* (Frankfort, 1995).

PAGE 215

Regional Employment in Travel and Tourism, 1984 and 1994. Ibid.; KDT, *Economic Impact of Kentucky's Tourism and Travel Industry, 1984 and 1985* (Frankfort, 1986).

Location Quotient: Tourism. KDT, *Economic Impact of Kentucky's Tourism and Travel Industry, 1994 and 1995.*

PAGE 216

Regional Travel Expenditures on Travel and Tourism, 1984 and 1994. KDT, *Economic Impact of Kentucky's Tourism and Travel Industry, 1984 and 1985;* ibid., *1994 and 1995.*

Travel Expenditures Per Capita. KDT, *Economic Impact of Kentucky's Tourism and Travel Industry, 1994 and 1995.*

PAGE 217

Top Twenty-Five Attractions. KDT, "Kentucky's Top Twenty-Five Attractions in 1994," unpublished.

Hotel, Motel, Resort, and Bed and Breakfast Rooms. KDT, *Economic Impact of Kentucky's Tourism and Travel Industry, 1994 and 1995.*

PAGE 218

State Parks and State Resort Parks. Kentucky Dept. of Parks, *Kentucky State Parks: The Nation's Finest* (Frankfort, 1994).

Home of Overnight Guests Visiting Kentucky's State Parks. Kentucky Dept. of Parks, "Statistical Information, Calendar Year 1993," unpublished.

PAGE 219

Federal Recreation Lands. KDT, *1994 Kentucky Travel Guide* (Louisville: Editorial Services, 1994); KDT, *Kentucky: Official Vacation Guide* (Frankfort, 1994); USGS map, *An Invitation to Enjoyment: Federal Recreation Lands of the United States* (Washington, D.C.: USGS, 1992).

PAGE 220

State Recreation Lands. KDT, *1994 Kentucky Travel Guide;* KDT, *Kentucky: Official Vacation Guide;* KDT, *Kentucky: A Traveller's Guide* (Frankfort, 1993); KDFWR, *A Guide to Public Wildlife Areas in Kentucky* (Frankfort, 1993); KSNPC, "Directory of Kentucky State Nature Preserves," unpublished, dated Jan. 1994; KSNPC, "Kentucky's State Nature Preserve System," unpublished map and pamphlet, n.d.

PAGE 221

Campsites. KDT, *Profile of Kentucky's Tourism and Travel Businesses: 1993* (Frankfort, 1993), pp. 81-88.

Historic Places. KDT, *1994 Kentucky Travel Guide;* KDT, *Kentucky: Official Vacation Guide;* KDT, *Kentucky: A Traveller's Guide.*

PAGE 222

Watchable Wildlife Areas. Same sources as above.

PAGE 223

Festivals and Shows. Louisville Convention and Visitors Bureau, *Louisville Visitor's Guide,* n.d.; publications of KDT: *1994 Kentucky Travel Guide; Kentucky: Official Vacation Guide; Kentucky: A Traveller's Guide; Kentucky Calendar of Events* (Frankfort, 1993); *The Complete Blue Grass from A to Z* (Frankfort, n.d.).

PAGE 224

Museums, Zoos, Planetariums, and Amusement Parks. Same sources as above.

PAGE 225

Horse-Related Attractions. Ibid.

PAGE 226

Change in the Number of Boat Registrations Issued. KDFWR, "Annual Boat Registrations, 1976-1993," unpublished.

Marinas and Boating Waters. KDT, "Marinas in Kentucky, 1993," unpublished; Kentucky Division of Water and the Rivers, Trails, and Conservation Assistance Program of the National Park Service, *Kentucky Rivers Assessment* (Atlanta: General Services Administration, 1992), esp. pp. 99-110.

PAGE 227

Fishing Waters. Prepared with the assistance of Lew Kornman, KDFWR, Northeastern Fishery District, Morehead; Brooks M. Burr and Melvin L. Warren Jr., *A Distributional Atlas of Kentucky Fishes,* Scientific and Technical Series no. 4 (Frankfort: KSNPC, 1986); KDFWR, *Kentucky Fish* (Frankfort, 1993).

PAGE 228

Total Number of Hunting and Fishing Licenses Sold. KDFWR, "Number and Type of Fishing and Hunting Licenses, 1971-1992," unpublished.

Types of Hunting and Fishing Licenses Sold. Ibid.; KDFWR, "Hunting and Fishing Licenses Sold, 1992-93," unpublished.

Hunting and Fishing Licenses Sold. KDFWR, "Hunting and Fishing Licenses Sold, 1992-93," unpublished.

PAGE 229

Deer Harvest. KDFWR, "Kentucky Deer Harvest, 1976-1992," by Roland L. Burns, unpublished.

Golf Courses. Kentucky Golf Association and Professional Golf Association, Louisville, "Kentucky Golf Association Member Clubs—1994" and "PGA of America Golf Facility Listing."

PAGE 230

Intercollegiate Sports. Charles T. Straughn II and Barbarasue Lovejoy Straughn, eds., *Lovejoy's College Guide,* 22d ed. (New York: Simon & Schuster/Macmillan, 1993), pp. 1101-36; *Barron's Profiles of American Colleges,* 19th ed. (Hauppauge, N.Y.: Barron's Educational Series, 1992), pp. 645-65.

Holders of Season Tickets for University of Kentucky Men's Basketball Season, 1993-94. Univ. of Kentucky Athletic Dept.

Holders of Season Tickets for University of Louisville Men's Basketball Season, 1993-94. Univ. of Louisville Athletic Dept.

PAGE 231

Sweet Sixteen High School Basketball Champions. Kentucky High School Athletic Association, table, "KHSAA State Tournament Finals," unpublished, 1995.

PAGE 232

Wagering at Kentucky's Horse Racing Tracks. Kentucky Racing Commission, *Forty-fourth Biennial Report* (Lexington, 1992-93).

Lottery Ticket Sales per Capita (4/4/89–12/29/93). Kentucky Lottery Corporation, unpublished data by county on sales and payouts of Kentucky lottery products.

Lottery Ticket Payouts Per Capita (4/4/89–10/30/93). Ibid.

OTHER SOURCE

Kleber, *Kentucky Encyclopedia.*

CHAPTER 11: THE POLITICAL LANDSCAPE

PAGE 236

United States Congressional Districts, 1972. Richard M. Scammon, and Alice V. McGillivray, *America Votes* (Washington, D.C.: Elections Research Center, Congressional Quarterly, 1981), p. 166.

United States Congressional Districts, 1991. Ibid. (1993), p. 206.

PAGE 237

Kentucky House of Representatives Districts, 1992. Map in Mary McKay Write, comp., *State Directory of Kentucky 1993* (Pewee Valley, Ky., 1993), p. 236.

Kentucky Senate Districts, 1992. Ibid., p. 235.

PAGE 238

Judicial Circuits, 1991. Ibid., p. 238.

Supreme Court Districts, 1991. Ibid., p. 239.

Kentucky Court System. Adapted from chart in "Justice in Our Commonwealth," compiled by Administrative Office of the Courts' Media and Public Information Office (Frankfort, 1996), p. 37.

PAGE 239

Voter Turnout, 1992 Presidential Election. Data provided by Roger A. Hunt, Dept. of Geography and Planning, Grand Valley State Univ., Allendale, Mich.

Voter Registration. Scammon and McGillivray, *America Votes* (1993), pp. 227-28.

Voter Turnout, May 26, 1992, Primary Election. Kentucky Secretary of State, *Primary and General Election Returns, 1992* (Frankfort, 1992), pp. 1-4.

Voter Turnout, November 3, 1992, General Election. Ibid.

PAGE 240

Presidential Election Scorecard. Scammon and McGillivray, *America Votes,* various vols., 1970-1993.

Presidential Election Trends, 1920-1996. Ibid., various vols., 1960-97.

Presidential Election Popular Vote, 1920-1996. Ibid.

PAGE 241

1968 Presidential Election. Ibid. (1970), pp. 140-41.

1972 Presidential Election. Ibid. (1973), pp. 146-47.

1976 Presidential Election. Ibid. (1977), pp. 144-45.

1980 Presidential Election. Ibid. (1981), pp. 167-68.

PAGE 242

1984 Presidential Election. Ibid. (1985), pp. 192-93.

1988 Presidential Election. Ibid. (1989), pp. 190-91.

1992 Presidential Election. Ibid. (1993), pp. 227-28.

1996 Presidential Election. *Lexington Herald-Leader,* Nov. 6, 1996, p. A3.

Vote for Independent Candidate Perot, 1992 Presidential Election. Scammon and McGillivray, *America Votes* (1993), pp. 227-28.

Vote for Independent Candidate Perot, 1996 Presidential Election. *Lexington Herald-Leader,* Nov. 6, 1996, p. A3.

PAGE 243

Vote for Republican Candidate Morton, 1962 Senatorial Election. Scammon and McGillivray, *America Votes* (1966), pp. 150-51.

Vote for Republican Candidate Cooper, 1966 Senatorial Election. Ibid. (1968), pp. 143-44.

PAGE 244

1990 Senatorial Election. Kentucky Secretary of State, *Official Primary and General Election Returns, 1990* (Frankfort, 1990), pp. 1-4.

1992 Senatorial Election. Kentucky Secretary of State, *Official Primary and General Election Returns, 1992* (Frankfort, 1992), pp. 1-3.

1996 Senatorial Election. Data provided by Carla Arnold, Kentucky State Board of Elections, Frankfort, Dec. 1996.

PAGE 245

1963 Gubernatorial Election. Scammon and McGillivray, *America Votes* (1968), pp. 141-42.

1967 Gubernatorial Election. Ibid. (1970), pp. 142-43.

1983 Gubernatorial Election. Ibid. (1985), pp. 194-95.

1987 Gubernatorial Election. Ibid. (1989), pp. 192-93.

1991 Gubernatorial Election. Ibid. (1993), pp. 227-28.

PAGE 246

1995 Gubernatorial Election. *Louisville Courier-Journal,* Nov. 8, 1995.

Birthplace of Kentucky's Governors. Robert Sobel and John Raimo, eds., *Biographical Directory of the Governors of the United States, 1789-1978* (Westport, Conn.: Meckler Books, 1978), vol. 2, *Iowa-Missouri;* John W. Raimo, ed., *Biographical Directory of the Governors of the United States, 1978-1983* (Westport, Conn.: Meckler Publishing, 1985); Lowell Harrison, ed., *Kentucky's Governors, 1792-1985* (Lexington: Univ. Press of Kentucky, 1985); Kleber, *Kentucky Encyclopedia;* Michael Barone and Grant Ujifusa with Richard Cohen, eds., *The Almanac of American Politics, 1996* (Washington, D.C.: National Journal, 1995); *State Yellow Book* 8, no. 1 (spring 1996): 160 (New York: Leadership Directories, 1996); *Who's Who in American Politics, 1995-96,* 15th ed., vol. 1, *Alabama-Montana,* pp. 1005-49 (New Providence, N.J.: R.R. Bowker, 1995).

PAGE 247

Constitutional Reform Amendment, 1966. Kentucky Secretary of State, *General Election Amendment Returns Report* (Frankfort, 1966), p. 484.

Judicial System Reform Amendment, 1975. Ibid. (1975), p. 491.

Legislative Reform Amendment, 1979. Ibid. (1979), n.p.

PAGE 248

State Lottery Amendment, 1988. Ibid. (1988), pp. 1-6.

Broad Form Deed Amendment, 1988. Ibid.

Elective Office Amendment, 1992. Ibid. (1993), pp. 1-3.

Charitable Lottery Amendment, 1992. Ibid.

CHAPTER 12: THE URBAN LANDSCAPE

PAGE 252

Metropolitan Area Populations. KCED, *1995 Deskbook*, p. 136.
Urban Population. Ibid., pp. 57-58.
Metropolitan Population. U.S. Bureau of the Census, *Statistical Abstract: 1994*, table 42, p. 38.
Urbanization, 1790-1990. U.S. data: Donald B. Dodd, comp., *Historical Statistics of the United States: Two Centuries of the Census, 1790-1990* (Westport, Conn.: Greenwood Press, 1993); Kentucky data: KCED, *1996 Deskbook*, table, p. 5.

PAGE 253

Kentucky's Metropolitan Areas. KCED, *1995 Deskbook*, p. 135.

PAGE 254

Classification of Incorporated Cities. Ibid., pp. 143-47.

PAGE 256

Urban Hierarchy Ordering Scheme. Compiled by Richard Ulack, Dept. of Geography, Univ. of Kentucky, 1996.
Urban Hierarchy, 1800. *Second Census of the United States, 1800*, ser. 2 (New York: Norman Ross Publishing, 1990), p. 2P.
Urban Hierarchy, 1840. Dept. of State, *Compendium of the Enumeration of the Inhabitants of the United States, 1840* (Washington, D.C., 1841), pp. 72-73.
Urban Hierarchy, 1880. Census Office, *Statistics of the Population of the United States at the Tenth Census, 1880* (Washington, D.C.: GPO, 1883), pp. 194-95.

PAGE 257

Urban Hierarchy, 1920. *U.S. Census, 1920, Population*, pp. 222-25.
Urban Hierarchy, 1960. *U.S. Census, 1960, Number of Inhabitants*, PC(1)-19A, pp. 23-24.
Urban Hierarchy, 1990. *U.S. Census, 1990: General Population Characteristics*, 1990 CP-1-19, pp. 2-5.

PAGE 261

Urban Expansion. "City of Lexington and Suburbs, Fayette County, Kentucky, 1950" compiled by B.L. Goldstein (Lexington: Chamber of Commerce); *U.S. Census, 1950*, vol. 2, pt. 17, p. 19; pt. 35, pp. 36-37; *U.S. Census, 1960*, vol. 1, pt. 19, p. 20; pt. 37, p. 39; *U.S. Census, 1970*, vol. 1, pt. 19, pp. 34-35; pt. 37, Ohio—Section 1, p. 62; *U.S. Census, 1980*, vol. 1, chap. A (1982), pt. 19, pp. 41, 44-45; *U.S. Census, 1990, Population and Housing Unit Counts*, pp. 12-13, 17.

PAGES 262-63

Kentucky's Major Metropolitan Areas. Census tract boundaries extracted from U.S. Bureau of the Census, TIGER/line 1992, Ohio (1) Adams-Ross computer file (CD-ROM), Ohio (2) Sandusky-Wyandot, Kentucky computer file (CD-ROM), and Illinois (2) St. Clair-Woodford, Indiana, computer file (CD-ROM) (Washington, D.C., 1993); census tract data from *U.S. Census, 1990*, STF—3A (1992).

PAGES 264-65

Urban Population. Same sources as above.

PAGES 266-67

Age of Housing. Ibid.

PAGES 268-69

Median House Value. Ibid.

PAGE 270

Selected Socioeconomic Characteristics. U.S. Bureau of the Census, *County and City Data Book, 1994;* vital data source: U.S. Bureau of the Census, *Statistical Abstract: 1995; U.S. Census, 1990*, Census Tracts and Block Numbering Areas, Louisville, KY-IN MSA, and Cincinnati, OH-KY-IN PMSA (Washington, D.C.: GPO).

PAGE 271

Black Population. Census tract boundaries extracted from U.S. Bureau of the Census, Data User Services Division, TIGER/line 1992, Ohio (1) Adams-Ross computer file (CD-ROM), Ohio (2) Sandusky-Wyandot, Kentucky computer file (CD-ROM), and Illinois (2) St. Clair-Woodford, Indiana, computer file (CD-ROM) (Washington, D.C., 1993); census tract data from *U.S. Census, 1990*, STF—3A (1992).

PAGE 272

Elderly Population. Same sources as above.

PAGE 273

Median Household Income. Ibid.

PAGE 274

College Graduates. Ibid.

PAGE 275

Louisville Commuter Shed, 1990. Kentucky State Data Center, Univ. of Louisville, *1990 Commuting Patterns, Kentucky* (Louisville, 1993).
Increase in the Number of Workers Commuting to Louisville, 1980-1990. Ibid.; *Commuting in Kentucky*, compiled by Kentucky State Data Center, Univ. of Louisville, for Kentucky Dept. of Economic Development (Frankfort, 1984).
Cross-State Commuting. Kentucky State Data Center, *1990 Commuting Patterns, Kentucky.*
Cincinnati Commuter Shed, 1990. Ibid.
Increase in the Number of Workers Commuting to Cincinnati, 1980-1990. Ibid.; Kentucky State Data Center, Commuting in Kentucky.

PAGE 276

Lexington Commuter Shed, 1990. Kentucky State Data Center, 1990 Commuting Patterns.
Increase in the Number of Workers Commuting to Lexington, 1980-1990. Ibid.; Kentucky State Data Center, Commuting in Kentucky.

CHAPTER 13: KENTUCKY IN THE FUTURE

PAGE 280

Change in Total Population, 1990-2020. Kentucky State Data Center, *Newsletter* 13, no. 1/2 (1995): 3-14.

Dominant Age Group Influence on Change, 1990-2020. Ibid.
Baseline Expenditure Projection by Function. Peter Schirmer, Michael T. Childress, and Charles C. Nett, *$5.8 Billion and Change: An Exploration of the Long-Term Budgetary Impact of Trends Affecting the Commonwealth* (Frankfort: LTPRC, 1996), p. xvi.

PAGE 281

Change in Population, Age 0 to 17, 1990-2020. Kentucky State Data Center, *Newsletter* 13, no. 1/2 (1995): 3-14.
Change in Population, Age 18 to 64, 1990-2020. Ibid.
Components of Population Change. Ibid.

PAGE 282

Kentucky's Policy Regions. Peter B. Meyer and Thomas S. Lyons, *Forecasting Kentucky's Environmental Futures* (Louisville: Kentucky Institute for the Environment and Sustainable Development, 1996), p. 30.
Estimated Gross State Product Gains, 1995-2025. Ibid., p. 49.
Counties at Risk from a Significant Tobacco Quota Decline. Information provided by KLTPRC.

PAGE 283

Energy Demand Changes, 1995-2025. Meyer and Lyons, *Forecasting Kentucky's Environmental Futures*, p. 68.

PAGE 284

National Labor Force Structure and Economic Development. Adapted from graph "Labor Force Structure and Economic Development" in *Human Geography in a Shrinking World*, ed. Ronald Abler et al. (North Scituate, Mass.: Duxbury Press, 1975), p. 49.
Adults Who Have Accessed the Internet. Information provided by KLTPRC and Univ. of Kentucky Survey Research Center, Lexington.
Change in Population, Age 65 and Older, 1990-2020. Kentucky State Data Center, *Newsletter* 13, no. 1/2 (1995): 3-14.

PAGE 285

Consolidated Counties. Gerald R. Webster, "The Spatial Reorganization of County Boundaries in Kentucky," *Southeastern Geographer* 24, no. 1 (May 1984): 14-29.

PAGE 286

Change in Annual Generation of Municipal Solid Waste, 1995-2025. Meyer and Lyons, *Forecasting Kentucky's Environmental Futures*, p. 73.
Change in Greenhouse Gas Emissions, 1995-2025. Ibid, p. 10.
Change in Ground-Level Emissions, 1995-2025. Ibid.

OTHER SOURCES

Michal Smith-Mello and Peter Schirmer. *The Context of Change: Trends, Innovations, and Forces Affecting Kentucky's Future* (Frankfort: LTPRC, 1994).

Ulack, Richard, Karl B. Raitz, and Hilary Lambert Hopper, eds. *Lexington and Kentucky's Inner Bluegrass Region*. Pathways in Geography Series no. 10. Indiana, Pa.: National Council for Geographic Information, 1994.

Sources of Illustrations

PHOTO CREDITS

Endpaper
Landsat TM data courtesy of Earth Observation Satellite Co., Lanham, Md., image enhancement by Mid-America Remote Sensing Center, Murray State Univ.

Frontispiece
Keith Mountain

Chapter 1. Kentucky: Its Setting
Page 1: Keith Mountain
Page 3: Keith Mountain
Page 4 top: Keith Mountain; bottom: Karl Raitz
Page 5: Keith Mountain
Page 8: flag and goldenrod: Kentucky Dept. of Travel; cardinal: Tom Barnes; Capitol: Richard Ulack
Page 12: Karl Raitz
Page 14: Karl Raitz

Chapter 2. The Natural Environment
Page 15: Karl Raitz
Page 17: Keith Mountain
Page 19: Landsat TM data courtesy of Earth Observation Satellite Co., Lanham, Md., image enhancement by Mid-America Remote Sensing Center, Murray State Univ.
Page 20: Keith Mountain
Page 21: Keith Mountain
Page 22: Keith Mountain
Page 23: Keith Mountain
Page 24: National Park Service
Page 28: Tom Barnes
Page 29: Keith Mountain
Page 30: Tom Barnes
Page 32: Keith Mountain
Page 34: Keith Mountain
Page 38 top: Keith Mountain; bottom: Landsat TM data courtesy of Earth Observation Satellite Co., Lanham, Md., image enhancement by Mid-America Remote Sensing Center, Murray State Univ.

Page 39: Keith Mountain
Page 41: Keith Mountain
Page 42 both photos: Tom Barnes
Pages 44-45 all photos by John MacGregor with the following exceptions: coyote: Roger W. Barbour; wild turkey: Lew Kornman; red-cockaded woodpecker: Jerome A. Jackson; bald eagle: Tom Barnes; little-wing pearlymussel: D. Biggins, U.S. Fish and Wildlife Service, courtesy of Brainard L. Palmer-Ball; Virginia spiraea: Julian Campbell; Short's goldenrod: Tom Barnes
Page 46 all photos: Tom Barnes

Chapter 3. Historical and Cultural Landscapes
Page 47: Keith Mountain
Page 49: Nancy O'Malley, Dept. of Anthropology, Office of Archaeological Research, Univ. of Kentucky
Page 50: Mary Powell, Museum of Anthropology, Univ. of Kentucky
Page 51: Jimmy A. Railey, by permission of the Kentucky Heritage Council, State Historic Preservation Office, Frankfort
Page 55: Gyula Pauer
Page 59 inset: Special Collections, M.I. King Library, Univ. of Kentucky
Page 60: Karl Raitz
Page 61: Karl Raitz
Page 62 top right: Keith Mountain
Page 67 both photos: Karl Raitz
Page 68: Kentucky Department of Travel
Page 69: Karl Raitz
Page 70: Special Collections, M.I. King Library, Univ. of Kentucky
Page 71: Karl Raitz
Page 73: Karl Raitz
Page 74 all photos: Karl Raitz
Page 75: Gyula Pauer

Chapter 4. Population
Page 79: Keith Mountain
Page 81: Keith Mountain
Page 82: Keith Mountain
Page 85: Karl Raitz
Page 89: Karl Raitz
Page 95: Karl Raitz
Page 96: Karl Raitz

Chapter 5. Social and Economic Characteristics
Page 101: Keith Mountain
Page 103: Keith Mountain
Page 104: Karl Raitz
Page 105: Karl Raitz
Page 106: Karl Raitz
Page 107: Keith Mountain
Page 108: Karl Raitz
Page 109: Karl Raitz
Page 112: Karl Raitz
Page 114: Karl Raitz
Page 116 all photos: Karl Raitz
Page 118: Keith Mountain
Page 120: Karl Raitz
Page 121: Karl Raitz
Page 123: Karl Raitz
Page 124: Karl Raitz
Page 126: Karl Raitz
Page 128: Keith Mountain
Page 130: Richard Ulack

Chapter 6. Mineral, Energy, and Timber Resources
All photos by Keith Mountain with the following exceptions:
Page 150: Porter Jarrard
Page 152: Janet Worne, *Lexington Herald-Leader,* Oct. 30,1991, p. A1

Chapter 7. Agriculture
All photos by Keith Mountain with the following exceptions:
Page 161 left: Karl Raitz; right: Richard Ulack
Page 163 bottom: Karl Raitz
Page 168 both photos: Karl Raitz
Page 169 left: Karl Raitz; center: Greg Perry, *Lexington Herald-Leader,* May 17, 1997
Page 172: Charles Bertram, *Lexington Herald-Leader,* Oct. 20, 1996, p. A1
Page 174: Frank Anderson, *Lexington Herald-Leader,* July 15, 1993

Chapter 8. Manufacturing
Page 175: Keith Mountain
Page 177: Keith Mountain

The Editors and Principal Contributors

RICHARD ULACK is professor and former chair of the Department of Geography at the University of Kentucky and the former Kentucky State Geographer. His research specialties include regional development, population and migration, urbanization, and tourism development. He has conducted extensive field research in the Appalachian region, the Philippines, and Fiji. He is co-editor (with Karl B. Raitz and Hilary Lambert Hopper) of *Lexington and Kentucky's Inner Bluegrass Region*, author of the *Atlas of Southeast Asia*, co-author (with Karl B. Raitz) of *Appalachia: A Regional Geography: Land, People, and Development*, and author or editor of many other publications.

KARL RAITZ is professor and chair of the Department of Geography at the University of Kentucky. His research interests include cultural and historical geography with a focus on interpretation of American landscapes. His numerous publications include *Appalachia: A Regional Geography: Land, People, and Development* (co-authored with Richard Ulack), *Rock Fences of the Bluegrass* (co-authored with Carolyn Murray-Wooley), *The Theater of Sport*, and *The National Road*.

GYULA PAUER is former director of the Center for Cartography and Geographic Information at the University of Kentucky. His training began in his native Hungary, from which he emigrated in 1956. He has been the cartographer for numerous atlases, maps, and other publications, including *Atlas of Southeast Asia, Historical Atlas of Political Parties in the U.S. Congress, Disease and Medical Care in the U.S.*, and *The Himalayan Kingdoms*.

DAVID A. HOWARTH is professor in the Department of Geography and Geosciences and Associate Provost and Dean of Undergraduate Studies at the University of Louisville. His research and teaching interests include climatology and meteorology, the hydrologic cycle, and statistical methods. His primary research has focused on water vapor transport in the southern hemisphere. His work has appeared in leading geographic and meteorologic journals. He is a former coordinator of the Kentucky Geographic Alliance.

RONALD L. MITCHELSON is professor and executive director of the Center for Community and Economic Development at Morehead State University. His research interests focus on economic development and the geography of movement, including aspects of transportation and communications. His research has been funded by the U.S. Department of State, Department of Commerce, Department of Transportation, Department of Agriculture, and the National Science Foundation. He has published widely in professional geographic journals and texts. In 1986 he was named Distinguished University Professor by the National Council of Geographic Education.

JOHN F. WATKINS is associate professor in the Department of Geography at the University of Kentucky. His research specialties emphasize population dynamics, including migration impact on local development, demographics of population aging, and life-course experience and migration theory. His work has covered several continents and most recently has focused on both Appalachia and the north central portion of the United States, on Southeast Asia, and on Scandinavia. He is co-editor (with Graham Rowles) of *Change in the Mountains* and of *Kentucky Atlas of the Elderly* and has published widely in development and gerontology journals.

RICHARD GILBREATH is director of the Center for Cartography and Geographic Information at the University of Kentucky. As a National Geographic Society intern and later contract employee, he produced maps for the *Picture Atlas of the World* on CD-ROM. He is currently principal cartographer on several publications, including the forthcoming text *Southeast Asia: Diversity and Development*.

DONNA GILBREATH is a freelance cartographer and graphic designer who has produced maps for books, journal articles, dissertations, and brochures. She is currently studying journalism at the University of Kentucky.

KEITH R. MOUNTAIN is an assistant professor with the Department of Geography and Geosciences at the University of Louisville, where he teaches courses in meteorology, statistics, and research methodology. His research interests lie in the fields of glaciology and geographic education. He has been involved in research projects in Peru, China, Antarctica, Greenland, Alaska, and British Columbia. He is also the coordinator for the Kentucky Geographic Alliance. He combines his career goals in geography with a long-term interest in aviation and photography.

Index